COMPUTER-
AIDED
MULTIVARIATE
ANALYSIS

COMPUTER-AIDED MULTIVARIATE ANALYSIS

A.A. Afifi
Virginia Clark
Professors of Biostatistics
and Biomathematics
University of California, Los Angeles

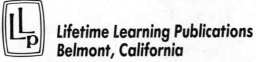
Lifetime Learning Publications
Belmont, California
A division of Wadsworth, Inc.
London, Singapore, Sydney, Toronto, Mexico City

Designer: Lois Stanfield
Copy Editor: Carol Beal
Illustrator: John Foster
Composition: Trigraph Inc.

Printed in the United States of America

1 2 3 4 5 6 7 8 9 10—87 86 85 84

Library of Congress Cataloging in Publication Data
Afifi, A.A.
    Computer-aided multivariate analysis.
    Bibliography: p.
    Includes index.
    1. Multivariate analysis—Data processing. I. Clark,
Virginia   II. Title.
QA278.A33 1984   519.5′35 02854   83-23882
ISBN 0-534-02786-5

CONTENTS

PREFACE

This book has been written for investigators, specifically behavioral scientists, biomedical scientists, econometricians, and industrial users who wish to perform *multivariate statistical analyses* on their data and understand the results. In addition, we believe that the book will be helpful to many statisticians who have been trained in conventional mathematical statistics (where applications of the techniques were not discussed) and who are now working as statistical consultants. Statistical consultants overall should find the book useful in assisting them in giving explanations to clients who lack sufficient background in mathematics.

We do not present mathematical derivations of the techniques in this book but, rather, rely on geometric and graphical arguments and on examples to illustrate them. The mathematical level has been kept deliberately low, with no mathematics beyond high-school level required. Ample references are included for those who wish to see the derivations of the results.

We have assumed that you have taken a basic course in statistics and are familiar with statistics such as the mean and the standard deviation. Also, tests of hypotheses are presented in several chapters and we assume you are familiar with the basic concept of testing a null hypothesis. Many of the computer programs utilize analysis of variance, and that part of the programs can only be understood if you are familiar with one-way analysis of variance.

Approach of the Book

The content and organizational features of this book are discussed in detail in Chapter 1, Section 1.4. Because no university-level mathematics is assumed, some topics often found in books on multivariate analysis have not been included. For example, there is no theoretical discussion of sampling from a multivariate normal distribution or the Wishart distribution, and the usual chapter on matrices has not been included. Also, we point out that the book is not intended to be a comprehensive text on multivariate analysis.

The choice of topics included reflects our preferences and experience as consulting statisticians. For example, we deliberately excluded multivariate analysis of variance (see, e.g., Anderson 1958; Morrison 1976) because we felt that it is not as commonly used as the other topics we describe. On the other hand, we would have liked to include the log-linear model for analyzing multivariate categorical data (see, e.g., Bishop, Fienberg, and Holland 1975; Upton 1978), but we decided against this for fear that doing so would have taken us far afield and would have added greatly to the length of the book.

The multivariate analyses have been discussed more as separate techniques than as special cases arising from some general framework. The advantage of the approach used here is that we can concentrate on explaining how to analyze a certain type of data by using output from readily available computer programs in order to answer realistic questions. The disadvantage is that the theoretical interrelationships among some of the techniques are not highlighted.

Uses of the Book

This book can be used as a text in an applied statistics course or in a continuing education course. We have used preliminary versions of this book in teaching behavioral scientists, epidemiologists, and applied statisticians. It is possible to start with Chapter 6 or 7 and cover the remaining chapters easily in one semester if the students have had a solid background in basic statistics. Two data sets are included that can be used for homework, and a set of problems is included at the end of each chapter except the first.

Computer Orientation

The original derivations for most of the current multivariate techniques were done over forty years ago. We feel that the application of these techniques to real life problems is now the "fun" part of this field. The presence of computer packages and the availability of computers have removed many of the tedious aspects of this discipline so that we can concentrate on thinking about the nature of the scientific problem itself and on what we can learn from our data by using multivariate analysis.

Because the multivariate techniques require a lot of computations, we have assumed that packaged programs will be used. Here we bring together discussions of data entry, data screening, data reduction, and data analysis aimed at helping you perform these functions, understand the basic methodology, and determine what insights into your data you can gain from multivariate analyses. Examples of control statements are given for various computer runs used in the text. Also included are discussions of the options available in the different statistical packages and how they can be used to achieve the desired output.

Acknowledgments

Our most obvious acknowledgment is due to those who have created the computer-program packages for statistical analysis. Many of the applications of multivariate analysis that are now prevalent in the literature are directly due to the development and availability of these programs. The efforts of the programmers and statisticians who managed and wrote the BMDP, SAS, SPSS, and other statistical packages have made it easier for statisticians and researchers to actually use the methods developed by Hotelling, Wilks, Fisher, and others in the 1930s or even earlier. While we have enjoyed longtime associations with members of the BMDP group, we also regularly use SAS and SPSS–X programs in our work, and we have endeavored to reflect this interest in our book. We recognize that other statistical packages are in use, and we hope that our book will be of help to their users as well.

We are indebted to Dr. Ralph Frerichs and Dr. Carol Aneshensel for the use of a sample data set from the Los Angeles Depression Study, which appears in many of the data examples. We wish to thank Dr. Roger Detels for the use of lung function data taken from households from the UCLA population studies of chronic obstructive respiratory disease in Los Angeles. We also thank *Forbes Magazine* for allowing us to use financial data from their publications.

We particularly thank Welden Clark and Nanni Afifi for reviewing drafts and adding to the discussion of the financial data. Helpful reviews and suggestions were also obtained from Dr. Mary Ann Hill, Dr. Roberta Madison, Mr. Alexander Kugushev, and several anonymous reviewers. Welden Clark has, in addition, helped with programming and computer support in the draft revisions and checking, and in preparation of the bibliographies and data sets. We would further like to thank Dr. Tim Morgan, Mr. Steven Lewis, and Dr. Jack Lee for their help in preparing the data sets. The BMDP Statcat (trademark of BMDP Statistical Software, Inc.) desktop computer with the UNIX (trademark of Bell Telephone Laboratories, Inc.) program system has served for text processing as well as further statistical analyses. The help provided by Jerry Toporek of BMDP and Howard Gordon of Network Research Corporation is appreciated.

In addition we would like to thank our copyeditor, Carol Beal, for her efforts in making the manuscript more readable both for the readers and for the typesetter. The major portion of the typing was performed by Mrs. Anne Eiseman; we thank her also for keeping track of all the drafts and teaching material derived in its production. Additional typing was carried out by Mrs. Judie Milton, Mrs. Indira Moghaddan, and Mrs. Esther Najera.

A.A. Afifi
Virginia Clark

COMPUTER-
AIDED
MULTIVARIATE
ANALYSIS

One | PREPARATION FOR ANALYSIS

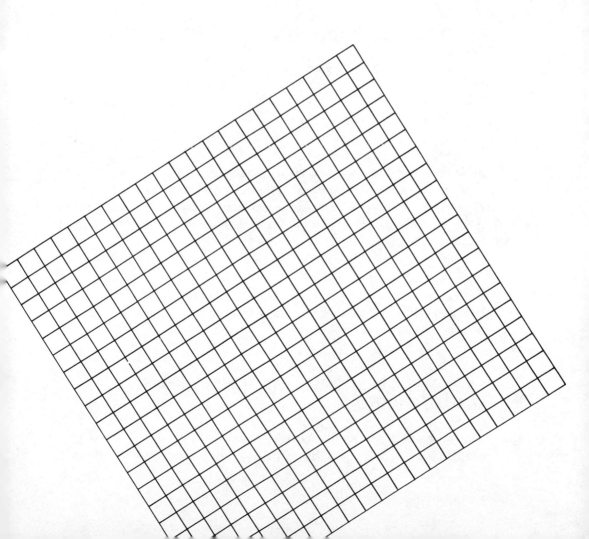

1 | WHAT IS MULTIVARIATE ANALYSIS?

1.1 HOW IS MULTIVARIATE ANALYSIS DEFINED?

The expression *multivariate analysis* is used to describe analyses of data that are multivariate in the sense that numerous observations or variables are obtained for each individual or unit studied. In a typical survey 30 to 100 questions are asked of each respondent. In describing the financial status of a company, an investor may wish to examine five to ten measures of the company's performance. Commonly, the answers to some of these measures are interrelated. The challenge of disentangling complicated interrelationships among various measures on the same individual or unit and of interpreting these results is what makes multivariate analysis a rewarding activity for the investigator. Often results are obtained that could not be attained without multivariate analysis.

In the next section of this chapter several studies are described in which the use of multivariate analyses is essential to understanding the underlying problem. Section 1.3 gives a listing and a very brief description of the multivariate analysis techniques discussed in this book. Section l.4 then outlines the organization of this book.

1.2 EXAMPLES OF STUDIES IN WHICH MULTIVARIATE ANALYSIS IS USEFUL

The studies described in the following subsections illustrate various multivariate analysis techniques. Some are used later in the book as examples.

Depression Study Example

The data for the depression study have been obtained from a complex, random, multiethnic sample of 1000 adult residents of Los Angeles County. The study was a *panel* or *longitudinal* design where the same respondents were interviewed four times between May 1979 and July 1980. About three-fourths of the respondents were reinterviewed for all four interviews. The field work for the survey was conducted by professional interviewers from the Institute for Social Science Research at UCLA.

This ongoing research is an epidemiological study of depression and help-seeking behavior among free-living (noninstitutionalized) adults. The major objectives are to provide estimates of the prevalence and incidence of depression and to identify causal factors and outcomes associated with this condition. The factors examined include demographic variables, life events stressors, physical health status, health care utilization, medication use, life-style, and social support networks. The major instrument used for classifying depression is the Depression Index (CESD) of the National Institute of Mental Health, Center for Epidemiologic Studies. A discussion of this index and the resulting prevalence of depression in this sample is given in Frerichs, Aneshensel, and Clark (1981).

The longitudinal design of the study offers advantages for assessing causal priorities since the time sequence allows us to rule out certain potential causal links. Nonexperimental data of this type cannot directly be used to establish causal relationships, but models based on an explicit theoretical framework can be tested to determine if they are consistent with the data. An example of such model testing is given in Aneshensel and Frerichs (1982).

Data from the first time period of the depression study are presented in Chapter 3. Only a subset of the factors measured on a sample of the respondents is included in order to keep the data set easily comprehensible. These data are used several times in subsequent chapters to illustrate some of the multivariate techniques presented in this book.

Bank Loan Study

The managers of a bank need some way to improve their prediction of which borrowers will successfully pay back a type of bank loan. They have data from the past of the characteristics of persons to whom the bank has lent money and the subsequent record of how well the person has repaid the loan.

Loan payers can be classified into several types: those who met all of the terms of the loan, those who eventually repaid the loan but often did not meet deadlines, and those who simply defaulted. They also have information on age, sex, income, other indebtedness, length of residence, type of residence, family size, occupation, and the reason for the loan. The question is, can a simple rating system be devised that will help the bank personnel improve their prediction rate and lessen the time it takes to approve loans? The methods described in Chapters 11 and 12 can be used to answer this question.

Chronic Respiratory Disease Study

The purpose of the ongoing respiratory disease study is to determine the effects of various types of smog on lung function of children and adults in the Los Angeles area. Because they could not randomly assign people to live in areas that had different levels of pollutants, the investigators were very concerned about the interaction that might exist between the locations where persons chose to live and their values on various lung function tests. The investigators picked four areas of quite different types of air pollution and are measuring various demographic and other responses on all persons over seven years old who live there. These areas were chosen so that they are close to an air-monitoring station.

 The researchers are taking measurements at two points in time and are using the change in lung function over time as well as the levels at the two periods as outcome measures to assess the effects of air pollution. The investigators have had to do the lung function tests by using a mobile unit in the field, and much effort has gone into problems of validating the accuracy of the field observations. A discussion of the particular lung function measurements used for one of the four areas can be found in Detels et al. (1975). In the analysis of the data, adjustments must be made for sex, age, height, and smoking status of each person.

 Over fifteen thousand respondents have been examined and interviewed in this study. The original data analyses were restricted to the first collection period, but now analyses include both time periods. This data set is being used to answer numerous questions concerning effects of air pollution, smoking, occupation, etc. on different lung function measurements. Studies of this type require multivariate analyses so that investigators can arrive at plausible scientific conclusions that could explain the resulting lung function levels.

A subset of this data set is included in Appendix B. Lung function and associated data are given for nonsmoking families for the father, mother, and up to three children ages 7–17.

Assessor Office Example

Local civil laws often require that the amount of property tax a homeowner pays be a percentage of the current value of the property. Local assessor's offices are charged with the function of estimating current value. Current value can be estimated by finding comparable homes that have been recently sold and using some sort of an average selling price as an estimate of the price of those properties not sold.

Alternatively, the sample of sold homes can indicate certain relationships between selling price and several other characteristics such as the size of the lot, the size of the livable area, the number of bathrooms, the location, etc. These relationships can then be incorporated into a mathematical equation used to estimate the current selling price from those other characteristics. Multiple regression analysis methods discussed in Chapters 7–10 can be used by many assessor's offices for this purpose (see Tchira 1973).

1.3 MULTIVARIATE ANALYSES DISCUSSED IN THIS BOOK

In this section a brief description of the major multivariate techniques covered in this book is presented. To keep the statistical vocabulary to a minimum, we illustrate the descriptions by examples.

Simple Linear Regression

A nutritionist wishes to study the effects of eating meat on the bone density of postmenopausal women. She can measure the bone density of the arm (radial bone), in grams per square centimeter, by using a noninvasive device. Women who are at risk of hip fractures because of too low a bone density will show low arm bone density also. The nutritionist intends to sample a group of elderly churchgoing women, some of whom eat very little meat but most of whom eat a normal amount. For women over 65 years of age, she will plot meat eaten per day (reported by the subjects) on the horizontal axis and arm bone density (measured) on the vertical axis. She expects the radial bone density to be lower in women who eat more meat. The nutritionist plans to fit a simple linear regression equation and test whether the slope of the regression line is zero. In this example a single outcome factor is being predicted by a single predictor factor.

Simple linear regression as used in this case would not be considered multivariate by some statisticians, but it is included in this book to introduce the topic of multiple regression.

Multiple Linear Regression

A manager is interested in determining which factors predict the dollar value of sales of the firm's personal computers. Aggregate data on population size, income, educational level, proportion of population living in metropolitan areas, etc. have been collected for 30 areas. As a first step, a multiple linear regression equation is computed, where dollar sales is the outcome factor and the other factors are considered as candidates for predictor factors. A linear combination of the predictor factors is used to predict the outcome or response factor.

Discriminant Function Analysis

A large sample of initially disease-free men over 50 years of age from a community has been followed to see who subsequently has a diagnosed heart attack. At the initial visit blood was drawn from each man, and numerous determinations were made from it, including serum cholesterol, phospholipids, and blood glucose. The investigator would like to determine a linear function of these and possibly other measurements that would be useful in predicting who would and who would not get a heart attack within ten years. That is, the investigator wishes to derive a classification (discriminant) function that would help determine whether or not a middle-aged man is likely to have a heart attack.

Logistic Regression

A television station staff has classified movies according to whether they have a high or low proportion of the viewing audience when shown. The staff has also measured factors such as the length and the type of story and the characteristics of the actors. Many of the characteristics are discrete yes-no or categorical types of data. The investigator may use logistic regression because some of the data do not meet the assumptions for statistical inference used in discriminant function analysis, but they do meet the assumptions for logistic regression. In logistic regression we derive an equation to estimate the probability of capturing a high proportion of the audience.

Principal Components Analysis

An investigator has made a number of measurements of lung function on a sample of adult males who do not smoke. In these tests each man is told to inhale deeply and then blow out as fast and as much as possible into a spirometer, which makes a trace of the volume of air expired over time. The maximum or forced vital capacity (FVC) is measured as the difference between maximum inspiration and maximum expiration. Also, the amount of air expired in the first second (FEV1), the forced midexpiratory flow rate (FEF 25–75), the maximal expiratory flow rate at 50% of forced vital capacity (V50), and other measures of lung function are calculated from this trace. Since all these measures are made from the same flow-volume curve for each man, they are highly interrelated. From past experience it is known that some of these measures are more interrelated than others and that they measure airway resistance in different sections of the airway.

The investigator performs a principal components analysis to determine whether a new set of measurements called principal components can be obtained. These principal components will be linear functions of the original lung function measurements and will be uncorrelated with each other. It is hoped that the first two or three principal components will explain most of the variation in the original lung function measurements among the men. Also, it is anticipated that some operational meaning can be attached to these linear functions that will aid in their interpretation. The investigator may decide to do future analyses on these uncorrelated principal components rather than on the original data. One advantage of this method is that often fewer principal components are needed than original variables. Also, since the principal components are uncorrelated, future computations and explanations can be simplified.

Factor Analysis

An investigator has asked each respondent in a survey whether he or she strongly agrees, agrees, is undecided, disagrees, or strongly disagrees with 15 statements concerning attitudes toward inflation. As a first step, the investigator will do a factor analysis on the resulting data to determine which statements belong together in sets that are uncorrelated with other sets. The particular statements that form a single set will be examined to obtain a better understanding of attitudes toward inflation. Scores derived from each set or factor will be used in subsequent analysis to predict consumer spending.

Canonical Correlation

A psychiatrist wishes to correlate levels of both depression and physical well-being from data on age, sex, income, number of contacts per month with family and friends, and marital status. This problem is different from the one posed in the multiple linear regression example because more than one outcome factor is being predicted. The investigator wishes to determine the linear function of age, sex, income, contacts per month, and marital status that is most highly correlated with a linear function of depression and physical well-being. After these two linear functions, called canonical variables, are determined, the investigator will test to see whether there is a statistically significant (canonical) correlation between scores from the two linear functions and whether a reasonable interpretation can be made of the two sets of coefficients from the functions.

Cluster Analysis

Investigators have made numerous measurements on a sample of patients who have been classified as being depressed. They wish to determine, on the basis of their measurements, whether these patients can be classified by type of depression. That is, is it possible to determine distinct types of depressed patients by performing a cluster analysis on patient scores on various tests?

Unlike the investigator of men who do or do not get heart attacks, these investigators do not possess a set of individuals whose type of depression can be known before the analysis is performed (see Andreasen and Grove 1982 for an example). Nevertheless, the investigators want to separate the patients into separate groups and to examine the resulting groups to see whether distinct types do exist and, if so, what their characteristics are.

1.4 ORGANIZATION AND CONTENT OF THE BOOK

This book is organized into three major parts. Part 1 (Chapters 1–5) deals with data preparation, entry, screening, transformations, and decisions about likely choices for analysis. Part 2 (Chapters 6–10) deals with regression analysis. Part 3 (Chapters 11–16) deals with a number of multivariate analyses. Statisticians disagree on whether or not regression is properly considered as part of multivariate analysis. We have tried to avoid this argument by including regression in the book, but as a separate part. Statisticians certainly agree that regression is an important technique for dealing with problems having multiple variables. In Part 2 on regression analysis we have included various topics, such as dummy variables, that are used in Part 3.

Chapters 2 through 5 are concerned with data preparation and the choice of what analysis to use. First, *variables* and how they are classified are discussed in Chapter 2. The next two chapters concentrate on the practical problems of getting data into the computer, getting rid of erroneous values, checking assumptions of normality and independence, creating new variables, and preparing a useful code book. The choice of appropriate statistical analyses is discussed in Chapter 5.

Readers who are familiar with handling data sets on computers could skip these initial chapters and go directly to Chapter 6. However, formal coursework in statistics often leaves an investigator unprepared for the complications and difficulties involved in real data sets. The material in Chapters 2–5 was deliberately included to fill this gap in preparing investigators for real world data problems.

For a course limited to multivariate analysis, Chapters 2–5 can be omitted if a carefully prepared data set is used for analysis. The depression data set, presented in Section 3.4, has been modified to make it directly usable for multivariate data analysis. Also, the lung function data presented in Appendix B can be used directly, although it has values that some investigators may question.

In Chapters 6–16 we follow a standard format. The topics discussed in each chapter are listed, followed by a discussion of when the technique is used. Then the basic concepts and formulas used are explained. Further interpretations and data examples follow, with topics chosen that relate directly to the technique. Finally, a summary of the available computer output that may be obtained from three widely used statistical package programs is presented, and examples of output from at least one of the packages are presented.

We recommend reading the material on regression in Chapters 6, 7, and 8 and the part of Chapter 10 up through Section 10.3 before proceeding with Chapters 11, 12, and 15, since some of the procedures used in the later chapters are explained in more detail in the regression chapters.

Some readers may prefer to study Chapter 15 directly after Chapter 10 since canonical correlation can be viewed as an extension of regression analysis. It has been placed later in the book because it is a less-often-used technique.

As much as possible, we tried to make each chapter self-contained. However, Chapters 11 and 12, on discriminant analysis and logistic regression, are somewhat interrelated, as are Chapters 13 and 14, covering principal components and factor analyses.

References for further information on each topic are given at the end of each chapter. A complete bibliography is also included at the end of the book. Most of the references at the ends of the chapters do require more mathematics than this book, but special emphasis has been placed on references that include examples. References requiring a strong mathematical background are preceded by an asterisk. If you wish primarily to learn the concepts involved in the multivariate techniques and are not as interested in performing the analysis, then a conceptual introduction to multivariate analysis can be found in Kachigan (1982).

We believe that the best way to learn multivariate analysis is to do it on data that the investigator is familiar with. No book can illustrate all of the features found in computer output for a real life data set. Learning multivariate analysis is similar to learning to swim: You can go to lectures, but the real learning occurs when you get into the water.

BIBLIOGRAPHY

Andreasen, N. C., and Grove, W. M. 1982. The classification of depression: Traditional versus mathematical approaches. *American Journal of Psychiatry* 139:45–52.

Aneshensel, C. S., and Frerichs, R. R. 1982. Stress, support, and depression: A longitudinal causal model. *Journal of Community Psychology* 10:363–376.

Detels, R.; Coulson, A.; Tashkin, D.; and Rokaw, S. 1975. Reliability of plethysmography, the single breath test, and spirometry in population studies. *Bulletin de Physiopathologie Respiratoire* 11:9–30.

Frerichs, R. R.; Aneshensel, C. S.; and Clark, V. A. 1981. Prevalence of depressions in Los Angeles County. *American Journal of Epidemiology* 113:691–699.

Kachigan, S. K. 1982. *Multivariate statistical analysis: A conceptual approach.* New York: Radius Press.

Tchira, A. A. 1973. Stepwise regression applied to a residential income valuation system. *Assessors Journal* 8:23–35.

2 | CHARACTERIZING DATA FOR FUTURE ANALYSES

2.1 WHAT WILL YOU LEARN FROM THIS CHAPTER?

From this chapter you will learn:

- The definition of the word *variable* (2.2).
- About the Stevens system for classification of variables (2.3).
- How variables are used in statistical analyses (2.4).

Chapter 2 is an introductory chapter in which concepts and vocabulary used later in the book are introduced.

2.2 DEFINING STATISTICAL VARIABLES

The word *variable* is used in statistically oriented literature to indicate a characteristic or a property that it is possible to measure. When we measure something, we make a numerical model of the thing being measured. We follow some rule for assigning a number to each level of the particular characteristic being measured. For example, height of a person is a variable. We assign a numerical value to correspond to each person's height. Two people who are equally tall are assigned the same numerical value. On the other hand, two people of different heights are assigned two different values. Measurements of a variable gain their meaning from the fact that there exists a unique correspondence between the assigned numbers and the levels of the property being measured. Thus two people with different assigned heights are not equally tall. Conversely, if a variable has the same assigned value for all individuals in a group, then this variable does not convey useful information about individuals in the group.

Physical measurements, such as height and weight, can be measured directly by using physical instruments. On the other hand, properties such as reasoning ability or the state of depression of a person must be measured indirectly. We might choose a particular intelligence test and define the variable "intelligence" to be the score achieved on this test. Similarly, we may define the variable "depression" as the number of positive responses to a series of questions. Although what we wish to measure is the degree of depression, we end up with a count of yes answers to some questions. These examples point out a fundamental difference between direct physical measurements and abstract variables.

Often the question of how to measure a certain property can be perplexing. For example, if the property we wish to measure is the cost of keeping the air clean in a particular area, we may be able to come up with a reasonable estimate, although different analysts may produce different dollar estimates. The problem becomes much more difficult if we wish to estimate the benefits of clean air.

On any given individual or thing we may measure several different characteristics. We would then be dealing with several variables, such as age, height, annual income, race, sex, and level of depression of a certain individual. Similarly, we can measure characteristics of a corporation, such as various financial measures. In this book we are concerned with analyzing data sets consisting of measurements on several variables for each individual in a given sample. We use the symbol P to denote the number of *variables* and the symbol N to denote the number of *individuals, observations, cases,* or *sampling units*.

2.3 HOW VARIABLES ARE CLASSIFIED: STEVENS'S CLASSIFICATION SYSTEM

In the determination of the appropriate statistical analysis for a given set of data, it is useful to classify variables by type. One method for classifying variables is by the degree of sophistication evident in the way they are measured. For example, we can measure height of people according to whether the top of their head exceeds a mark on the wall: If yes, they are tall; and if no, they are short. On the other hand, we can also measure height in centimeters or inches. The latter technique is a more sophisticated way of measuring height. As a scientific discipline advances, measurements of the variables with which it deals become more sophisticated.

Various attempts have been made to formalize variable classification. A commonly accepted system is that proposed by *Stevens* (1951). In this system measurements are classified as *nominal, ordinal, interval*, or *ratio*. In deriving his classification, Stevens characterized each of the four types by a transformation that would not change a measurement's classification. In the subsections that follow, rather than discuss the mathematical details of these transformations, we present the practical implications for data analysis.

Nominal Variables

With *nominal variables* each observation belongs to one of several distinct categories. The categories are not necessarily numerical, although numbers may be used to represent them. For example, "sex" is a nominal variable. An individual's sex is either male or female. We may use any two symbols, such as M and F, to represent the two categories. In computerized data analysis, numbers are used as the symbols since many computer programs are designed to handle only numerical symbols. Since the categories may be arranged in any desired order, any set of numbers can be used to represent them. For example, we may use 0 and 1 to represent males and females, respectively. We may also use 1 and 2 to avoid the use of zero since some programs do not distinguish zeros from blanks. Any two other numbers can be used as long as they are used consistently.

An investigator may rename the categories, thus performing a numerical operation. In so doing, the investigator must preserve the uniqueness of each category. Stevens expressed this last idea as a "basic empirical operation" that preserves the category to which the observation belongs. For example, two males must have the same value on the variable "sex," regardless of the two numbers chosen for the categories. Table 2.1 summarizes these ideas and presents further examples. Nominal variables with more than two categories, such as race or religion, may present special challenges to the multivariate data analyst. Some ways of dealing with these variables are presented in Chapter 10.

TABLE 2.1. Stevens's Measurement System

Type of Measurement	Basic Empirical Operation	Examples
Nominal	Determination of equality of categories	Company names Race Religion Basketball players' numbers
Ordinal	Determination of greater than or less than (ranking)	Hardness of minerals Socioeconomic status Rankings of wines
Interval	Determination of equality of differences between levels	Temperature, in degrees Fahrenheit Calendar dates
Ratio	Determination of equality of ratios of levels	Height Weight Density Difference in time

Ordinal Variables

Categories are used for *ordinal variables* as well, but there also exists a known order among them. For example, in the Mohs Hardness Scale minerals and rocks are classified according to ten levels of hardness. The hardest mineral is diamond and the softest is talc (see Pough 1960). Any ten numbers may be used to represent the categories, as long as they are ordered in magnitude. For instance, the integers 1 to 10 would be natural to use. On the other hand, any sequence of increasing numbers may also be used. Thus the basic empirical operation defining ordinal variables is whether one observation is greater than another. For example, we must be able to determine whether one mineral is harder than another. Hardness can be easily tested by noting which mineral can scratch the other. Note that for most ordinal variables there is an underlying continuum being approximated by artificial categories. For example, in the above hardness scale fluorite is defined as having a hardness of 4, and calcite, 3. However, there is a range of hardness between these two numbers not accounted for by the scale.

Interval Variables

An *interval variable* is a special ordinal variable in which the differences between successive values are always the same. For example, the variable "temperature," in degrees Fahrenheit, is measured on the interval scale since the difference between 12° and 13° is the same as the difference between 13° and 14° or the difference between any two successive temperatures. In contrast, the Mohs Hardness Scale does not satisfy this condition since the intervals between successive categories are not necessarily the same. The scale must satisfy the basic empirical operation of preserving the equality of intervals.

Ratio Variables

Ratio variables are interval variables with a natural point representing the origin of measurement, i.e., a natural zero point. For instance, height is a ratio variable since zero height is a naturally defined point on the scale. We may change the unit of measurement (e.g., centimeters to inches), but we would still preserve the zero point and also the ratio of any two values of height. Temperature in degrees Fahrenheit is not a ratio variable since we may choose the zero point arbitrarily, thus not preserving ratios.

There is an interesting relationship between interval and ratio variables. For example, although time of day is measured on the interval scale, the length of a time period is a ratio variable since it has a natural zero point.

Other Classifications

Other methods of classifying variables have also been proposed (Coombs 1964). We mention, in particular, that variables may be classified as discrete or continuous.

A variable is called *continuous* if it can take on any value in a specified range. Thus the height of an individual may be 70 in. or 70.4539 in. Any numerical value in a certain range is a conceivable height.

A variable that is not continuous is called *discrete*. A discrete variable may take on only certain specified values. For example, counts are discrete variables since only zero or positive integers are allowed. In fact, all nominal and ordinal variables are discrete. Interval and ratio variables can be continuous or discrete. This latter classification carries over to the possible distributions assumed in the analysis. For instance, the normal distribution is often used to describe the distribution of continuous variables.

Statistical analyses have been developed for various types of variables. In Chapter 5 a guide to selecting the appropriate descriptive measures and multivariate analyses will be presented. The choice depends on how the variables are used in the analysis, a topic that is discussed next.

2.4 HOW VARIABLES ARE USED IN DATA ANALYSIS

The type of data analysis required in a specific situation is also related to the way in which each variable in the data set is used. Variables may be used to measure outcomes or to explain why a particular outcome resulted. For example, in the treatment of a given disease a specific drug may be used. The outcome may be a discrete variable classified as "cured" or "not cured." Also, the outcome may depend on several characteristics of the patient such as age, genetic background, and severity of the disease. These characteristics are sometimes called *explanatory variables*. Equivalently, we may call the outcome the *dependent variable* and the characteristics the *independent variables*. The latter terminology is very common in statistical literature. This choice of terminology is unfortunate in that the "independent" variables do not have to be statistically independent of each other. Indeed, these independent variables are usually interrelated in a complex way. Another disadvantage of this terminology is that the common connotation of the words implies a causal model, an assumption not needed for the multivariate analyses described in this book. In spite of these drawbacks, the widespread use of these terms forces us to adopt them.

In other situations the dependent variable may be treated as a continuous variable. For example, in household survey data we may wish to relate monthly expenditure on cosmetics per household to several explanatory or independent variables such as the number of individuals in the household, their sex, and household income.

In some situations the roles that the various variables play are not obvious and may also change, depending on the question being addressed. Thus a data set for a certain group of people may contain observations on their sex, age, diet, weight, height, and blood pressure. In one analysis we may use weight as a dependent variable with height, sex, age, and diet as the independent variables. In another analysis blood pressure might be the dependent variable with weight and the other variables considered as independent variables.

In certain exploratory analyses all the variables may be used as one set with no regard to whether they are dependent or independent. For example, in the social sciences a large number of variables may be defined initially, followed by attempts to combine them into a smaller number of summary variables. In this analysis the original variables are not classified as dependent or independent. The summary variables may later be used to possibly explain certain outcomes or dependent variables. In Chapter 5 multivariate analyses described in this book will be characterized by situations in which they apply according to the types of variables analyzed and the roles they play in the analysis.

2.5 EXAMPLES OF CLASSIFYING VARIABLES

In the depression data example several variables are measured on the nominal scale: sex, marital status, employment, and religion. The general health scale is an example of an ordinal variable. Income and age are both ratio variables. No interval variable is included in the data set. A listing and code book for this data set are given in Chapter 3.

One of the questions that may be addressed in analyzing this data set is, What are the factors related to the degree of psychological depression of a person? The variable "cases" may be used as the dependent or outcome variable since an individual is considered a case if his or her score on the depression scale exceeds a certain level. "Cases" is an ordinal variable, although it can be considered nominal because it has only two categories. The independent variable could be any or all of the other variables (except ID and measures of depression). Examples of analyses without regard to variable roles are given in Chapters 13 and 14, using the variables C_1 to C_{20} in an attempt to summarize them into a small number of components or factors.

Sometimes, the Stevens classification system is difficult to apply, and two investigators could disagree on a given variable. For example, there may be disagreement about the ordering of the categories of a socioeconomic status variable. Thus the status of blue-collar occupations with respect to the status of certain white-collar occupations might change over time or from culture to culture. So such a variable might be difficult to justify as an ordinal variable, but we would be throwing away valuable information if we used it as a nominal variable. Despite these difficulties, the Stevens system is useful in making decisions on appropriate statistical analysis, as will be discussed in Chapter 5.

SUMMARY

In this chapter statistical variables were defined. Their types and the roles they play in data analysis were discussed.

These concepts can affect the choice of analyses to be performed, as will be discussed in Chapter 5. We point out that the SPSS-X manual uses the Stevens classification as a guide for the selection of statistical analyses*.

*SPSS-X is the trademark of SPSS Inc. of Chicago, Illinois, for its proprietary computer software.

BIBLIOGRAPHY

Churchman, C. W., and Ratoosh, P., eds. 1959. *Measurement: Definition and theories*. New York: Wiley.

*Coombs, C. H. 1964. *A theory of data*. New York: Wiley.

Ellis, B. 1966. *Basic concepts of measurement*. London: Cambridge University Press.

Pough, F. H. 1960. *Field guide to rocks and minerals*. 3rd ed. Boston: Houghton Mifflin.

Stevens, S. S. 1951. *Handbook of experimental psychology*. New York: Wiley.

Torgerson, W. S. 1958. *Theory and methods of scaling*. New York: Wiley.

*References preceded by an asterisk require strong mathematical background.

PROBLEMS

2.1 Classify the following types of data by using Stevens's measurement system: decibels of noise level, father's occupation, parts per million of an impurity in water, density of a piece of bone, rating of a wine by one judge, net profit of a firm, and score on an attitude test.

2.2 In a survey of users of a walk-in mental health clinic, data have been obtained on sex, age, household roster, race, educational level (number of years of school), family income, reason for coming to the clinic, symptoms, and scores on screening examination. The investigator wishes to determine what variables affect whether or not coercion by the family, friends, or a governmental agency was used to get the patient to the clinic. Classify the data according to Stevens's measurement system. What would you consider to be possible independent variables? Dependent variables? Do you expect the independent variables to be independent of each other?

2.3   For the chronic respiratory study data presented in Appendix B, classify each variable according to the Stevens scale and according to whether it is discrete or continuous. Pose two possible research questions and decide on the appropriate dependent and independent variables.

2.4   From a field of statistical application (perhaps your own field of specialty), describe a data set and repeat the procedures described in Problem 2.3.

3 PREPARING FOR DATA ANALYSIS

3.1 WHAT WILL YOU LEARN FROM THIS CHAPTER?

From this chapter you will learn about the features of some of the major statistical computing software programs, how to get data into the computer, and how to do preliminary data screening. In particular, you will learn:

▶ About the most used statistical package programs (3.2).

▶ About the advantage of using SAS, SPSS–X, or BMDP and about their manuals (3.2).

▶ How to code data so that they can be easily entered into the computer (3.3).

▶ What to include in a code book (3.3).

▶ What variables are available in the depression data set (3.4).

▶ How to find erroneous data (3.5).

3.2 STATISTICAL PACKAGE PROGRAMS

In this book we are concerned mainly with analyzing data on several variables simultaneously. Such multivariate analyses usually require a large amount of computations, thus almost precluding hand calculations. We therefore assume that the analyses described in this book will be performed by a high-speed digital computer through the use of a packaged program. The development of packaged programs has simplified the computations greatly and increased the usability of multivariate statistical analyses. Such packages present a reliable, standardized way of performing the computations. Thus it is usually preferable to use a tested package to perform standard statistical analyses rather than to attempt to write your own programs from scratch.

We assume that you have access to one or more statistical packages. Table 3.1 presents a list of five of the major program packages. All five packages include preliminary screening and some multivariate analysis programs.

Not all packages have been adapted for use on all computers. For example, SAS is available for IBM and IBM–compatible computers and for Digital Equipment Corporation's VAX, Data General's Eclipse, and Prime minicomputers. But for some years these major statistical packages have been available on large mainframe computers. Versions of them have also been available on some medium-scale minicomputers. Most recently, the BMDP programs have become available on desktop computers using the Motorola 68000 microprocessors. Neffendorf (1983) presents a list of other statistical packages available for microcomputers. If you are interested in implementing a package on a particular machine, obtain a reference manual from the sources listed in Table 3.1 and contact the developer listed in the manual.

TABLE 3.1. List of Some Multivariate Statistical Package Programs

Name	Brief Description	Reference Manual
BMDP	General-purpose statistical package	W. J. Dixon, *BMDP Statistical Software* (1964 Westwood Blvd., Los Angeles, Calif. 90025)
MINITAB	Instructional package, batch or interactive, useful for screening and preliminary analysis	T. A. Ryan, Jr., et al., Minitab Reference Manual (Boston: Duxbury Press)
OSIRIS	General-purpose statistical package	*Osiris User's Manual* (Ann Arbor: Institute for Social Research, University of Michigan)
SAS	General-purpose statistical and data management package	SAS Institute, Inc., *SAS User's Guide*, (Cary, N. C. 27511)
SPSS–X	General-purpose statistical package	N. H. Nie, *SPSS-X* (SPSS Inc., 444 North Michigan Ave., Chicago, Ill. 60611)

In the following subsections we describe the three packages we use extensively in this book. We also discuss their manuals and their relative advantages.

Packages Used in This Book

In this book we will be making specific references to programs or procedures from three packages: BMDP*, SAS®**, and SPSS–X. These three packages are widely used, and courses are frequently offered to explain their capabilities. In Chapters 6 through 16 we include tables that highlight specific programs that perform the analyses discussed in each chapter. Each package, however, is capable of performing many more analyses than the ones we cite.

Our discussion applies, for the most part, to the following editions:

▸ BMDP, 1983.

▸ SPSS–X, 1983.

▸ *SAS Supplemental Library User's Guide*, 1980;
SAS Changes and Enhancements, Technical Report
p–115, 1981; and *SAS User's Guide: Statistics*,
1982.

Both BMDP and SPSS–X adopt the philosophy of offering a number of comprehensive programs, each with its own options and variants for performing portions of the analyses. The SAS package, on the other hand, offers a large number of procedures, each with a specific function to perform. The user of SAS specifies a sequence of such procedures to form a complete analysis. Some of the SAS procedures also include options. When referring to SAS procedures, we only mention the specific procedure for that analysis. The SAS package provides, however, a rich assortment of options by combining its various procedures.

User's Manuals

These three computer packages are large systems, with manuals that go through repeated updates at unpredictable intervals. Further, different com-

*BMDP™ is the registered trademark of BMDP Statistical Software, Inc.
**SAS® is the registered trademark of SAS Institute, Inc., Cary, NC 27511, USA.

puter centers with different hardware configurations may use slightly different versions of the programs and do not always have the latest releases in
operation. It makes sense to find out which release is available on the
computer you intend to use and to get a manual to match that release.

When you are learning how to use a package for the first time, there is no
substitute for reading the manuals. For any of the packages the examples
reproduced in the manuals are often very helpful. However, some advice from
a person experienced in the use of a given package may save you several
agonizing hours.

Whenever alternative options are available for performing a specific analysis, the package performs one of these options automatically if not instructed
otherwise. These options are called the *default options*. In this book we often
recommend which options you should use. If you wish to become familiar
with a package, begin by using the default options. Further insight into the
packages' capabilities may be gained by later specifying other options and
comparing the resulting outputs.

Advantages of Each Package

Each of the three packages has a special strength in one area. The SPSS–X,
for example, is designed specifically to handle questionnaire and survey data
arising in the social sciences. The clarity of its manual is also a decided
advantage. The SPSS–X offers users an understandable and comprehensive
cleanup and data management system.

The BMDP package pays special attention to the technical aspects of the
statistical procedures and emphasizes graphical output that is useful in
checking assumptions. It incorporates some recent developments in exploratory data analysis (Tukey 1977). Also, it includes programs of particular
relevance to the biomedical sciences.

The SAS package is the most sophisticated of the three packages in terms
of *data management capabilities*. Many investigators use SAS to set up
their data files and then call one of the procedures of SAS or a program from
another package to perform the analysis. SAS has also incorporated programs developed by users into its system, resulting in a wide variety of
procedures.

3.3 DATA ENTRY AND CODING

In this book we are concerned with analyzing data sets already collected rather than collecting new data. In this section we discuss preliminary steps in preparing existing data for analysis by the computer. The discussion here is very brief because of the differences in capabilities and configurations of computer centers. If you intend to make use of a particular computer facility, you should make an effort to become familiar with the preferred forms of data entry for that facility.

Preparing Data for the Computer

In general, a computer can read data supplied in the form of one or more of the following: cards, magnetic tapes, floppy disks, and hard disks. The format in which the data appear varies depending on the computer and the program used. All of the major packages accept the so-called *card image format*. In this format the data for each individual observation or case are arranged in the form of one or more cards, each having 80 columns. Each variable occupies one or more columns, called a *field*.

For example, suppose that an investigator has data on ten mice. Each mouse is given a number from 1 to 10, called the ID. Two variables are weight (in grams) and age (in days); a third variable indicates whether the mouse has a particular disease. The latter variable is coded as 1 = diseased and 2 = not diseased. In Figure 3.1a the original *data sheet* is given with data on the first three mice. The data can then be transferred onto a *code sheet*, as shown in Figure 3.1b, in preparation for keypunching. Note that the decimal points for weight are not necessarily shown on the code sheet. The place of the decimal point can be indicated to the computer by a *format statement*.

In situations where the data for each each observation occupy more than one card, a field in each card should be assigned to the card number. Sometimes, the column assignment is indicated on the data collection forms themselves, so that code sheets are not necessary.

The data are transferred from the code sheet into a *computer-readable form* (cards, tapes, or disks) by using keypunch machines or terminals. The choice depends on the computer facility available and on the user's prefer-

FIGURE 3.1. Illustration of Data, Code Sheet, Cards, and Code Book for a Small Data Set

A. DATA SHEET

ID	Age (Days)	Weight (Grams)	Diseased (1 = Yes 2 = No)
1	430	285.3	2
2	272	291.5	1
3	514	307.1	2
.	.	.	.
.	.	.	.
.	.	.	.

B. CODE SHEET

	ID		AGE			WEIGHT			DISEASE							
	1	2	3	4	5	6	7	8	9	10	11	12	13	14	15	16	17
1	0	1	4	3	0	2	8	5	3	2							
2	0	2	2	7	2	2	9	1	5	1							
3	0	3	5	1	4	3	0	7	1	2							
4		.		.						.							
5		.		.						.							
6		.		.						.							

D. CODE BOOK

Variable Number	Variable Location (Column)	Variable Name	Description
1	1–2	ID	ID number from 1 to 10
2	3–5	WEIGHT	Weight in grams (XXX.X)
3	6–9	AGE	Age in days
4	10	DISEASE	Disease status (1 = yes, 2 = no)

C. CARDS

035143071?

027229151

014302853?

GLOBE NO. 1   STANDARD FORM 5081

27

ence. Some investigators prefer keeping punched cards as a backup form of the data even if tapes or disks are used. Keypunch machines such as the IBM 120 allow easy verification of the entered data—i.e. keypunching the number twice to detect possible errors. Verification is advisable for all data when possible; as a minimum, ID and card number should always be verified. Figure 3.1c illustrates the punched cards for the first three cases of the mice study.

On the other hand, data sets could be entered via the terminal to save time. Terminal use is particularly advisable if interactive-checking, verification, and editing programs are available. For instance, form-handling routines on the terminal allow the user to lay out data entry forms on the display screen to improve the efficiency of data entry. Then during entry operations the user is warned when data do not fall within acceptable values.

The following suggestions may be helpful for data entry:

1. Whenever possible, code information in numbers, not letters. Very few packaged programs are designed to deal with alphabetical letters (many SAS, BMDP4D, and nonmathematical SPSS–X procedures, such as COUNT or DATA LIST, can handle the alphabet, though).

2. Code information in the most detailed form you ever intend to use. You can use the computer to aggregate the data in coarser groupings later. For example, it is better to record age at last birthday than to record ten-year age groups.

3. For medium or large data sets, have data entered and verified by professional operators.

4. Use several cards (or lines) per case rather than overcondense all the information on one.

5. If data are put on cards, make a duplicate of them as a backup copy. It is easy to lose or destroy a few cards; these cards can easily be replaced if a duplicate copy is made. If editing is done later on tapes or disks, then the cards should also be corrected.

6. For each variable, use a code to indicate missing values. This code should be a value outside the

acceptable limits of the variable. A blank may be
used as a missing value code if a blank is
appropriate for the package used.

Finally, an important and often forgotten part of data analysis is to create
a data *code book*. The code book should contain a description and information
on the location of each variable. It serves as a guide and record for all users
of the data set and for future documentation of the results. In the mice
example the code book could be as brief as the table illustrated in Figure
3.1d. If desired, it may also include other information useful in interpreting
the data. For example, the acceptable range of values for each variable can
be included. For questionnaire data the complete question, or a shortened
version of it, and possible answers may be part of the description.

Computer Statements

Once the data set is in computer-readable form, it is then possible to use one
of the packages to analyze it. The specific steps involved in getting the
computer to perform the desired analysis are outlined in the package manuals. In this book we only cover these details for specific examples; for your
facility and packages, refer to the appropriate manuals or consultants if
available.

In brief, the following statements are required by the computer:

1. *System control statements* to inform the computer
   of the account number being used, request use of the
   desired package, and perhaps indicate the form and
   the location of the data set.

2. *Package control statements* specifying which
   options or procedures of the package are to be used.

3. Other *data definition statements* specifying the
   variables to be used, their location and format, and
   the number of cases.

4. *Data transformation statements* that allow the user
   to transform the input data prior to analysis. If
   transformation is desired, such statements must be
   supplied along with those mentioned above. A
   detailed discussion of data transformations is given
   in Chapter 4.

5. *File statements* that instruct the computer to use the output of one program as input for other programs. This statement is usually achieved by creating a file containing the output data and saving it for future use.

Examples will be given in each chapter to illustrate the use of these statements.

Sometimes, the data constitute a set created for a special purpose. In that case few modifications, if any, are needed to prepare the data for use in conjunction with a packaged computer program. More often, the data set is more complex, with multiple possibilities for analysis. Complex data result, for example, from questionnaire studies or from long ongoing projects where many variables are collected from a large number of individuals. In those situations data subsets are usually created especially for specific analyses. The management of such data sets and subsets can be a complex process requiring specialized expertise. But packaged programs usually include some capabilities that may be helpful for data management. As noted earlier, SAS and SPSS–X pay special attention to this aspect; OSIRIS and BMDP also have some data management options. If you are interested in further reading on this subject, see Flores (1977), Augenstein and Tenenbaum (1979), or Capron and Williams (1982).

3.4 DATA EXAMPLE: LOS ANGELES DEPRESSION STUDY

In this section we discuss a data set that will be used in several of the succeeding chapters to illustrate multivariate analyses. The depression study itself is described in Chapter 1.

The data given in this chapter are from a subset of 294 respondents randomly chosen from the original 1000. This subset of the observations is large enough to provide a good illustration of the statistical techniques but small enough to save computer costs. Only data from the first time period are included. Variables were chosen so that they would be easily understood and would be sensible to use in the multivariate statistical analyses described in Chapters 6 through 16.

The code book, the variables used, and the data set are described below.

Code Book

Table 3.2 contains a code book for the depression data set. In the first column the variable number is listed, since that is often the simplest way to refer to the variables in the computer. The location of each variable is listed next for the set of cards that were used in the computer runs. A variable name is given next, and this name is used in later data analyses. These names were chosen to be eight characters or less in length so that they could be used by all three packaged programs. It is helpful to choose variable names that are easy to remember but are short enough to reduce entry time when running programs; then use them consistently.

Finally, a description of the outcome of each variable is given in the last column of Table 3.2. For nominal or ordinal data the numbers used to code each answer are listed. For ratio data such as age the units used are included. Note that income is given in thousands of dollars per year for the household; thus an income of 15 would be $15,000 per year.

Depression Variables

The 20 items used in the depression scale are variables 9–28 and are named C1, C2,..., C20. (The wording of each item is given later in the text, in Table 13.1.) Each item was written on a card and the respondent was asked to tell the interviewer the number that best describes how often he or she felt or behaved this way during the past week. Thus respondents who answered item C2, "I felt depressed," could respond 0 through 3, depending on whether this particular item applied to them rarely or none of the time (less than 1 day: 0), some or a little of the time (1–2 days: 1), occasionally or a moderate amount of the time (3–4 days: 2), or most or all of the time (5–7 days: 3).

Most of the items are worded in a negative fashion, but items C8 through C11 are positively worded. For example, C8 is "I felt that I was as good as other people." For positively worded items the scores are *reflected*; that is, a score of 3 is changed to be 0, 2 is changed to 1, 1 is changed to 2, and 0 is changed to 3. In this way, when the total score of all 20 items is obtained by summation of variables C1 through C20, a large score indicates a person who is depressed. This sum is the 29th variable, named CESD.

Persons whose CESD score is greater than or equal to 16 are classified as

TABLE 3.2. Code Book for Depression Data

Variable Number	Variable Location (Columns)	Variable Name	Description
1	6–8	ID	Identification number from 1 to 294
2	9	SEX	1 = male; 2 = female
3	10–11	AGE	Age in years at last birthday
4	12	MARITAL	1 = never married; 2 = married; 3 = divorced; 4 = separated; 5 = widowed
5	13	EDUCAT	1 = less than high school; 2 = some high school; 3 = finished high school; 4 = some college; 5 = finished bachelor's degree; 6 = finished master's degree; 7 = finished doctorate
6	14	EMPLOY	1 = full time; 2 = part time; 3 = unemployed; 4 = retired; 5 = houseperson; 6 = in school; 7 = other
7	15–16	INCOME	Thousands of dollars per year
8	17	RELIG	1 = Protestant; 2 = Catholic; 3 = Jewish; 4 = none; 5 = other
9–28	18–37	C1–C20	"Please look at this card and tell me the number that best describes how often you felt or behaved this way during the past week." 20 items from depression scale (already reflected; see text) 0 = rarely or none of the time (less than 1 day); 1 = some or a little of the time (1–2 days); 2 = occasionally or a moderate amount of the time (3–4 days); 3 = most or all of the time (5–7 days)
29	38–39	CESD	Sum of C1–C20; 0 = lowest level possible; 60 = highest level possible
30	40	CASES	0 = normal; 1 = depressed, where depressed is $CESD \geq 16$
31	41	DRINK	Regular drinker? 1 = yes; 2 = no
32	42	HEALTH	General health? 1 = excellent; 2 = good; 3 = fair; 4 = poor
33	43	REGDOC	Have a regular physician? 1 = yes; 2 = no
34	44	TREAT	Has a doctor prescribed or recommended that you take medicine, medical treatments, or change your way of living in such areas as smoking, special diet, exercise, or drinking? 1 = yes; 2 = no
35	45	BEDDAYS	Spent entire day(s) in bed in last two months? 0 = no; 1 = yes
36	46	ACUTEILL	Any acute illness in last two months? 0 = no; 1 = yes
37	47	CHRONILL	Any chronic illnesses in last year? 0 = no; 1 = yes

depressed since this value is the common cutoff point used in the literature (see Frerichs, Aneshensel, and Clark 1981). These persons are given a score of 1 in variable 30, the CASES variable. The particular depression scale employed here was developed for use in community surveys of noninstitutionalized respondents (see Comstock and Helsing 1976; Radloff 1977).

Data Set

As can be seen by examining the code book given in Table 3.2, demographic data (variables 2–8), depression data (variables 9–30), and general health data (variables 32–37) are included in this data set. Variable 31, drinking habits, was included so that it would be possible to determine if an association exists between drinking and depression. Frerichs et al. (1981) have already noted a lack of association between smoking and scores on the depression scale.

The actual data for the 294 respondents are listed in Table 3.3. The headings at the top of Table 3.3 correspond to the variable names given in Table 3.2. For example, the first respondent is a female, aged 68, who is widowed, has had some high school education, is retired with a current income of $4000 per year, and is a Protestant. She has a CESD score of 0, so she would not be considered depressed. She is not a regular drinker and considers herself in good health. She has a regular physician who prescribed or recommended a treatment to her. She has had no bed days or acute illnesses in the last two months, but she has had at least one chronic illness. Similar statements could be made about each of the 294 respondents by reading the data in Table 3.3.

3.5 PRELIMINARY DATA SCREENING

Prior to any data analysis some initial *screening* of the data set should be performed. For small to medium data sets it is useful to obtain a list of the data and critically examine it. If the number of observations is small, a careful examination of each listed value is feasible. These values can be compared with the original data sheets for any possible entry errors. For larger data sets this comparison may be so tedious that it is not useful. However, a cursory examination of the pattern of the listing may reveal that the keypuncher shifted the data by one or more columns.

TABLE 3.3. Depression Data

OBS	ID	SEX	AGE	MARITAL	EDUC	EMPLOY	INCOME	RELIG	C1	C2	C3	C4	C5	C6	C7	C8	C9	C10	C11	C12	C13	C14	C15	C16	C17	C18	C19	C20	CESD	CASES	DRINK	HEALTH	REGDOC	TREAT	BEDDAYS	ACUTEILL	CHRONILL
1	1	2	68	5	2	4	4	1	0	0	0	0	0	0	0	0	0	0	0	0	0	0	0	0	0	0	0	0	0	0	2	2	1	1	0	0	1
2	2	1	58	3	4	1	15	1	0	0	0	0	0	0	0	0	0	0	0	0	0	0	0	0	2	0	2	0	4	0	1	1	1	0	0	0	1
3	3	2	45	2	3	1	28	1	0	0	0	0	1	0	0	3	0	0	0	0	0	0	0	0	0	0	0	0	4	0	2	2	1	1	0	0	0
4	4	2	50	3	3	1	9	1	0	0	0	0	0	0	0	0	0	0	1	0	0	0	0	0	0	0	0	0	5	0	2	1	1	0	0	0	1
5	5	2	33	4	3	1	35	1	0	0	0	0	0	0	0	0	0	0	0	0	0	0	0	0	0	0	0	0	6	0	2	2	1	0	1	0	0
6	6	1	24	2	3	5	11	1	0	1	0	0	0	0	0	0	0	0	0	0	0	0	0	0	0	0	0	0	7	0	2	1	1	0	0	1	0
7	7	2	58	2	2	4	9	1	0	0	0	0	0	0	0	0	0	0	0	0	0	0	0	0	0	0	0	0	15	1	2	3	1	0	0	0	0
8	8	1	22	1	3	1	23	2	1	0	0	0	0	0	0	0	0	0	0	0	1	0	0	0	0	1	1	1	10	1	1	1	1	0	0	0	1
9	9	2	47	5	3	3	35	4	0	0	0	0	0	0	0	0	0	0	0	0	0	0	0	0	0	0	0	0	16	1	1	3	1	0	0	0	0
10	10	1	30	2	2	2	25	4	0	0	0	0	0	0	0	0	0	0	0	0	0	0	0	0	0	0	0	0	10	1	2	3	1	0	0	0	0
11	11	2	20	1	3	4	24	1	0	0	0	0	0	0	0	0	0	0	0	0	0	0	0	0	0	0	0	0	8	0	1	3	1	0	0	0	1
12	12	2	57	4	3	1	28	2	1	0	0	0	0	0	0	0	0	0	0	0	0	0	0	0	0	0	0	0	4	0	1	2	1	0	0	0	0
13	13	2	39	1	3	1	13	1	0	0	0	0	0	0	0	0	3	0	0	0	0	0	0	0	0	0	0	0	8	0	2	3	1	0	0	0	0
14	14	2	61	5	2	1	16	1	0	0	0	0	0	0	0	0	0	0	0	0	0	0	0	0	0	0	0	0	4	0	2	3	1	0	0	0	0
15	15	2	23	1	4	3	8	2	0	3	0	2	0	0	3	0	0	0	0	0	0	0	0	0	0	0	0	0	21	1	2	2	1	0	0	0	1
16	16	1	21	2	2	1	19	1	0	0	3	0	1	0	0	0	0	2	0	0	0	0	0	0	0	0	0	0	22	1	2	2	1	0	0	0	0
17	17	2	55	5	6	4	15	2	0	0	0	0	0	0	0	0	0	0	0	0	0	0	0	0	0	0	0	0	6	0	2	2	1	0	0	0	1
18	18	2	26	2	2	1	9	1	0	0	0	0	0	0	0	0	0	0	0	0	0	0	0	0	0	0	0	0	0	0	2	2	1	0	0	0	0
19	19	2	64	3	3	1	35	2	0	0	0	0	0	0	0	0	0	0	0	0	0	0	0	0	0	0	0	0	3	0	2	2	1	0	0	0	0
20	20	2	44	2	3	3	7	1	2	0	2	0	0	0	0	0	0	0	0	0	0	0	0	0	0	0	0	0	11	0	2	1	1	0	0	0	1
21	21	2	25	2	3	4	19	2	2	2	0	0	3	3	0	0	0	2	0	2	2	0	0	3	0	0	0	0	42	1	2	3	1	0	0	0	1
22	22	1	72	2	3	5	6	1	1	0	0	0	3	0	0	0	0	0	0	0	0	0	0	0	0	0	0	0	10	1	1	3	1	0	0	0	1
23	23	2	61	4	3	4	13	2	0	0	0	0	1	0	0	0	0	0	0	0	0	0	0	0	0	0	2	0	12	1	2	3	1	0	0	0	1
24	24	1	43	2	5	2	19	1	2	0	0	2	3	3	0	0	3	0	0	0	0	0	0	0	0	0	0	0	69	1	2	5	1	0	0	0	0
25	25	2	52	2	4	1	15	2	1	0	0	0	0	0	0	0	0	0	0	0	0	0	0	0	0	0	0	0	28	1	2	4	1	1	0	0	1
26	26	1	23	2	3	1	20	2	1	0	0	0	0	0	0	0	0	0	0	0	0	0	0	0	0	0	0	0	17	1	2	3	1	0	0	0	1
27	27	2	73	3	3	5	19	4	0	0	0	0	0	0	0	0	0	0	0	0	0	0	0	0	0	0	0	0	11	0	2	2	1	0	0	0	0
28	28	2	34	1	3	2	45	4	0	0	0	0	0	0	0	0	0	0	0	0	0	0	0	0	0	0	1	1	80	1	2	3	1	0	0	0	0
29	29	2	34	2	4	1	23	4	0	0	0	0	0	0	0	0	0	0	0	0	0	0	0	0	0	0	0	0	40	1	2	3	1	0	0	0	1
30	30	1	47	2	2	1	15	1	0	0	0	0	0	0	0	0	0	0	0	0	0	0	0	0	0	0	0	0	14	1	2	3	1	0	0	0	0
31	31	1	31	2	2	1	19	1	1	1	0	0	1	0	0	0	0	1	0	2	1	1	1	1	3	1	1	0	46	1	2	4	1	1	0	0	0
32	32	1	60	5	5	1	15	1	0	0	0	0	0	0	0	0	0	0	0	0	0	0	0	0	0	0	0	0	8	0	2	2	1	0	0	0	1
33	33	1	35	2	3	2	23	1	0	0	0	0	0	0	0	0	0	0	0	0	0	0	0	0	0	0	0	0	0	0	1	2	1	0	0	0	0
34	34	2	56	2	7	4	11	4	1	1	1	0	1	0	0	0	0	0	0	0	0	0	0	0	0	0	1	0	2	0	2	2	1	0	0	0	0
35	35	1	33	5	6	2	23	2	0	0	0	0	0	0	0	1	0	0	0	0	0	0	0	0	0	0	0	0	17	1	2	3	1	0	0	0	0
36	36	2	35	2	4	1	28	4	0	0	0	0	0	0	0	0	0	0	0	0	0	0	0	0	0	0	0	0	5	0	2	1	1	0	0	0	0
37	37	2	59	2	3	1	23	4	0	0	0	0	0	0	0	0	0	0	0	0	0	0	0	0	0	0	0	0	9	0	2	2	1	0	0	0	0
38	38	2	42	2	2	1	29	4	0	0	0	0	0	0	0	0	0	0	0	0	0	0	0	0	0	0	0	0	0	0	1	1	1	0	0	0	0
39	39	1	19	1	3	1	35	1	1	0	0	0	1	0	0	0	2	0	0	0	0	0	0	0	0	0	0	0	24	1	1	3	1	0	0	1	0
40	40	2	32	2	5	1	35	1	0	0	0	0	0	0	0	0	0	0	0	0	0	0	0	0	0	2	2	0	13	1	2	2	1	0	0	0	0
41	41	1	57	2	6	1	55	4	0	0	0	1	0	0	0	0	3	2	1	0	0	0	0	0	0	0	2	3	12	1	2	6	1	1	0	0	1

TABLE 3.3. Depression Data (*Continued*)

OBS	ID	SEX	AGE	MAR-ITAL	EDUC	EMPLOY	INCOME	REL-G	C1	C2	C3	C4	C5	C6	C7	C8	C9	C10	C11	C12	C13	C14	C15	C16	C17	C18	C19	C20	CESD	CASES	DRINK	HEALTH	REGDOC	TREAT	BEDDAYS	ACUTE-ILL	CHRON-ILL
50	50	2	30	3	4	7	15	4	0	0	1	2	1	2	0	0	0	1	1	0	1	0	1	1	0	0	0	0	9	0	2	3	1	2	1	0	1
51	51	1	77	3	2	4	5	4	0	1	0	1	0	0	0	0	1	0	0	0	0	0	1	0	1	0	0	0	5	0	1	2	1	2	0	0	1
52	52	2	24	1	4	2	25	2	0	0	0	3	0	0	0	0	1	2	0	1	1	0	0	0	0	0	1	0	10	0	2	2	1	2	0	0	0
53	53	2	31	3	4	4	7	2	0	1	1	0	0	0	0	0	0	0	0	0	1	0	0	1	0	1	0	0	8	0	1	3	1	2	0	0	0
54	54	1	81	2	2	4	23	1	0	0	0	0	0	0	0	0	0	0	0	0	0	0	0	0	0	0	0	0	0	0	1	2	1	2	0	0	0
55	55	2	66	2	2	4	6	1	0	0	0	0	0	2	0	0	0	0	0	0	0	0	0	0	0	0	1	0	8	0	2	2	1	2	0	0	1
56	56	2	26	1	5	4	20	1	0	1	3	1	1	1	0	1	3	0	3	0	0	0	0	0	0	0	0	0	4	0	1	3	1	1	0	0	0
57	57	2	48	2	2	2	23	1	0	1	0	0	0	0	0	2	2	0	0	0	0	0	0	0	0	0	0	0	0	0	1	2	1	2	0	0	0
58	58	1	55	2	3	4	15	1	0	0	0	0	0	0	0	0	0	0	0	0	0	0	0	0	0	0	0	0	0	0	1	4	1	2	0	0	0
59	59	2	59	4	3	4	28	3	0	0	0	0	0	0	0	0	0	0	0	1	0	0	0	0	0	0	1	0	14	0	2	2	1	2	0	0	0
60	60	2	20	5	3	4	6	2	0	1	3	0	1	1	0	2	2	0	0	0	0	0	0	0	0	0	0	0	40	1	1	4	2	2	0	1	0
61	61	1	68	5	3	4	35	1	0	0	1	0	1	0	0	1	0	0	0	0	0	0	0	0	0	1	0	0	8	0	1	2	1	1	0	1	0
62	62	1	55	1	3	4	23	1	0	0	0	0	0	0	0	0	0	0	0	0	0	0	0	0	0	0	2	1	18	0	2	2	1	1	0	0	0
63	63	2	34	2	4	1	9	1	0	0	0	0	0	2	0	0	0	0	0	0	0	0	0	0	0	0	0	0	26	1	2	2	1	2	0	0	0
64	64	2	58	2	2	1	28	1	3	2	3	0	1	1	0	0	0	0	0	0	0	0	0	0	0	0	0	0	12	0	1	4	1	2	0	0	0
65	65	1	60	4	5	3	11	3	0	1	1	1	3	0	0	1	0	1	0	0	0	0	0	0	0	0	0	0	8	0	1	2	1	2	1	1	0
66	66	2	34	3	3	4	18	2	1	1	0	0	0	0	0	0	1	1	1	0	0	0	0	0	0	0	0	0	7	0	2	3	2	2	0	1	1
67	67	1	68	5	2	4	45	3	0	0	0	1	0	0	0	1	1	0	1	0	1	0	0	0	0	0	0	0	9	0	1	2	1	2	0	0	0
68	68	2	70	5	4	3	13	3	0	0	0	0	0	0	0	0	0	0	0	0	0	0	0	0	0	0	0	0	5	0	1	3	1	2	0	0	1
69	69	2	24	1	4	3	7	1	0	1	3	0	2	0	0	0	0	0	0	0	0	0	0	0	0	0	0	0	0	0	2	3	1	1	0	0	0
70	70	1	32	2	7	4	17	1	0	0	1	0	0	2	0	0	0	0	0	0	0	0	0	0	0	0	0	0	21	1	1	1	1	2	0	0	0
71	71	2	26	1	4	4	12	2	0	0	0	0	0	0	0	0	0	0	0	0	0	0	0	0	0	0	0	0	16	0	2	3	1	2	0	0	0
72	72	2	65	5	4	3	12	3	0	0	0	0	0	0	3	0	3	1	0	0	0	0	0	0	0	0	0	0	3	0	2	2	1	2	0	0	0
73	73	1	71	3	3	4	65	1	2	2	0	0	0	0	0	0	0	1	0	0	0	0	0	0	0	0	0	0	1	0	1	2	1	2	0	0	1
74	74	2	20	5	3	4	1	1	0	0	0	0	0	0	0	0	0	0	0	0	0	0	0	0	0	0	0	0	7	0	1	1	1	2	0	0	1
75	75	1	70	1	2	3	11	3	0	0	0	2	0	0	0	0	1	0	0	0	0	0	0	0	0	0	0	0	42	1	1	3	1	2	0	0	1
76	76	2	83	2	3	4	7	1	0	0	0	0	0	0	0	0	0	0	0	0	0	0	0	0	0	0	0	0	18	0	2	4	1	2	0	0	0
77	77	1	67	3	2	3	13	1	1	1	2	2	0	3	0	2	3	2	2	0	0	0	0	0	0	0	0	0	17	0	2	2	1	2	1	0	0
78	78	1	24	2	2	3	28	1	1	1	1	1	1	0	0	0	0	0	0	0	0	0	0	0	0	0	0	0	5	0	1	2	1	2	0	0	0
79	79	2	71	2	6	4	19	1	0	0	0	0	2	0	0	0	0	0	0	0	0	0	0	0	0	0	0	0	50	1	3	2	1	1	0	0	0
80	80	2	66	5	3	1	2	3	0	0	0	0	0	0	2	2	3	2	1	0	0	0	0	0	0	0	0	0	26	1	2	3	1	2	0	0	0
81	81	2	59	4	5	5	8	1	1	0	0	1	0	0	0	0	1	0	0	0	0	0	0	0	0	0	0	0	8	0	1	2	1	2	0	0	0
82	82	1	48	3	3	5	19	3	0	0	1	0	0	0	0	0	0	0	0	0	0	0	0	0	0	0	0	0	7	0	2	2	1	1	0	0	0
83	83	2	65	2	3	4	29	1	0	1	0	0	0	0	0	2	0	0	0	0	0	0	0	0	0	0	0	0	4	0	1	2	1	1	0	0	0
84	84	2	43	5	3	4	22	3	0	0	0	0	0	0	0	0	0	0	0	0	0	0	0	0	0	0	0	0	5	0	1	2	1	2	0	0	1
85	85	2	38	4	6	3	9	3	0	0	0	0	0	0	0	0	0	0	0	0	0	0	0	0	0	0	0	0	4	0	2	2	1	2	0	0	0
86	86	1	22	3	5	4	8	1	0	0	0	0	0	0	0	0	0	0	0	0	0	0	0	0	0	0	0	0	13	0	3	3	1	1	0	0	1
87	87	2	23	1	7	3	18	3	0	0	0	0	1	0	0	0	1	0	0	0	0	0	0	0	0	0	0	0	6	0	1	2	1	1	0	0	0
88	88	1	59	1	3	4	15	1	1	1	0	0	0	0	0	0	0	0	0	0	0	0	0	0	0	0	0	0	3	0	2	2	1	2	0	1	0
89	89	2	65	2	2	5	19	1	0	0	0	0	1	1	0	0	0	0	0	0	0	0	0	0	0	0	0	0	9	0	1	1	1	2	0	0	1
90	90	2	21	3	2	2	65	3	0	1	0	0	0	0	0	0	1	0	0	0	0	0	0	0	0	0	0	0	3	0	1	3	1	1	0	0	0
91	91	2	82	5	4	2	8	1	0	0	0	0	0	3	0	0	0	0	0	1	0	0	0	0	0	0	0	0	8	0	2	2	1	2	0	0	0
92	92	2	54	1	5	4	7	2	1	1	0	0	0	1	0	0	0	0	0	0	0	0	0	0	0	0	0	0	9	0	2	2	1	1	0	0	1
93	93	2	64	5	2	1	8	1	0	0	0	0	1	0	0	1	1	1	1	0	0	0	0	0	0	0	0	0	2	0	1	2	1	2	0	0	1
94	94	2		4	4	4		3	0	1	0	0	0	0	0	0	0	0	0	0	0	0	0	0	0	0	0	0		0	1	3	1	2	0	0	0
95	95	2		5	2	2		1	0	0	0	0	0	0	0	0	0	0	0	0	0	0	0	0	0	0	0	0	12	0	2	2	1	2	0	0	1
96	96	2		5	4	1		1	1	0	0	0	0	0	0	0	0	0	0	0	0	0	0	0	0	0	0	0		0	1	2	1	1	0	0	1
97	97	2		4	2	2		1	0	0	1	0	0	0	0	1	1	0	0	0	0	0	0	0	0	0	0	0		0	1	3	1	2	0	0	0
98	98	1		2	3	2		2	3	0	0	1	1	0	0	0	0	0	0	0	0	0	0	0	0	0	0	0	15	0	1	2	1	1	0	0	0

TABLE 3.3. Depression Data (*Continued*)

OBS	ID	SEX	AGE	MARITAL	EDUC	EMPLOY	INCOME	RELIG	C1	C2	C3	C4	C5	C6	C7	C8	C9	C10	C11	C12	C13	C14	C15	C16	C17	C18	C19	C20	CESD	CASES	DRINK	HEALTH	REGDOC	TREAT	BEDDAYS	ACUTEILL	CHRONILL
99	99	1	72	5	2	4	11	1	0	2	3	0	2	0	2	0	3	0	2	0	0	3	0	0	2	0	3	0	25	1	1	2	1	1	0	0	0
100	100	1	52	2	5	4	45	3	0	0	3	0	2	2	0	0	0	0	0	1	0	0	0	1	0	0	0	5	0	1	1	1	2	0	0	1
101	101	2	21	1	5	2	19	3	1	0	0	0	1	0	0	1	0	0	0	0	0	0	0	1	1	0	0	6	0	1	1	1	2	0	0	1
102	102	1	37	3	3	2	8	1	0	1	0	1	2	0	0	1	0	1	0	0	0	0	0	0	0	0	0	8	0	1	3	2	2	0	0	1
103	103	2	60	2	6	2	65	4	0	0	2	0	1	0	0	0	0	0	0	0	0	0	0	1	1	1	0	17	0	1	2	1	2	0	0	1
104	104	1	32	3	3	2	35	4	1	1	0	2	2	1	0	0	0	1	0	0	0	0	0	0	1	0	0	15	1	2	3	1	2	1	0	0
105	105	2	27	2	5	6	13	3	1	0	0	0	0	3	0	0	0	1	0	0	0	0	0	1	0	2	1	16	0	1	3	2	2	0	0	0
106	106	1	28	1	3	2	7	1	0	1	2	0	0	0	0	0	0	0	0	0	0	0	0	0	0	0	0	4	0	1	2	1	2	0	0	0
107	107	1	24	4	6	1	18	4	1	1	0	1	1	1	1	1	0	0	0	0	1	0	0	0	1	1	0	13	0	1	2	1	1	0	1	0
108	108	2	20	1	5	4	23	3	0	0	0	1	0	0	0	0	0	1	0	0	0	0	0	0	0	0	0	8	0	1	2	2	2	0	1	0
109	109	2	18	1	4	1	27	1	0	0	2	0	1	0	0	0	0	1	0	0	0	0	0	0	0	1	0	6	0	1	2	2	2	1	0	1
110	110	1	24	1	2	2	9	3	0	1	0	1	1	1	0	0	1	0	1	0	0	0	0	0	0	0	0	27	1	1	3	1	2	0	0	1
111	111	2	79	2	3	5	15	2	0	0	0	0	0	0	0	0	0	0	0	1	0	0	0	0	0	0	0	1	0	1	2	1	2	0	1	1
112	112	1	50	1	3	1	15	3	1	2	0	0	2	3	0	0	2	1	3	0	0	0	0	3	0	2	3	37	1	1	3	1	2	0	0	1
113	113	2	51	3	5	1	8	2	0	0	0	1	0	0	0	0	0	1	0	0	0	0	0	0	0	0	0	7	0	1	1	1	2	0	0	1
114	114	1	50	1	2	1	9	3	0	0	0	0	1	3	3	0	0	0	0	0	0	0	0	0	0	0	0	23	1	1	3	1	2	0	0	0
115	115	2	52	1	7	4	65	3	0	0	0	1	0	0	0	0	0	0	0	0	0	0	0	0	3	0	0	6	0	1	1	2	1	0	0	0
116	116	2	52	2	5	5	19	2	0	0	3	0	0	0	0	0	3	0	3	0	0	0	0	0	0	0	0	6	0	1	3	2	2	0	0	1
117	117	2	28	3	5	1	24	3	0	1	2	0	1	3	0	2	1	2	2	0	3	0	0	3	0	0	0	26	1	1	3	1	2	0	1	1
118	118	1	70	5	5	3	7	3	0	0	0	0	0	0	0	0	0	0	0	0	0	0	0	0	0	1	0	12	0	1	1	1	2	0	0	0
119	119	2	62	2	5	4	15	4	0	0	0	0	0	1	0	0	0	0	0	0	0	0	0	0	1	0	0	6	0	1	1	1	2	0	0	0
120	120	2	83	2	2	3	9	3	0	0	0	0	0	0	0	0	0	3	0	3	0	0	0	0	0	0	0	39	2	1	3	1	2	0	0	0
121	121	1	46	2	3	4	19	1	0	0	0	2	0	2	0	0	0	2	0	0	0	0	0	0	0	0	0	29	0	1	2	1	1	0	0	0
122	122	2	18	1	5	5	2	4	0	0	3	0	0	1	0	0	0	1	1	0	0	0	0	3	0	0	0	19	0	1	3	1	2	0	0	1
123	123	2	25	5	2	1	44	2	0	0	0	0	0	0	0	0	0	0	1	2	0	0	0	0	0	0	0	13	0	1	3	1	2	0	0	1
124	124	2	78	5	2	1	9	3	0	0	3	0	0	0	0	0	0	0	3	0	0	0	0	0	0	0	0	9	0	1	3	2	2	0	0	0
125	125	1	71	2	3	1	7	3	1	0	0	0	0	0	0	0	0	0	0	0	0	0	0	0	0	0	0	6	0	1	3	1	2	0	0	1
126	126	2	73	1	4	3	51	2	0	0	3	2	0	3	0	0	0	1	0	0	0	0	0	0	0	0	0	21	0	1	3	2	2	0	0	0
127	127	2	60	3	4	2	34	4	0	0	0	2	0	2	0	0	0	0	0	0	0	0	0	0	0	0	0	29	0	1	2	1	2	0	1	1
128	128	1	42	1	4	4	9	2	0	0	2	0	2	0	0	0	3	1	0	0	0	0	0	1	3	0	0	7	0	1	2	2	2	0	0	1
129	129	1	60	5	6	2	51	3	0	2	3	0	0	0	0	0	0	0	0	0	0	0	0	0	0	0	0	1	0	1	1	1	2	0	0	1
130	130	2	51	1	3	5	17	2	0	0	3	0	0	0	3	0	0	1	0	0	0	0	0	0	0	0	0	4	0	1	3	1	2	0	1	1
131	131	1	78	3	5	4	8	3	0	0	3	0	0	0	0	0	0	0	0	0	0	0	0	0	0	0	0	3	0	1	3	1	2	0	0	1
132	132	1	34	3	3	1	13	2	0	0	3	2	0	0	0	0	0	0	3	0	0	0	0	0	0	0	0	27	1	1	2	1	2	0	0	1
133	133	2	33	3	4	4	35	4	0	0	2	0	0	2	0	0	0	0	0	0	0	0	0	0	0	0	0	20	0	1	3	2	2	0	0	1
134	134	2	48	2	5	5	15	4	0	0	0	0	2	0	0	0	3	3	2	0	0	0	0	0	2	0	0	1	0	1	2	1	2	0	1	1
135	135	2	22	2	3	1	45	3	0	0	0	0	0	0	0	0	0	0	0	0	0	0	0	0	0	0	0	38	1	1	2	1	2	0	0	1
136	136	1	34	3	6	1	42	3	0	0	0	0	0	0	0	0	0	0	0	0	0	0	0	2	0	0	0	13	0	1	3	1	2	0	1	1
137	137	2	29	1	5	1	29	2	0	0	3	0	0	0	3	0	0	0	0	0	0	0	0	0	2	3	3	2	0	1	2	2	2	0	0	0
138	138	1	26	1	5	1	9	2	1	0	0	1	0	0	0	2	0	1	0	0	0	0	0	0	0	0	0	1	0	1	2	1	1	0	0	1
139	139	1	24	1	5	4	11	4	0	0	0	0	2	0	0	0	0	2	0	0	0	0	0	0	0	0	0	9	0	1	2	1	1	0	0	0
140	140	1	27	1	4	2	45	2	0	0	0	0	3	0	0	0	0	0	0	0	0	0	0	0	3	0	0	2	0	1	2	1	2	0	0	1
141	141	2	40	2	6	5	20	4	2	3	2	1	2	0	0	0	1	1	1	0	0	0	0	1	0	3	3	0	0	1	3	1	2	0	0	0
142	142	2	32	2	2	1	28	2	0	1	1	0	3	1	0	0	1	0	0	0	0	0	0	0	0	0	0	0	0	1	2	1	2	0	0	0
143	143	2	33	2	3	1	35	4	0	0	3	0	2	0	0	0	0	0	0	0	0	0	0	0	0	0	3	0	0	1	2	1	2	0	1	0
144	144	1	53	2	4	1	28	2	1	0	0	0	0	0	0	0	0	0	0	0	0	0	0	0	0	0	0	0	0	1	2	1	1	0	1	1
145	145	1	31	3	3	1	35	2	0	3	3	0	3	0	0	0	0	2	0	0	0	0	0	0	0	0	0	0	0	1	2	1	1	0	1	0
146	146	1	53	2	4	5	11	2	1	1	0	0	1	3	0	0	0	0	1	0	0	0	0	3	0	0	0	12	0	1	2	1	2	0	0	0
147	147	2	19	2	3	1	11	1	1	2	0	1	1	3	0	0	0	1	0	0	0	0	0	0	0	0	0	19	0	1	2	1	1	1	1	1

TABLE 3.3. Depression Data (Continued)

OBS	ID	SEX	AGE	MARITAL	EDUC	EMPLOY	INCOME	RELIG	C1	C2	C3	C4	C5	C6	C7	C8	C9	C10	C11	C12	C13	C14	C15	C16	C17	C18	C19	C20	CESD	CASES	DRINK	HEALTH	REGDOC	TREAT	BEDDAYS	ACUTEILL	CHRONILL
148	148	2	23	1	4	1	9	2	0	0	0	0	0	2	0	1	1	0	0	0	1	0	0	0	0	0	1	0	5	0	1	1	1	2	0	1	0
149	149	1	27	3	3	1	19	2	0	0	0	0	1	1	0	0	1	0	1	1	1	1	1	0	0	0	0	0	9	0	0	2	1	2	0	0	0
150	150	1	57	4	3	1	28	1	0	0	0	0	1	1	0	0	0	1	0	1	0	2	1	1	1	1	1	1	28	1	0	3	2	2	0	0	0
151	151	2	26	3	4	2	27	2	2	0	0	1	0	0	0	0	0	0	0	1	1	1	1	1	1	0	1	0	18	0	0	2	1	2	0	0	1
152	152	1	31	5	5	1	13	1	0	0	0	1	0	0	0	0	0	1	0	0	1	1	1	1	0	0	0	0	1	0	0	2	2	2	0	0	1
153	153	2	74	1	3	4	5	1	0	0	0	0	0	0	0	0	0	0	1	0	1	1	1	1	0	0	0	0	0	0	0	3	2	2	0	0	1
154	154	2	24	1	5	1	8	2	0	0	0	0	0	0	0	1	0	0	0	0	0	1	1	1	1	0	0	0	4	0	0	2	1	2	0	1	0
155	155	2	21	1	3	3	11	1	0	0	0	0	0	0	0	0	0	0	0	1	0	1	1	1	0	0	0	0	3	0	0	3	1	2	0	0	0
156	156	1	22	1	3	1	19	1	0	0	0	0	0	0	0	0	0	0	1	0	1	1	1	1	1	0	0	0	2	0	0	2	1	2	0	1	1
157	157	2	22	2	6	1	16	1	0	0	0	0	0	0	0	0	0	0	0	0	1	1	1	1	1	1	0	0	3	0	0	3	1	2	0	1	0
158	158	2	28	2	5	1	35	1	0	0	0	0	0	0	0	0	0	0	0	0	1	1	1	1	1	0	0	0	4	0	0	2	1	2	0	1	0
159	159	2	49	2	6	2	32	1	0	0	0	0	0	0	0	0	0	0	0	0	1	1	1	1	0	0	0	0	5	0	0	3	1	2	0	0	1
160	160	2	47	2	5	4	65	2	0	0	0	0	0	0	0	0	0	0	0	0	1	1	1	1	0	0	0	0	1	0	0	2	1	2	0	0	1
161	161	2	32	2	7	2	42	1	0	0	0	0	0	0	0	0	0	0	0	0	1	1	1	1	0	0	0	0	1	0	0	2	1	2	0	0	1
162	162	2	36	3	3	1	6	2	0	0	0	0	0	0	0	0	0	0	0	0	1	1	1	1	1	0	0	0	9	0	0	2	1	2	0	0	0
163	163	2	45	2	4	1	36	1	0	0	0	0	0	0	0	0	0	0	0	0	1	1	1	1	0	0	0	0	0	0	0	3	1	2	0	0	0
164	164	1	43	2	7	4	45	2	0	0	0	0	0	0	0	0	0	0	0	0	1	1	1	1	0	0	0	0	0	0	0	2	1	2	0	1	0
165	165	1	22	1	6	1	55	3	0	0	0	0	0	0	0	0	0	0	0	0	1	1	1	1	0	0	0	0	0	0	0	2	1	2	0	0	0
166	166	1	20	2	5	2	19	3	0	0	0	0	0	0	0	0	0	0	0	0	1	1	1	1	0	0	0	0	3	0	0	2	1	2	0	0	0
167	167	2	36	3	3	1	18	1	0	1	2	1	1	1	2	1	1	1	1	1	1	2	2	2	1	1	1	3	0	0	3	1	2	0	0	0
168	168	1	59	1	2	4	9	1	0	0	0	0	0	0	0	0	0	0	0	0	1	1	1	1	0	0	2	0	0	0	0	3	1	2	0	0	0
169	169	1	57	2	2	4	6	1	0	0	0	0	0	1	0	0	0	1	0	1	1	1	1	0	0	0	3	6	0	0	3	1	2	0	0	1
170	170	2	23	1	4	1	23	1	0	0	0	2	0	0	0	0	0	1	0	1	1	1	1	0	1	0	6	0	0	0	2	1	2	0	1	0
171	171	1	74	1	3	1	7	1	0	0	0	2	0	0	0	0	0	1	0	2	2	2	2	2	0	0	20	0	0	0	2	2	2	0	1	0
172	172	1	29	1	3	4	13	1	0	0	0	0	1	0	0	0	0	1	0	1	1	1	2	1	1	2	27	1	0	3	1	2	0	0	1
173	173	2	65	1	2	1	15	1	0	1	0	0	0	0	1	0	0	0	0	1	1	1	1	1	0	0	7	0	0	3	1	2	0	0	0
174	174	1	50	5	3	4	5	1	0	0	0	2	0	0	0	0	0	1	0	1	1	1	1	0	0	0	15	0	0	3	2	2	0	0	1
175	175	2	51	3	3	2	9	1	0	0	0	3	2	2	0	0	0	0	0	1	1	2	1	1	0	0	33	1	0	3	1	2	0	1	0
176	176	2	70	2	2	1	19	1	0	0	0	0	0	0	0	0	0	0	0	1	1	1	1	0	0	0	8	0	0	3	2	2	0	0	1
177	177	2	57	2	3	2	6	1	0	0	3	0	0	2	1	0	3	1	0	1	1	2	3	1	1	1	1	0	0	2	1	2	0	0	1
178	178	2	45	1	2	2	23	1	0	0	0	0	0	0	0	0	2	1	0	1	1	2	3	1	1	1	16	1	1	3	2	2	0	0	0
179	179	2	36	1	4	1	73	2	0	0	2	0	0	0	0	1	0	0	0	1	1	1	1	0	0	0	16	1	0	3	1	2	0	1	0
180	180	1	25	1	5	4	15	1	1	1	3	1	3	0	0	3	3	2	3	2	2	3	3	3	1	1	39	1	1	4	1	2	1	0	0
181	181	1	58	2	3	2	19	2	0	0	0	0	0	0	0	0	0	0	0	1	1	1	1	0	0	0	9	0	0	3	2	2	0	0	1
182	182	1	59	1	4	3	6	2	0	0	2	2	0	2	0	1	0	0	0	1	1	1	1	0	0	0	8	0	0	3	1	2	0	0	0
183	183	2	26	1	3	1	7	1	1	1	0	0	0	0	0	0	0	0	0	1	1	1	1	0	0	0	3	0	0	2	1	2	0	0	0
184	184	2	21	1	2	1	23	1	0	0	0	0	0	1	1	1	0	0	0	1	1	1	1	0	0	0	2	0	0	2	1	2	0	0	0
185	185	2	83	5	3	4	7	1	1	0	2	0	3	2	0	3	0	1	0	0	0	1	0	0	0	1	2	0	0	4	1	2	0	0	0
186	186	2	34	1	4	1	13	1	1	0	1	0	0	2	1	0	1	0	0	1	1	2	2	1	1	2	1	0	0	2	1	2	0	0	0
187	187	2	42	1	5	4	15	1	0	1	2	0	3	1	0	3	3	0	0	0	0	3	3	0	3	2	39	1	0	3	1	2	0	0	1
188	188	2	42	3	2	1	9	1	0	0	0	0	0	0	0	0	0	0	0	0	1	1	1	1	0	0	9	0	0	2	1	2	0	0	0
189	189	2	24	1	3	1	19	1	0	0	0	0	0	0	0	0	0	0	0	0	1	1	1	1	0	0	0	8	0	0	2	1	2	0	0	1
190	190	2	80	5	2	4	6	1	0	0	0	0	0	0	0	0	0	0	0	0	0	1	1	1	0	0	0	0	0	0	2	2	2	0	0	0
191	191	2	18	1	2	3	2	2	0	1	2	0	0	0	0	0	0	0	0	0	0	1	1	1	0	0	0	3	0	0	2	1	2	0	0	1
192	192	2	58	5	3	2	45	1	0	0	0	0	0	0	0	0	0	0	0	0	0	1	1	1	0	0	0	2	0	0	3	1	2	0	0	1
193	193	2	54	2	2	5	4	1	1	0	1	0	0	0	0	1	0	0	0	0	0	1	1	1	0	0	0	2	0	0	2	1	2	0	0	1
194	194	1	47	3	2	4	24	1	0	1	0	0	0	0	0	0	0	0	0	0	0	1	1	1	0	0	0	1	0	0	3	1	2	0	0	0
195	195	1	61	2	2	6	—	1	0	1	0	0	0	0	0	0	0	0	0	0	0	1	1	1	0	0	0	1	0	0	3	1	2	0	0	0
196	196	1	18	1	2	6	—	2	0	0	0	0	0	0	0	0	0	0	0	0	0	1	1	1	0	0	0	5	0	1	2	1	2	1	1	0

TABLE 3.3. Depression Data (*Continued*)

OBS	ID	SEX	AGE	MARITAL	EDUC	EMPLOY	INCOME	REL-IG	C1	C2	C3	C4	C5	C6	C7	C8	C9	C10	C11	C12	C13	C14	C15	C16	C17	C18	C19	C20	CESD	CASES	DRINK	HEALTH	REGDOC	TREAT	BEDDAYS	ACUTE-ILL	CHRON-ILL
197	197	2	51	2	5	1	45	1	0	0	0	0	0	0	0	0	0	0	0	1	0	0	1	0	0	0	1	1	4	0	1	1	1	2	0	0	0
198	198	2	28	2	2	1	26	2	0	0	0	0	0	0	0	0	0	0	0	0	0	0	0	0	0	0	0	0	0	0	1	1	1	2	0	0	0
199	199	2	44	2	3	1	29	1	0	0	0	0	0	0	0	0	0	0	0	0	0	0	0	0	0	0	0	0	0	0	2	2	1	2	0	0	1
200	200	1	50	3	1	1	19	1	0	0	0	0	0	0	0	0	0	0	0	0	0	0	3	0	0	0	0	0	8	0	1	3	1	1	1	0	1
201	201	2	33	3	3	1	17	1	0	0	0	0	0	0	0	0	0	0	0	0	0	0	0	0	0	0	0	0	3	0	1	2	2	1	0	0	1
202	202	2	53	2	2	1	9	1	0	0	0	0	0	0	0	0	0	0	0	0	0	0	0	0	0	0	0	0	2	0	2	2	1	1	0	0	1
203	203	2	26	3	4	1	19	1	0	0	0	0	0	0	0	0	0	0	0	0	0	0	0	0	0	0	0	0	2	0	1	1	1	1	0	0	0
204	204	2	58	5	3	1	6	1	0	0	0	0	0	0	0	0	0	0	0	0	0	0	0	0	0	0	0	0	1	0	2	3	1	2	0	0	0
205	205	2	67	5	3	7	8	1	0	0	0	0	0	0	0	0	0	0	0	0	0	0	0	0	0	0	0	0	3	0	1	1	1	2	0	0	0
206	206	2	49	2	3	1	28	1	0	0	0	0	0	0	0	0	0	0	0	0	0	0	0	0	0	0	0	0	0	0	1	1	1	1	0	1	0
207	207	1	42	2	2	5	23	1	1	1	1	0	1	1	1	0	1	1	0	0	0	0	0	0	3	0	0	1	13	1	2	1	1	2	0	0	1
208	208	2	61	2	2	5	23	1	0	0	0	0	0	0	1	0	0	0	0	0	0	0	0	0	0	0	0	0	1	0	1	1	1	1	0	0	1
209	209	1	23	2	3	2	28	1	0	0	0	0	0	0	0	0	0	0	0	0	0	0	0	0	0	0	0	0	0	0	2	2	1	1	0	1	0
210	210	1	47	2	2	7	8	2	0	0	0	0	0	0	0	0	0	0	0	0	0	0	0	0	0	0	0	0	3	0	1	2	1	2	0	0	0
211	211	2	32	5	2	5	5	2	0	3	3	3	0	3	3	3	3	3	0	0	0	0	0	0	0	0	0	0	0	0	2	2	1	2	0	1	0
212	212	2	59	2	2	1	15	1	0	0	0	0	0	0	0	0	0	0	0	0	0	0	0	0	0	0	0	0	1	0	2	2	1	1	0	0	0
213	213	2	25	2	2	5	11	1	0	0	0	0	0	0	0	0	0	0	0	0	0	0	0	0	0	0	0	0	3	0	1	1	1	1	0	0	0
214	214	1	74	5	3	4	19	2	0	0	1	2	1	1	0	0	3	0	0	1	0	0	0	0	0	0	0	0	0	0	2	2	1	2	0	0	0
215	215	2	30	2	3	4	35	1	0	0	0	0	0	0	0	0	0	0	0	0	0	0	0	0	0	0	0	0	4	0	1	1	1	2	0	0	0
216	216	2	59	2	2	2	11	1	0	0	0	0	0	0	0	0	0	0	0	0	0	0	0	0	0	0	0	0	2	0	2	2	1	2	0	0	0
217	217	2	46	5	3	1	45	1	0	0	0	0	0	0	0	0	0	0	0	0	0	0	0	0	0	0	0	0	6	0	1	1	1	2	0	0	1
218	218	2	51	2	3	1	65	2	0	0	0	0	0	2	2	1	0	1	0	0	0	0	0	0	0	0	0	0	6	0	2	2	1	2	0	0	0
219	219	1	25	2	3	5	55	1	0	0	0	0	0	0	0	0	0	0	0	0	0	0	0	0	0	0	0	0	3	0	2	1	1	2	0	0	1
220	220	2	18	1	2	2	28	2	0	0	0	0	0	0	0	0	0	0	0	0	0	0	0	0	0	0	0	0	9	0	1	2	1	2	0	0	0
221	221	2	52	2	4	5	13	2	0	0	0	0	0	2	2	0	0	0	0	0	0	0	0	0	0	0	0	0	3	0	2	2	1	2	0	0	0
222	222	2	37	5	4	2	37	4	0	0	0	0	0	0	3	0	0	0	0	0	0	0	0	0	0	0	0	0	9	0	1	1	1	2	0	0	0
223	223	2	42	2	4	5	13	1	0	0	0	0	0	0	0	0	0	0	0	0	0	0	0	0	0	0	0	0	3	0	2	2	1	2	0	0	0
224	224	1	43	5	5	5	7	1	1	1	1	1	1	1	2	1	1	1	1	1	1	1	2	1	2	2	1	1	29	1	2	3	1	2	0	0	0
225	225	2	63	1	3	2	15	6	0	0	0	0	0	0	0	0	0	0	0	0	0	0	0	0	0	0	1	0	1	0	2	2	1	2	0	0	0
226	226	1	89	2	5	2	8	2	3	3	2	0	2	0	2	0	2	2	0	0	0	0	0	0	0	0	0	0	21	1	1	4	1	2	0	0	1
227	227	2	43	2	3	5	55	2	0	0	0	0	0	0	0	0	0	0	0	0	0	0	0	0	0	0	0	0	9	0	1	1	1	2	0	0	0
228	228	2	60	2	7	2	7	1	0	0	0	0	0	2	2	0	0	0	0	0	0	0	0	0	0	0	0	0	9	0	2	2	1	2	0	0	0
229	229	2	22	2	5	1	15	1	3	0	0	0	0	0	0	0	0	0	0	0	0	0	0	0	0	0	0	0	0	0	2	2	1	2	0	0	0
230	230	1	40	2	4	1	8	2	0	0	0	0	0	0	0	0	0	0	0	0	0	0	0	0	0	0	0	0	1	0	1	2	1	2	0	0	0
231	231	2	42	2	4	2	19	2	0	0	0	0	0	0	0	0	0	0	0	0	0	0	0	0	0	0	0	0	0	0	2	2	1	2	0	0	0
232	232	2	19	1	3	4	45	4	0	0	0	0	0	0	0	0	0	0	0	0	0	0	0	0	0	0	0	0	1	0	2	1	1	2	0	0	0
233	233	2	30	2	5	2	65	4	0	0	0	0	0	1	1	0	0	0	0	0	0	0	0	0	0	0	0	0	5	0	2	2	1	2	0	0	0
234	234	1	19	2	7	4	55	6	0	0	0	1	2	2	2	0	0	0	0	0	0	0	0	0	0	0	0	0	28	1	2	3	1	2	0	0	0
235	235	2	68	2	5	5	11	2	3	0	0	0	0	0	0	0	0	0	0	0	0	0	0	0	0	0	0	0	4	0	1	1	1	2	0	0	0
236	236	1	63	2	4	5	55	2	0	0	0	0	0	0	0	0	0	0	0	0	0	0	0	0	0	0	0	0	1	0	2	3	1	2	0	0	0
237	237	2	32	1	3	2	13	6	0	0	0	1	0	1	3	0	3	0	0	0	3	0	0	0	0	0	0	0	35	1	2	3	1	3	0	1	0
238	238	2	57	2	3	1	7	2	3	0	0	0	0	0	0	0	0	0	0	0	0	0	0	0	0	0	0	0	1	0	2	2	1	1	0	0	1
239	239	2	36	2	3	2	19	2	0	0	0	2	0	1	1	2	0	0	0	0	0	0	0	0	0	0	0	0	13	0	2	2	1	2	0	0	1
240	240	2	23	2	4	4	45	2	0	0	0	0	0	0	0	0	0	0	0	0	0	0	0	0	0	0	0	1	1	0	2	2	1	2	0	0	1
241	241	2	36	2	3	5	37	3	0	0	0	2	1	0	0	0	0	0	1	1	0	2	0	0	0	0	0	0	4	0	2	3	1	2	0	0	1
242	242	2	40	2	3	2	23	3	0	0	0	2	0	0	0	0	0	0	0	0	0	0	0	0	0	0	0	0	7	0	1	2	1	1	0	0	1
243	243	2	23	1	3	4	19	1	0	0	0	0	0	0	0	0	0	0	0	0	0	0	0	0	0	0	0	0	0	0	1	1	1	1	0	0	1
244	244	2	40	1	3	3	7	1	0	0	0	0	0	0	0	0	0	0	0	0	0	0	0	0	0	0	0	0	0	0	2	2	1	1	0	0	1
245	245	2	54	5	2	5	7	1	0	0	0	0	0	0	0	0	0	0	0	0	0	0	0	0	0	0	0	0	0	0	1	2	1	1	0	0	1

TABLE 3.3. Depression Data (*Continued*)

OBS	ID	SEX	AGE	MARITAL	EDUC	EMPLOY	INCOME	RELIG	C1	C2	C3	C4	C5	C6	C7	C8	C9	C10	C11	C12	C13	C14	C15	C16	C17	C18	C19	C20	CESD	CASES	DRINK	HEALTH	REGDOC	TREAT	BEDDAYS	ACUTEILL	CHRONILL
246	246	1	27	1	3	1	15	4	0	1	0	0	0	2	1	1	1	1	0	0	0	0	0	1	2	0	0	2	12	0	2	1	2	1	1	1	0
247	247	2	49	3	3	1	15	2	0	0	2	1	0	1	0	0	0	1	0	0	0	0	0	0	1	0	0	4	0	1	2	1	1	0	0	1
248	248	2	69	2	3	2	28	2	0	0	0	0	0	0	0	0	0	0	0	0	0	0	0	0	0	0	0	3	0	1	2	1	1	0	0	0
249	249	2	35	5	2	2	7	4	0	0	0	0	0	0	0	0	0	1	0	0	0	0	0	0	0	0	0	7	0	1	2	1	2	0	0	1
250	250	2	60	3	3	4	9	2	1	0	0	1	0	0	0	0	3	1	0	0	0	2	0	0	3	0	0	1	2	0	2	2	1	2	1	1	1
251	251	2	70	5	2	4	9	4	0	0	0	0	0	0	0	0	0	0	0	0	0	0	0	0	0	0	0	3	0	1	2	1	2	0	1	1
252	252	2	83	5	2	2	15	2	0	0	0	0	0	0	0	0	1	1	0	0	0	0	0	0	0	0	0	7	0	1	3	1	1	0	0	0
253	253	2	79	3	2	1	5	2	0	0	0	0	0	0	0	0	1	1	0	0	0	0	0	0	0	0	0	3	0	1	2	1	1	0	0	0
254	254	2	28	3	4	2	12	1	0	0	0	0	0	0	0	0	0	0	0	0	0	0	0	0	0	0	0	4	0	1	2	1	2	0	0	0
255	255	2	29	2	4	1	13	1	0	0	0	0	0	1	0	0	0	1	0	1	1	1	1	0	0	0	0	6	0	1	3	1	2	0	0	1
256	256	2	51	3	5	2	15	4	0	0	0	1	0	1	0	0	0	0	0	0	0	0	0	0	0	0	0	2	0	1	1	1	2	1	0	0
257	257	2	30	2	2	5	11	4	0	0	0	0	0	0	0	0	2	0	0	0	0	0	0	0	0	0	0	1	0	1	1	1	2	0	0	0
258	258	1	57	2	2	4	28	3	1	0	0	0	0	0	0	0	0	0	0	0	0	0	0	0	0	0	0	7	0	1	4	1	1	0	0	0
259	259	2	37	2	3	1	13	3	0	0	0	0	0	0	0	0	0	0	0	1	0	0	0	0	0	0	0	8	0	1	1	1	2	0	0	0
260	260	1	55	2	4	1	35	4	0	0	0	3	2	1	0	0	0	0	0	0	0	0	1	1	0	0	0	5	0	1	3	1	1	0	0	0
261	261	2	34	2	5	2	15	1	0	0	0	0	0	0	0	0	0	0	0	0	0	0	0	0	0	0	0	3	0	1	2	1	2	0	0	0
262	262	2	48	2	5	1	11	2	0	0	0	0	1	1	1	0	0	0	0	0	0	0	0	0	0	0	0	4	0	1	3	1	2	0	0	1
263	263	2	22	2	3	4	7	1	0	0	0	0	0	0	0	0	0	0	0	0	0	0	0	0	0	0	0	7	0	1	1	1	2	0	0	0
264	264	2	62	2	4	2	8	6	0	0	0	0	0	3	2	0	0	0	0	0	3	0	0	0	0	0	0	15	1	1	3	1	2	0	0	0
265	265	2	43	2	3	1	23	4	0	0	0	1	0	0	0	0	0	0	0	0	0	0	0	2	1	0	0	13	0	1	4	1	2	0	0	0
266	266	2	23	2	3	2	9	3	0	0	0	0	0	0	0	0	0	0	0	0	0	0	0	0	0	0	0	3	0	1	2	1	1	0	0	1
267	267	2	36	3	4	1	15	4	0	0	0	0	0	0	0	1	0	0	0	1	0	0	0	0	0	0	0	8	0	1	1	1	1	0	0	0
268	268	1	32	2	5	1	23	1	0	0	0	0	0	0	0	0	0	0	0	0	0	0	0	0	0	1	0	12	0	1	4	1	2	0	0	1
269	269	2	26	2	5	1	65	3	0	0	0	1	0	2	0	2	0	0	0	0	0	0	0	0	0	0	0	0	0	1	2	1	1	0	1	0
270	270	2	20	3	3	1	13	1	1	0	0	0	0	0	0	0	0	0	0	0	0	0	0	0	0	0	0	8	0	1	1	1	2	0	0	0
271	271	2	42	2	4	1	11	1	0	0	0	0	0	0	0	0	0	0	0	0	0	0	0	0	0	0	0	3	0	1	2	1	2	0	0	0
272	272	2	35	2	5	1	17	3	0	0	0	0	1	0	0	0	0	0	0	0	0	0	0	0	0	0	0	0	0	1	1	1	2	0	0	0
273	273	1	77	2	5	5	8	1	0	0	0	0	0	0	0	0	2	0	0	0	0	0	1	0	0	0	0	0	0	1	3	1	2	0	0	1
274	274	1	32	3	5	1	23	5	0	0	0	0	0	0	0	0	0	0	0	0	0	0	0	0	0	0	0	3	0	1	3	1	2	0	0	1
275	275	2	62	2	3	1	35	2	0	0	0	1	2	1	0	2	2	0	0	0	0	1	0	0	0	0	0	0	0	1	3	1	1	0	0	1
276	276	2	62	2	5	1	45	2	0	0	0	0	0	0	0	0	0	0	0	0	0	0	0	1	0	0	0	4	0	1	4	1	2	0	0	0
277	277	2	30	2	3	1	35	3	0	0	0	0	0	0	0	0	0	0	0	0	0	0	0	0	0	0	0	2	0	1	2	1	2	0	0	0
278	278	1	83	2	4	5	28	1	0	0	0	0	0	0	0	2	2	0	0	1	0	0	0	0	0	1	0	13	0	1	4	1	1	0	0	0
279	279	2	37	2	3	1	14	3	0	0	0	0	0	0	0	0	0	0	0	0	0	0	0	0	0	0	0	0	0	1	2	1	1	0	1	0
280	280	2	42	2	4	1	13	2	0	0	1	0	0	1	0	0	0	0	0	0	0	0	0	0	0	0	0	0	0	1	3	1	2	1	0	0
281	281	2	56	2	4	1	35	3	0	0	0	0	0	0	0	0	0	0	0	0	0	0	0	0	0	0	0	2	0	1	4	1	1	0	0	1
282	282	2	61	2	4	1	28	2	0	0	0	0	0	0	0	0	0	0	0	0	0	0	0	3	0	0	0	13	0	1	4	1	2	0	0	1
283	283	1	19	2	3	1	14	2	1	2	1	3	0	2	0	3	0	3	3	0	0	3	3	0	0	0	2	0	0	1	1	1	2	0	0	1
284	284	2	49	2	2	1	23	4	0	0	0	2	0	0	0	0	0	0	0	0	0	0	0	0	0	0	0	0	0	1	3	1	1	0	0	0
285	285	1	45	2	3	1	23	3	0	0	0	0	0	0	3	0	0	0	0	0	0	0	0	0	0	0	0	3	0	1	3	1	2	0	0	0
286	286	2	43	2	3	1	55	3	0	0	0	0	0	0	0	0	0	0	0	0	0	0	0	0	0	0	0	9	0	1	4	1	2	0	0	1
287	287	2	58	1	5	1	55	3	0	0	0	0	0	0	0	0	0	0	0	0	0	0	0	0	0	0	0	2	0	1	4	1	2	0	0	0
288	288	1	45		4	1	28	1	0	0	0	3	2	3	3	3	3	3	0	3	0	3	3	2	3	3	3	47	1	1	4	1	2	0	0	0
289	289	2	64	2	5	2	23	3	0	0	0	3	0	0	0	0	0	0	0	0	0	0	0	0	0	0	0	0	0	1	4	1	1	0	0	1
290	290	2	43	2	3	1	9	3	0	0	0	0	0	2	0	3	3	3	3	0	3	0	0	0	0	0	0	3	0	1	3	1	2	0	0	1
291	291	2	45	2	5	1	28	3	0	0	0	0	0	0	0	0	0	0	0	0	0	0	0	0	0	0	0	9	0	1	2	1	1	0	0	1
292	292	1	43	1	6	5	23	3	0	0	0	0	0	0	0	0	0	0	0	0	0	0	0	0	0	0	0	2	0	1	4	1	2	0	0	1
293	293	1	64	2	3	1	55	1	0	0	0	0	0	0	0	0	0	0	0	0	0	0	0	0	0	0	0	0	0	1	4	1	1	0	0	1
294	294	2	58	1	3	4	28	1	1	0	0	2	1	0	0	0	1	1	1	2	0	1	1	2	0	0	2	10	0	1	3	2	1	0	0	1

Initial Screening Using Frequencies

For any data set, frequency count programs provide a means of identifying certain types of errors. For example, a numerical value may be erroneously punched as a letter. In this case you can use an *alphanumeric* frequency count procedure, i.e., a procedure that allows variables whose values may be expressed as numbers or letters. This procedure enables you to detect such an error even if it is missed in examining the listing. Numerical values outside the possible range of the variable are also detectable in this manner. For example, we performed the SPSS–X procedure FREQUENCIES on the variables SEX and MARITAL of the depression data set by using the commands listed in Table 3.4. For illustrative purposes we deliberately added a case whose sex is coded as 3 and whose MARITAL is 8, both outside the permissible ranges. Note that this case increased the sample size to 295.

The functions to be performed by SPSS–X are specified by a series of statements called *commands* that are either actual 80-column punched cards or card images entered on a computer terminal. An example of commands for the SPSS-X program run is shown in Table 3.4.

First, *data definition commands* are supplied. They are used to describe the data set, its location on the cards or in the data file, and the number and the names of the variables. The DATA LIST command says that there are only two variables, sex and marital status. It also specifies that they are located in columns 9 and 12, respectively, as was shown in the code book example in Table 3.2. The VALUE LABELS command defines the coding used for each variable. A BEGIN DATA command, followed by the data, and an END DATA command are included because the data are entered as lines in the SPSS-X instructions.

The data definition commands are followed by a *task definition*, or procedure, command, which tells the computer what statistical analysis to run. In this example the FREQUENCIES procedure was run with the sex and marital variables. Most procedures offer one or more options called subcommands, as listed in the manual. In this case the histogram subcommand was used. Also, all statistics offered by the procedure were requested. This choice is a good one for some problems. However, if the data set is very large and computation costs are a consideration, only the desired statistics should be requested.

A FINISH command is used to tell the computer that everything has been read in.

TABLE 3.4. Examples of Commands for SPSS-X FREQUENCIES Program

```
TITLE SCREENING SEX AND MARITAL STATUS
DATA  LIST /SEX 9 MARITAL 12
VALUE  LABELS SEX 1 'MALE' 2 'FEMALE'/
              MARITAL 1 'NEVER MARRIED' 2 'MARRIED'
               3 'DIVORCED' 4 'SEPARATED' 5 'WIDOWED'
FREQUENCIES VARIABLE = SEX MARITAL
              /HISTOGRAM
        /STATISTICS  ALL
BEGIN DATA


{DATA}



END DATA
FINISH
```

Other computer programs such as SAS or BMDP also ask for the same information, but the form of the statements is often different. (Examples using SAS and BMDP are given elsewhere in this book.) The user always has to tell the computer where the data are and what to do.

Careful reading of the manuals is essential, because very minor errors in input can cause a program to give incorrect answers or not to run at all. The manual explanations are sometimes difficult to follow, but the examples are almost always helpful. If you are not familiar with a given package, taking a short course or finding a colleague to help in the early attempts may be useful.

Figure 3.2 shows part of the output for the variable MARITAL. This graphical presentation was obtained by specifying/HISTOGRAM in the procedure FREQUENCIES. Figure 3.2 indicates the number of observations corresponding to each value (code). As was shown in Table 3.2, the permissible codes for MARITAL are 1, 2, 3, 4, and 5. Therefore, it is immediately seen that the value 8 is an error. This error could also be noted from the fact that

FIGURE 3.2. SPSS–X Output of Depression Data Variable MARITAL with One Erroneous Observation Added with the Numerical Value of 8

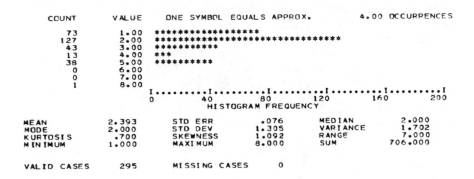

COUNT	VALUE	ONE SYMBOL EQUALS APPROX.			4.00 OCCURRENCES
73	1.00	******************			
127	2.00	********************************			
43	3.00	***********			
13	4.00	***			
38	5.00	**********			
0	6.00				
0	7.00				
1	8.00				

```
                    I.........I.........I.........I.........I.........I
                    0        40        80       120       160       200
                            HISTOGRAM FREQUENCY
```

MEAN	2.393	STD ERR	.076	MEDIAN	2.000
MODE	2.000	STD DEV	1.305	VARIANCE	1.702
KURTOSIS	.700	SKEWNESS	1.092	RANGE	7.000
MINIMUM	1.000	MAXIMUM	8.000	SUM	706.000

VALID CASES	295	MISSING CASES	0

the printed maximum of 8 is outside the acceptable limits. In other data sets the printed minimum and range may serve the same purpose. Note that for this nominal variable the only statistic appropriate for describing location is the mode, since there is no inherent order to the possible values. More on this subject will be given in Chapter 5.

The frequency table or histogram can also be useful for other purposes. For example, if it is noted that a particular code occurs only rarely, the investigator may wish to delete this code altogether or collapse it with other categories prior to future analyses.

Detection of Outliers

Outliers are observations that appear to be inconsistent with the remainder of the data set (Barnett and Lewis 1978). One method for the detection of outliers is to print out the maximum, minimum, and range of each variable. Some investigators will decide on reasonable maximum and minimum values on the basis of their knowledge of the variables and include this information in the code book, which ensures that different users of the data will exclude the same observations. It is also a sensible procedure when an investigator has the packaged programs run by a statistical clerk or an employee who is not familiar with the nature of the variables. If some observations exceed the maximum or minimum, then these observations are excluded. Programs such as BMDP1D can be used to print out the cases with outliers. If possible, outliers should be checked against the results recorded on the original data forms to see if some numerical or transcription error was made. Replacing an outlier with its correct value keeps that observation from being treated like a missing value.

Often observations are obtained that seem quite high or low but are not impossible. These values are the most difficult ones for most persons doing analyses to cope with. Should they be removed or not? Statisticians differ in their opinions, from the "if in doubt, throw it out" point of view to the view point that it is unethical to remove an outlier for fear of biasing the results. Using observations that are clearly in error seems unreasonable, but discarding observations without a convincing explanation can reduce the investigator's confidence in the results. The investigator may wish to eliminate gross outliers from the results but report them along with the statistical analysis. Another possibility is to run the analyses twice, both with the outliers and without them, to see if they make an appreciable difference in the results. Most investigators would hesitate to report that they rejected a null hypothesis if the removal of an outlier would result in the hypothesis not being rejected.

The presence of several outliers may indicate that the data are not normally distributed. In Chapter 4 methods are presented for detection of nonnormal distributions and possible transformations to use to obtain normal-appearing distributions from them.

A review of format tests for detection of outliers is given in Dunn and Clark (1974) and in Barnett and Lewis (1978). To make the formal tests, you must assume normality of the data. Some of the formal tests are known to be quite sensitive to nonnormality and should only be used when you are convinced that this assumption is reasonable. Often a level of alpha of 0.10 or 0.15 is used for the testing if it is suspected that outliers are not that unusual. Smaller values of alpha can be used if outliers are thought to be rare.

Each of the three statistical packages we consider in this book provides methods for excluding observations falling beyond specified limits. Similarly, observations equal to the missing value code could be excluded from the analysis. You should determine from the manual the exact way in which the program or procedure handles incomplete cases. For example, the program may offer the option of deleting a case in which any variable is missing. On the other hand, it may be possible to use the nonmissing values on one variable although the case has missing values on other variables. If such options are offered, it is advisable to carefully select the one you prefer.

Logical Relationships

Some data checking can also be based on inherent logical relationships among the variables in the data set. For example, if the data contain the two variables SEX (male, female) and PREGNANCY (yes, no), then a pregnant male would be an obvious error. The investigator is advised to search for such logical relationships and use them to detect possible errors. In this example a two-way frequency table will show the number of pregnant males. If that number is greater than zero, then those cases can be identified by special features of the package used. Transformations, discussed in Chapter 4, may also be helpful.

Once the cases with errors are identified, a decision must be made to either correct the errors, if possible, or delete the erroneous value(s) or the whole case. Most investigators prefer to check with the original records or respondent to replace the erroneous values. If the correct values are obtained, the investigator may repunch the wrong cards or reenter the data. In some cases it is preferable to keep the original data file intact and use transformations to replace the erroneous values (as shown Chapter 4). A new file containing the corrected values may be created for use in future analyses.

SUMMARY

Five packaged statistical programs were mentioned in this chapter. Three of these—BMDP, SAS, and SPSS–X—were described in some detail. They will be used extensively in the remainder of the book. Data coding and entry were discussed, with some practical advice given. An example of the use of a statistical package for outlier detection was given.

It is assumed that you will use a statistical package for performing the analyses. The preliminary steps discussed here can be time-consuming and may require assistance from other professionals. It is often helpful to ask a colleague to review what you have done before data entry, because it is often hard to find your own errors.

For investigators who do a great deal of data analysis, it may be advisable to obtain a specialized program for data screening. One such program is the NORC recode program available from the National Opinion Research Center at the University of Chicago.

It is also advisable to spend some time reading the manual of the selected package, including the section on data preparation. How variables are named and labeled varies slightly from one package to another, and it is sensible to use names that comply with the package you intend to use and that are easy to remember and type.

BIBLIOGRAPHY

Aneshensel, C. S.; Frerichs, R. R.; and Clark, V. A. 1981. Family roles and sex differences in depression. *Journal of Health and Social Behavior* 22:379–393.

Augenstein, M. J., and Tenenbaum, A. M. 1979. *Data structures and PL/1 programming*. Englewood Cliffs, N.J.: Prentice-Hall.

Barnett, V., and Lewis, T. 1978. *Outliers in statistical data*. New York: Wiley.

Berk, K. W., and Francis, I. S. 1978. A review of the manuals for BMDP and SPSS. *Journal of the American Statistical Association* 73:65–71.

Capron, H.L., and Williams, B.K. 1982. *Computers and data processing*. Menlo Park, Calif.: Benjamin/Cummings.

Comstock, G. W., and Helsing, K. J. 1976. Symptoms of depression in two communities. *Psychological Medicine* 6:551–563.

Dixon, W. J., ed. 1983. *BMDP statistical software 1983*. Berkeley: University of California Press.

Dunn, O. J., and Clark, V. A. 1974. *Applied statistics: Analysis of variance and regression*. New York: Wiley.

Flores, I. 1977. *Data structure and management*. 2nd ed. Englewood Cliffs, N.J.: Prentice-Hall.

Frerichs, R.R.; Aneshensel, C. S.: and Clark, V. A. 1981. Prevalence of depression in Los Angeles County. *American Journal of Epidemiology* 113:69–99.

Frerichs, R.; Aneshensel, C. S.; Clark, V. A.; and Yokopenic, P. 1981. Smoking and depression: A community survey. *American Journal of Public Health* 71:637–640.

Helwig, J. T., and Council, K. A., eds. 1979. *SAS user's guide*. Raleigh, N.C.: SAS Institute, Inc., P.O. Box 10066.

Hill, M. A. 1982. *BMDP user's digest*. 2nd ed. Los Angeles: BMDP Statistical Software, Inc., P.O. Box 24 A 26.

Hull, C. H., and Nie, N. H. 1981. *SPSS update 7–9*. New York: McGraw-Hill.

Muller, M. E. 1978. A review of the manuals for BMDP and SPSS (followed by comments by several authors). *Journal of the American Statistical Association* 73:71–80 (80–98).

Neffendorf, H. 1983. Statistical packages for microcomputers: A listing. *American Statistician* 37:83–86.

Nie, N. H.; Hull, C. H.; Jenkins, J. G.; Steinbrenner, K.; and Bent, D. H. 1975. *SPSS*. New York: McGraw-Hill.

———. 1983. *SPSS–X user's guide*. New York: McGraw-Hill.

Radloff, L. S. 1977. The CES–D scale: A self-report depression scale for research in the general population. *Applied Psychological Measurement* 1:385–401.

Ray, A. A., ed. 1982*a*. *SAS user's guide: Basics*. Cary, N.C.: SAS Institute, Inc., Box 8000.

———. 1982*b*. *SAS user's guide: Statistics*. Cary, N.C.: SAS Institute, Inc., Box 8000.

SAS. 1981. *SAS 79.5 changes and enhancements*. Technical Report P–115. Cary, N.C.: SAS Institute, Inc., Box 8000.

Schucany, W. R.; Shannon, B. S.; and Minton, P. D. 1972. A survey of statistical packages. *Computer Surveys* 4:65–79.

Tukey, J. W. 1977. *Exploratory data analysis*. Reading, Mass: Addison-Wesley.

PROBLEMS

3.1   Using a packaged program, compute a two-way table of income versus employment status from the depression data. From the data in this table, decide if there are any adults whose income is unusual considering their employment status. Are there any adults whose income is unusual considering their age or education?

3.2   List the programs from SAS that you would be likely to use to screen interval or ratio data. Repeat for SPSS and BMDP.

3.3   Describe the person in the depression data set who has the highest total CESD score.

3.4   Make a code book for a data set you have.

3.5   Using a packaged program, compute histograms for mothers' and fathers' heights and weights from the lung function data set given in Appendix B. Describe any cases that you consider potential outliers.

3.6   For the chronic respiratory disease study, use the information given in Section 1.2 and Appendix B to create a code book for the data set. If you were to create a more complete code book, what other information would you require?

3.7 For the data set given in Appendix B, use a packaged computer program to obtain a frequency count for the variables in columns 5, 19, and 33. Do you notice anything unusual?

3.8 For the data set given in Appendix B, use a packaged computer program to find out how many families have one child, two children, and three children between the ages of 7 and 18.

3.9 For the data set given in Appendix B, produce a two-way table of sex of child 1 versus sex of child 2 (for families with at least two children). Comment.

4 DATA SCREENING AND DATA TRANSFORMATION

4.1 WHAT WILL YOU LEARN FROM THIS CHAPTER?

From this chapter you will learn:

▸ How to combine information obtained in separate questions (4.2).

▸ How to exclude cases that have erroneous values (4.2).

▸ How to assess the effect of commonly used transformations (4.3).

▸ What a normal probability plot looks like for various distributions (4.4).

▸ How to assess whether data are normally distributed (4.4).

▸ How to transform data to data that are normally distributed (4.4).

▸ How to assess whether the data are independent (4.5).

Each computer package offers the user the capability to create new variables and modify or delete existing ones. In this chapter we discuss how these data modification operations can be used for further screening of the data and for preparing the data for analysis.

4.2 REDEFINITION OF DATA

In this section we describe how to use variable modification statements to replace certain values, to collapse categories of a variable, and to combine two or more variables into one.

Replacing Erroneous and Missing Values

In Chapter 3 we presented some methods for detecting errors in the data. To replace an erroneous value, the user may specify, for example, that the erroneous observation on some variable should be replaced with a particular value. This procedure would be a substitute for repunching a card or reentering the data on a terminal. Alternatively, the case may be identified by specifying the values of some variables. For instance, suppose that an individual has a value of age = 80 years, height = 74 in., and social security number = 123–45–6789. Suppose also that the correct value of age is 50 years. The computer may then be instructed to find the case whose age is 80 and replace it with 50. Note that if so instructed, the computer will replace *each* value of age that is 80 with 50. Thus the user should be sure that the case is identified uniquely so that additional errors may be avoided. Therefore it may be preferable in the above case to instruct the computer to find the case with the appropriate social security number and then do the correction. A case can be identified by specifying the values of more than one variable, and a list of cases fitting this description may be requested from the computer as a further check on the operations carried out.

In certain situations an investigator may wish to replace each missing value by another value. The technique described above can achieve this result. Packaged programs include special features to assist the user in identifying patterns of missing data and possibly performing special analyses for incomplete cases.

As will be discussed in later chapters, most multivariate analyses require complete data on the variables used in the analysis. To obtain such a data set, the user is advised to examine the patterns of missing and erroneous data in order to select the variables and the cases to be included. Most statisticians agree on the following guidelines:

> **1.** If a variable is missing in a high proportion of the cases, then that variable should be deleted.

> **2.** If a case is missing variables that you wish to analyze, then that case should be deleted.

Following these guidelines will give you a complete data set.

Packaged programs offer easy ways to exclude variables and cases with specified properties, such as a case with any missing observation. But you should be aware of some peculiarities of the statistical packages. For example, suppose that ten variables are read in the data set but only five of them are used in a certain analysis. Some programs would eliminate a case with a missing value on any of the original ten variables, even though it might be complete on the five variables used. This error can be avoided by specifying that only the five appropriate variables should be used from the data set. Further discussion of handling missing values in multivariate analysis is given in Section 10.2.

Collapsing Values

Another use of data modification statements is to collapse possible values or categories of a variable. Thus if a particular response in a questionnaire variable is rarely or never mentioned by a respondent, the investigator may wish to combine that response with another or possibly delete it as a response. Packages offer methods for achieving either of these possibilities. To this end, it is advisable to plan the order of the responses so that responses that may have to be collapsed are next to each other.

Collapsing can also be done for two or more variables. For example, suppose that variable 1 is sex (male, female) and question 2 is pregnancy (yes, no), to be answered only if the respondent is a female. Question 2 would thus be missing for all male respondents. It is possible to instruct the computer to create a *new* variable with the following values: 0 if male, 1 if female and not pregnant, 2 if female and pregnant. This new variable is then added to the list of variables and may be used in subsequent analyses. The same technique can be used in dealing with questionnaires constructed so that a particular response to one question may lead to a series of other questions.

Creating New Variables

The user may also create new variables that are arithmetic functions of existing variables. For instance, the user may modify all of the values of a variable by performing an arithmetic operation such as changing length from inches to centimeters or taking the square root or adding a constant. New variables can also be created as the result of arithmetic operations on two or more variables.

Each package summarizes the available transformations in its manual. The use of transformations to attempt to satisfy distributional assumptions will be presented in Section 4.4. In the next section the arithmetic relationships between functions of variables will be illustrated.

4.3 COMMON TRANSFORMATIONS

In the analysis of data it is often useful to transform certain variables before performing the analyses. Examples of such uses are found in the next section and in Chapter 6. In this section we present some of the most commonly used transformations. If you are familiar with this subject, you may wish to skip to the next section.

To develop a feel for transformations, let us examine a plot of transformed values versus the original values of the variable. To begin with, a plot of values of a variable X against itself produces a 45° diagonal line going through the origin, as shown in Figure 4.1a. One of the most commonly performed transformations is taking the *logarithm* (log) to the base 10. Recall that the logarithm Y of a number X satisfies the relationship $X = 10^Y$. That is, the logarithm of X is the power Y to which 10 must be raised in order to produce X. As shown in Figure 4.1b, the logarithm of 10 is 1 since $10 = 10^1$. Similarly, the logarithm of 1 is 0 since $1 = 10^0$, and the logarithm of 100 is 2 since $100 = 10^2$. Other values of logarithms can be obtained from tables of common logarithms or from a hand calculator with a log function. All statistical packages discussed in this book allow the user to make this transformation as well as others.

Note that an increase in X from 1 to 10 increases the logarithm from 0 to 1, that is, an increase of one unit. Similarly, an increase in X from 10 to 100 increases the logarithm also by one unit. For larger numbers it takes a great increase in X to produce a small increase in log X. Thus the logarithmic transformation has the effect of stretching small values of X and condensing large values of X. Note also that the logarithm of any number less than 1 is negative, and the logarithm of a value of X that is less than or equal to 0 is not defined. In practice, if negative or zero values of X are possible, the

FIGURE 4.1.   Plots of Selected Transformations of Variable X

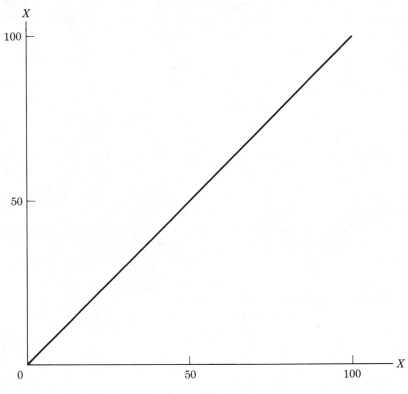

a. Plot of X Versus X

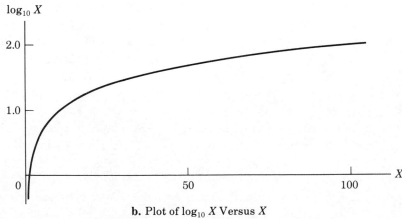

b. Plot of $\log_{10} X$ Versus X

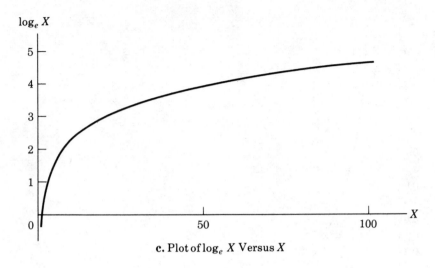

c. Plot of $\log_e X$ Versus X

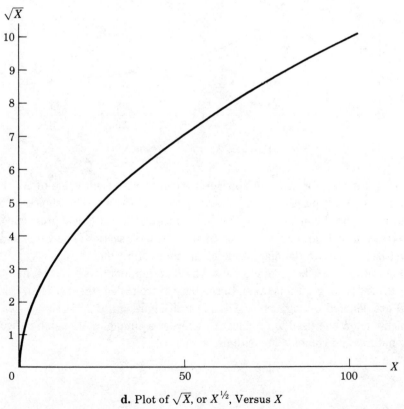

d. Plot of $\sqrt{X}$, or $X^{1/2}$, Versus X

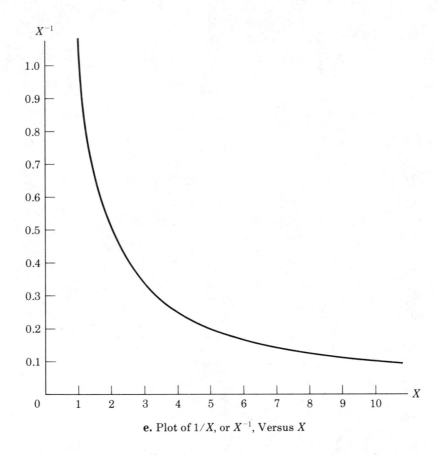

e. Plot of $1/X$, or X^{-1}, Versus X

investigator may first add an appropriate constant to each value of X, thus making them all positive prior to taking the logarithms. The choice of the additive constant can have an important effect on the statistical properties of the transformed variable, as will be seen in the next section. The value added must be larger than the magnitude of the minimum value of X.

Logarithms can be taken to any base. A familiar base is the number $e = 2.7183\ldots$ Logarithms taken to the base e are called *natural logarithms* and are denoted by $\log_e$ or ln. The natural logarithm of X is the power to which e must be raised to produce X. There is a simple relationship between the natural and common logarithms, namely,

$$\log_e X = 2.3026 \log_{10} X$$

$$\log_{10} X = 0.4343 \log_e X$$

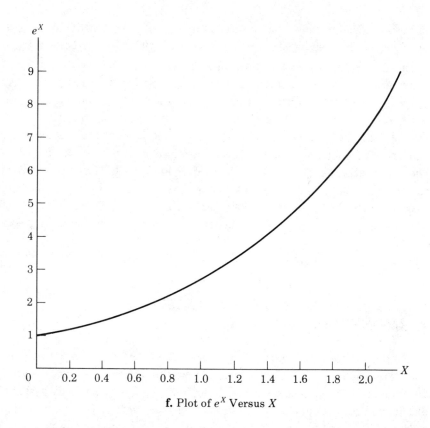

f. Plot of e^X Versus X

Figure 4.1c shows that a graph of $\log_e X$ versus X has the same shape as $\log_{10} X$, with the only difference being in the vertical scale; i.e., $\log_e X$ is larger than $\log_{10} X$ for $X > 1$ and smaller than $\log_{10} X$ for $X < 1$. The natural logarithm is used frequently in theoretical studies because of certain appealing mathematical properties.

Another class of transformations is known as *power transformations*. For example, the transformation X^2—i.e., X raised to the power of 2—is used frequently in statistical formulas such as computing the variance. The most commonly used power transformations are the square root—i.e., X raised to the power $\frac{1}{2}$—and the inverse $1/X$—i.e., X raised to the power −1. Figure 4.1d shows a plot of the square root of X versus X. Note that this function is also not defined for negative values of X. Compared with taking the logarithm, taking the square root also progressively condenses the values of X as X increases. However, the degree of condensation is not as severe as in the case of logarithms. That is, the square root of X tapers off slower than $\log X$, as can be seen by comparing Figure 4.1d with Figure 4.1b or 4.1c.

Unlike $\sqrt{X}$ and $\log X$, the function $1/X$ decreases with increasing X, as shown in Figure 4.1e. To obtain an increasing function, you may use $-1/X$.

Finally, exponential functions are also sometimes used in data analysis. An *exponential function* of X may be thought of as the antilogarithm of X. For example, the antilogarithm of X to the base 10 is 10 raised to the power X. Similarly, the antilogarithm of X to the base of e is e raised to the power X. The exponential function e^X is illustrated in Figure 4.1f. The function 10^X has the same shape but increases faster than e^X. Both have the opposite effect of taking logarithms; i.e., they increasingly stretch the larger values of X. Additional discussion of these interrelationships can be found in Tukey (1977).

4.4 ASSESSING THE NEED FOR AND SELECTING A TRANSFORMATION

In the theoretical development of statistical methods some assumptions are usually made regarding the distribution of the variables being analyzed. Often the form of the distribution is specified. The most commonly assumed distribution for continuous observations is the *normal*, or Gaussian, *distribution*. Although the assumption is sometimes not crucial to the validity of the results, some check of normality is advisable in the early stages of analysis. In this section methods for choosing a transformation to induce normality are presented.

Transformations to Induce Normality: Graphical Methods

Graphical methods present an appealing option for checking normality. Histograms for this purpose can be obtained from SAS, SPSS–X, and BMDP programs. If the population distribution is normal, the sample histogram should resemble the famous bell-shaped Gaussian curve. However, assessing the degree of normality or lack thereof is sometimes difficult to do by simply looking at the histograms, so a different graphical aid is required. A useful tool for this purpose is the *normal probability plot*. Such a plot can be obtained as the optional output of program BMDP5D, and from the SAS UNIVARIATE procedure using the PLOT option.

One axis of the probability plot shows values of X, and the other shows expected values, Z, of X if its distribution were exactly normal. The computation of these expected Z values is discussed in Johnson and Wichern (1982). Equivalently, this graph is a plot of the *cumulative distribution* found in the data set, with the vertical axis adjusted to produce a straight line if the data followed an exact normal distribution. Thus if the data were from a normal distribution, the normal probability plot should approximate a straight line, as shown in Figure 4.2a for a normal distribution with zero mean. In this graph values of the variable X are shown on the horizontal axis, and values of the expected normal values appear on the vertical axis. The Z values correspond to those usually listed in standard distribution tables.

The remainder of Figure 4.2 illustrates various other distributions with their corresponding normal probability plots. The purpose of these probability plots is to help you associate the appearance of the probability plot with the shape of the histogram or frequency distribution. Figures 4.2b and 4.2f both show symmetric distributions with more observations in the tails, i.e., heavier tails than those of the normal distribution. The corresponding normal probability plot is an inverted S. Taking the inverse of the data may produce an approximately normal distribution in this case (Figure 4.1f), although in other cases transformations are of little help (e.g., Figure 4.1b).

On the other hand, taking the logarithm of the data illustrated in Figure 4.2c would induce normality. The distribution before the transformation is called the *lognormal distribution*. This distribution is encountered in numerous real life applications. It can be recognized from its normal probability plot, whose curvature resembles a quarter of a circle. In this case we can create a new variable by taking logarithms to either base 10 or base e.

The investigator should request the normal probability plot of the transformed data. If the plot is still curved in the same direction, then subtracting a positive number that is smaller than the minimum X value before taking logarithms may straighten out the remaining curvature. You are encouraged to experiment with various constants to find the one producing the nearest plot to a straight line. If, on the other hand, taking logarithms produces a plot with the curvature in the opposite direction, as shown in Figure 4.2d, then adding a positive constant prior to taking logarithms may

FIGURE 4.2. Plots of Hypothetical Histograms and the Resulting Normal Probability
Plots from Those Distributions

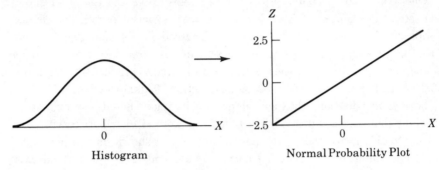

a. Normal Distribution

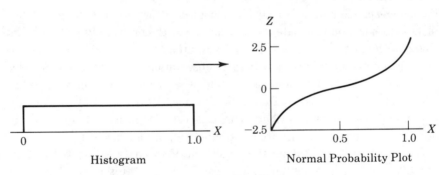

b. Rectangular Distribution

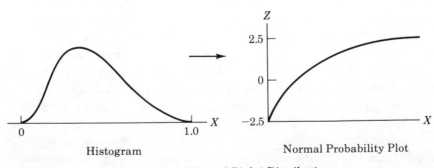

c. Lognormal (Skewed Right) Distribution

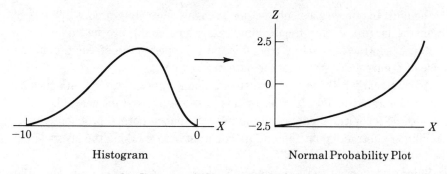

Histogram   Normal Probability Plot

d. −Lognormal (Skewed Left) Distribution

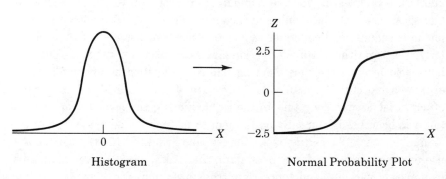

Histogram   Normal Probability Plot

e. 1/Normal Distribution

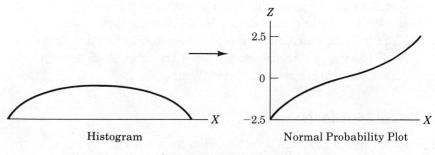

Histogram   Normal Probability Plot

f. High-Tail Distribution

be helpful. In some cases subtracting each X value from a constant larger than the largest X and then taking logarithms could also be helpful. For instance, subtracting each X from 0 is the appropriate transformation in the case of Figure 4.2d, where all the data are negative.

As noted earlier, the inverse transformation is appropriate for data such as those illustrated in Figure 4.2e. In this case the histogram has tails lower than those for normal data, resulting in an S-shaped normal probability plot. Adding or subtracting an appropriate constant before taking the inverse may be helpful in this case as well.

Some transformations are commonly used in certain situations. Thus the square root transformation is used with Poisson-distributed variables. Such variables represent counts of events occurring randomly in space or time with a small probability such that the occurrence of one event does not affect another. Examples include the number of cells per unit area on a slide of biological material, the number of incoming phone calls per hour in a telephone exchange, and counts of radioactive material per unit of time. Similarly, the logarithmic transformation has been used on variables such as household income, the time elapsing before a specified number of Poisson events have occurred, systolic blood pressure of older individuals, and many other variables with long tails to the right. Finally, the inverse transformation has been used frequently on the length of time it takes a person or an animal to complete a certain task. Bennett and Franklin (1954) illustrate the use of transformations for counted data, and Tukey (1977) gives numerous examples of variables and appropriate transformations for them.

It should be noted that not every distribution can be transformed to a normal distribution. For example, variable 29 in the depression data set is the sum of the results from the 20-item depression scale. The mode (the most commonly occurring score) is at zero, thus making it virtually impossible to transform these CESD scores to a normal distribution. However, if a distribution has a single mode in the interior of the distribution, then it is usually not too difficult to find a transformation producing a distribution that appears to be normal. Ideally, you should search for a transformation that has the proper scientific meaning in the context of your data.

Statistical Tests for Normality

It is also possible to do formal statistical tests for normality. For example, the SAS UNIVARIATE procedure will compute a Shapiro-Wilks W statistic if the number of observations is less than 50. This test is well suited to small sample sizes (see Dunn and Clark 1974). The null hypothesis of normality is rejected for small values of W. The program prints out the value of W and the associated probability (p value) for testing the hypothesis that the data came from a normal distribution. If the p value is small, then the data may not be normally distributed.

For larger N the program prints a modified version of the Kolmogorov-Smirnov D statistic and approximate p values. Here the null hypothesis is rejected for large values of D or a small p value. Afifi and Azen (1979) discuss the characteristics of this test.

The BMDP2D program computes the skewness of a variable along with its standard error. *Skewness* is a measure of how nonsymmetric a distribution is. If the data are normally or symmetrically distributed, then the computed skewness will be close to zero. The numerical value of the ratio of the skewness to its standard error can be compared with tabled normal Z values, and normality would be rejected if the value of the ratio is large. Positive skewness indicates that the distribution has a long tail to the right and probably looks like Figure 4.2c. (Note that it is important to remove outliers or erroneous observations prior to performing this test.) The SAS KSLTEST procedure also computes the Kolmogorov-Smirnov D statistic and the sample skewness along with a test statistic to test whether the data are normally distributed. Note that the SAS definition of skewness differs slightly from that found in most textbooks.

Very little is known about what levels of α should be chosen to compare with the p value obtained from formal tests of normality. So the sense of increased preciseness gained by performing a formal test over examining a plot is somewhat of an illusion. From the normal probability plot you can both decide whether the data are normally distributed and get a suggestion about what transformation to use.

Assessing the Need for a Transformation

In general, transformations are more effective in inducing normality when the standard deviation of the untransformed variable is large relative to the mean. If the standard deviation divided by the mean is less than $\frac{1}{4}$, then the transformation may not be necessary. In deciding whether to make a transformation, you may wish to perform the analysis with and without the proposed transformation. Examining the results will frequently convince you that the conclusions are not altered after making the transformation. In this case it is preferable to present the results in terms of the most easily interpretable units. And it is often helpful to conform to the customs of the particular field of investigation.

Sometimes, transformations are made to simplify later analyses rather than to approximate normal distributions. For example, it is known that FEV1 (forced expiratory volume in 1 s) and FVC (forced vital capacity) decrease in adults as they grow older. (See Section 1.3 for a discussion of these variables.) Some researchers will take the ratio FEV1/FVC and work with it because this ratio is less dependent on age. Using a variable that is independent of age can make analyses of a sample including adults of varying ages much simpler. In a practical sense, then, the researcher can use the transformation capabilities of the computer program packages to create new variables to be added to the set of initial variables rather than only modify and replace them.

If transformations alter the results, then you should select the transformation that makes the data conform as much as possible to the assumptions. If a particular transformation is selected, then all analyses should be performed on the transformed data, and the results should be presented in terms of the transformed values. Inferences and statements of results should reflect this fact.

4.5 ASSESSING INDEPENDENCE

Measurements on two or more variables collected from the *same* individual are not expected to be, nor are they assumed to be, independent of each other. On the other hand, independence of observations collected from *different* individuals or items is an assumption made in the theoretical derivation of most multivariate statistical analyses. This assumption is crucial to the validity of the results, and violating it may result in erroneous conclusions. Unfortunately, little is published about the quantitative effects of various degrees of nonindependence. Also, few practical methods exist for checking whether the assumption is valid.

In situations where the observations are collected from people, it is frequently safe to assume independence of observations collected from different individuals. Dependence could exist if a factor or factors exist to affect all of the individuals in a similar manner with respect to the variables being

FIGURE 4.3. Graphs of Hypothetical Data Sequences Showing Lack of Independence

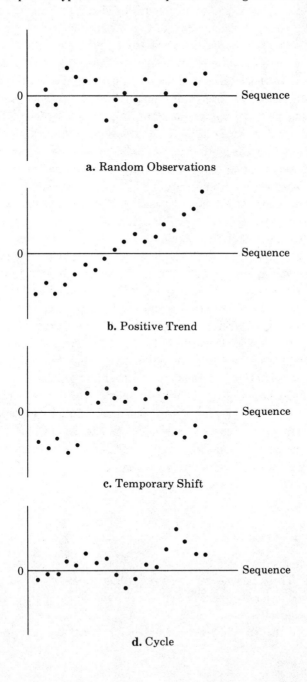

a. Random Observations

b. Positive Trend

c. Temporary Shift

d. Cycle

measured. For example, political attitudes of adult members of the same household cannot be expected to be independent. Inferences that assume independence are not valid when drawn from such responses. Similarly, biological data on twins or siblings are usually not independent.

Data collected in the form of a sequence either in time or in space can also be dependent. In those cases it is useful to plot the data in the appropriate sequence and search for trends or changes over time. This search may be done by using package scatter diagram programs such as BMDP6D, SAS PLOT, and SPSS–X SCATTERGRAM procedures. Values of the variable being considered appear on the vertical axis, and values of time, location, or observation identification number (ID) appear on the horizontal axis. If the plot resembles that shown in Figure 4.3a, little reason exists for suspecting lack of independence or nonrandomness. The data in that plot were, in fact, obtained as a sequence of random standard normal numbers. In contrast, Figure 4.3b shows data that exhibit a positive trend. Figure 4.3c is an example of a temporary shift in the level of the observations, followed by a return to the initial level. This result may occur, for example, in laboratory data when equipment is temporarily out of calibration or when a substitute technician, following different procedures, is used temporarily. Finally, a common situation occurring in some business or industrial data is the existence of seasonal cycles, as shown in Figure 4.3d.

Some formal tests for the randomness of a sequence of observations are given in Brownlee (1965) or Bennett and Franklin (1954). One such test is presented in Chapter 6 of this book in the context of regression and correlation analysis. If you are dealing with series of observations you may also wish to study the area of forecasting and time series analysis. Some books on this subject are included in the bibliography at the end of this chapter and in the bibliography following Chapter 6.

SUMMARY

In this chapter we showed that data screening involves several interrelated components. These include decisions regarding the necessity of transformations and how to handle missing values (also see Chapter 10). Screening the data for outliers (Sections 3.5 and 6.6) and deciding how to treat them is another chore that must be carried out before data analysis. Missing data, errors, and outliers may constitute a large portion of the data set, depending on the nature of the project. Another aspect of data screening is checking the assumptions underlying the analyses selected. In particular, we discussed in this chapter some methods for checking normality and independence or randomness.

No special order of these tasks is best for all situations. However, most investigators would delete obvious errors from the data before any other data screening is performed. They may then check the assumption of normality, attempt some transformation, and recheck the data for outliers. Other combinations of data-screening activities are usually dictated by the nature of the problem.

We wish to emphasize that data screening could be the most time-consuming and costly portion of data analysis. Investigators should not underestimate this aspect. If the data are not screened properly, much of the analysis may have to be repeated, resulting in an unnecessary waste of time and resources. Once the data have been carefully screened, the investigator will then be in a position to select the appropriate analysis to answer specific questions. In Chapter 5 we present a guide to the selection of the appropriate data analysis.

BIBLIOGRAPHY

Afifi, A. A., and Azen, S. P., 1979. *Statistical analysis: A computer oriented approach*, 2nd ed. New York: Academic Press.

Bennett, C. A., and Franklin, N. L., 1954. *Statistical analysis in chemistry and the chemical industry*, New York: Wiley.

*Box, G. E. P., and Jenkins, G. M., 1950. *Time series analysis: forecasting and control*, San Francisco: Holden-Day.

Brownlee, K. A., 1965. *Statistical theory and methodology in science and engineering*, 2nd ed. New York: Wiley.

Dixon, W. J., and Massey, F. J., 1983. *Introduction to statistical analysis*, 4th ed. New York: McGraw-Hill.

Dunn, O. J., and Clark, V. A., 1974. *Applied statistics: Analysis of variance and regression*, New York: Wiley.

Johnson, R. A., and Wichern, D. W., 1982. *Applied multivariate statistical analysis*, Englewood Cliffs, N.J.: Prentice-Hall.

McCleary, R., and Hay, R. A., 1980. *Applied time series analysis for the social sciences*, Beverly Hills: Sage.

Mage, D. T., 1982. An objective graphical method for testing normal distribution assumptions using probability plots, *American Statistician* 36:116–120.

*Montgomery, D. C., and Johnson, L. A., 1976. *Forecasting and time series analysis*, New York: McGraw-Hill.

Ostrom, C. W., 1978. *Time series analysis: Regression techniques*, Beverly Hills: Sage.

Tukey, J. W. 1977. *Exploratory data analysis*, Reading, Mass.: Addison-Wesley.

PROBLEMS

4.1   Combine the results from the following two questions into a single variable:
a. Have you been sick during the last two weeks?

Yes, go to (b)_____
No   _____

b. How many days were you sick? _____

4.2   Take the logarithm to the base 10 of the income variable in the depression data set. Compare the histogram of income with the histogram of log(income). Also, compare the normal probability plots of income and log(income).

4.3   Repeat Problem 4.2, taking the square root of income.

4.4   Generate a set of 100 random normal deviates by using a computer program package. Display a histogram and normal probability plot of these values. Square these numbers by using transformations. Compare histograms and normal probability plots of the logarithms and the square roots of the set of squared normal deviates.

4.5   Take the logarithm of the CESD score plus 1 and compare the histograms of CESD and log(CESD + 1). (A small constant must be added to CESD because CESD can be zero.)

4.6   The accompanying data are from the New York Stock Exchange Composite Index for the August 9 through September 17, 1982, period. Run a program to plot these data in order to assess the lack of independence of successive observations. (Note that the data cover five days per week, except for a holiday on Monday, September 6.) This time period saw an unusually rapid rise in stock prices (especially for August), coming after a protracted falling market. Compare this time series with prices for the current year.

Day	Month	Index
9	Aug.	59.3
10	Aug.	59.1
11	Aug.	60.0
12	Aug.	58.8
13	Aug.	59.5
16	Aug.	59.8
17	Aug.	62.4
18	Aug.	62.3
19	Aug.	62.6
20	Aug.	64.6
23	Aug.	66.4
24	Aug.	66.1
25	Aug.	67.4
26	Aug.	68.0
27	Aug.	67.2
30	Aug.	67.5
31	Aug.	68.5
1	Sept.	67.9
2	Sept.	69.0
3	Sept.	70.3
6	Sept.	—
7	Sept.	69.6
8	Sept.	70.0
9	Sept.	70.0
10	Sept.	69.4
13	Sept.	70.0
14	Sept.	70.6
15	Sept.	71.2
16	Sept.	71.0
17	Sept.	70.4

4.7 Obtain a normal probability plot of the data set given in Problem 4.6. Suppose that you had been ignorant of the lack of independence of these data and had treated them as if they were independent samples. Assess whether they are normally distributed.

4.8 Obtain normal probability plots of mothers' and fathers' weights from the lung function data set in Appendix B. Discuss whether or not you consider weight to be normally distributed in the population from which this sample is taken.

4.9 For the data set in Appendix B, create a new variable; AGEDIFF = (age of child 1) − (age of child 2), for families with at least two children. Produce a frequency count of this variable. Are there any negative values? Comment.

5 SELECTING APPROPRIATE ANALYSES

5.1 WHAT WILL YOU LEARN FROM THIS CHAPTER?

From this chapter you will learn how to use the information on independent and dependent variables and Stevens's classification system to assist you in selecting analyses for your data. In particular, you will learn:

▶ How to decide what descriptive measures should be used (5.3).
▶ How to decide which measure of central tendency and dispersion should be used (5.3).
▶ How to determine which multivariate analyses presented in Chapters 7–16 fit your data (5.4).

5.2 WHY SELECTION OF ANALYSES IS OFTEN DIFFICULT

There are two reasons why deciding what descriptive measures or analyses to perform and report is often difficult for an investigator with real life data. First, in statistics textbooks statistical methods are presented in a logical order from the viewpoint of learning statistics but not from the viewpoint of doing data analysis by using statistics. Most texts are either mathematical statistics texts or are imitations of them with the mathematics simplified or left out. Also, when learning statistics for the first time, the student often finds mastering the techniques themselves tough enough without worrying about how to use them in the future. The second reason is that real life data often contain mixtures of types of data, which makes the choice of analysis somewhat arbitrary. Two trained statisticians presented with the same set of data will often opt for different ways of analyzing the set, depending on what assumptions they are willing to take into account in the interpretation of the analysis.

Acquiring a sense of when it is safe to ignore assumptions is difficult both to learn and to teach. Here, for the most part, an empirical approach will be suggested. For example, it is often a good idea to perform several different analyses, one where all the assumptions are met and one where some are not, and compare the results. The idea is to use statistics to obtain insights into the data and to determine how the system under study works.

One point to keep in mind is that the examples presented in many statistics books are often ones the authors have selected after a long period of working with a particular technique. Thus they usually are "ideal" examples, designed to suit the technique being discussed. This feature makes learning the technique simpler but does not provide insight into its use in typical real life situations. In this book we will attempt to be more flexible than standard textbooks so that you will gain experience with commonly encountered difficulties.

In the next section suggested graphical and descriptive statistics are given for several types of data collected for analysis. Note, however, that these suggestions should not be applied rigidly in all situations. They are meant to be a framework for assisting the investigator in analyzing and reporting the data.

5.3 APPROPRIATE STATISTICAL MEASURES UNDER STEVENS'S CLASSIFICATION

In Chapter 2 Stevens's system of classifying variables into nominal (naming results), ordinal (determination of greater than or less than), interval (equal intervals between successive values of a variable), and ratio (equal intervals and a true zero point) was presented. In this section this system is used to obtain suggested descriptive measures. Table 5.1 shows appropriate graphical and computed measures for each type of variable. It is important to note that the descriptive measures are appropriate to that type of variable listed on the left *and to all below it*. Note also that Σ signifies addition, Π multiplication, and P percentile in Table 5.1.

TABLE 5.1. Descriptive Measures Depending Upon Stevens's Scale

Classification	Graphical Measures	Measures of the Center of a Distribution	Measures of the Variability of a Distribution
Nominal	Bar graphs Pie charts	Mode	Binomial or multinomial variance
Ordinal	Histogram	Median	Range $P_{75} - P_{25}$
Interval	Histogram with areas measurable	Mean = $\overline{X}$	Standard deviation = S
Ratio	Histogram with areas measurable	Geometric mean = $\left(\prod_{i=1}^{N} X_i \right)^{1/N}$ Harmonic mean = $N / \sum_{i=1}^{N} 1/X_i$	Coefficient of variation = $S/\overline{X}$

Measures for Nominal Data

For nominal data, the order of the numbers has no meaning. For example, in the depression data set the respondent's religion was coded 1 = Protestant, 2 = Catholic, 3 = Jewish, and 4 = Other. Any other four distinct numbers could be chosen, and their order could be changed without changing the empirical operation of equality. The measures used to describe this data should *not* imply a sense of order.

Suitable graphical measures for nominal data are bar graphs or pie charts. Both of these graphical measures are available from SPSS–X (BARCHART and PIECHART commands) and SAS (CHART procedure). These graphical measures will show the proportion of respondents who have each of the four responses to the religion question. The length of each bar represents the proportion for bar graphs, and the size of the piece of the pie represents the proportion for the pie charts.

The *mode*, or outcome of a variable that occurs most frequently, is the only appropriate measure of the center of the distribution. For a variable such as sex, where only two outcomes are available, the variability, or variance, of the proportion of cases who are male (female) can be measured by the *binomial variance*, which is

$$\text{Estimated Variance of } p = \frac{p(1-p)}{N}$$

where p is the proportion of respondents who are males (females). If more than two outcomes are possible, then the variance of the ith proportion is given by

$$\text{Estimated Variance of } p_i = \frac{p_i(1-p_i)}{N}$$

Measures for Ordinal Data

For ordinal variables order or ranking does have relevance, and so more descriptive measures are available. In addition to the pie charts and bar graphs used for nominal data, histograms can now be used. The area under the histogram still has *no* meaning because the intervals between successive numbers are not necessarily equal. For example, in the depression data set a general health question was asked and later coded 1 = excellent, 2 = good, 3 = fair, and 4 = poor. The distance between 1 and 2 is not necessarily equal to the distance between 3 and 4 when these numbers are used to identify answers to the health question.

An appropriate measure of the center of the distribution is the median. Roughly speaking, the *median* is the value of the variable that half the respondents exceed and half do not. The *range*, or largest minus smallest value occurring, is a measure of how variable or disperse the distribution is. Another measure that is sometimes reported is the difference between two *percentiles*. For example, sometimes the 5th percentile (P_5) will be subtracted from the 95th percentile (P_{95}). The 5th percentile is the value of the variable that divides the total sample such that 5% of the respondents are below and 95% are above this value. Some investigators prefer to report the *interquartile range*, $P_{75} - P_{25}$, or the *quartile deviation*, $(P_{75} - P_{25})/2$.

Histograms are available from SPSS–X, SAS, or BMDP. The BMDP2D computes the median, the range, and the quartile deviation. The UNIVARIATE procedure of SAS provides the median, the range, and percentiles P_{75}, P_{25}, and $P_{75} - P_{25}$, as well as six other percentiles. Any percentile can be obtained from the PCTL procedure. The SPSS–X CONDESCRIPTIVE provides the median and the range. The percentiles for a particular outcome can be obtained from the SPSS–X FREQUENCIES subprogram.

Measures for Interval Data

For interval data the full range of descriptive statistics generally used are available to the investigator. This set includes graphical measures such as histograms, with the area under the histogram now having meaning. The well-known *mean* and *standard deviation* can now be used also. These descriptive statistics are part of the output in most programs and hence are easily obtainable.

Measures for Ratio Data

The additional measures available for ratio data are seldom used. The *geometric mean* (GM) is sometimes used when the log transformation is used, since

$$\log \text{GM} = \frac{\sum\limits^{N} \log X}{N}$$

It is also used when computing the mean of a process where there is a constant rate of change. For example, suppose a rapidly growing city has a population of 2500 in 1970 and a population of 5000 according to the 1980 census. An estimate of the 1975 population (or halfway between 1970 and 1980) can be estimated as

$$\text{GM} = \left(\prod^{2} X_i\right)^{1/2} = (2500 \times 5000)^{1/2} = 3525$$

The *harmonic mean* (HM) is the reciprocal of the arithmetic mean of the reciprocals of the data. It is used for obtaining a mean of rates when the quantity in the numerator is fixed. For example, if an investigator wishes to analyze distance per unit of time that N cars require to run a fixed distance, then the harmonic mean should be used.

The *coefficient of variation* can be used to compare the variability of distributions that have different means. It is a unitless statistic.

The BMDP1S provides the geometric and the harmonic means, and BMDP1D prints out the coefficient of variation. In SAS the SUMMARY and the MEANS procedures, among others, provide the coefficient of variation. In SPSS–X, the harmonic mean is available in the ONEWAY program.

Stretching Assumptions

In the data analyses given in the next section, it is the ordinal variables that often cause confusion. Some statisticians treat them as if they were nominal data, often splitting them into two categories if they are dependent variables or using the dummy variables described in Section 10.3 if they are independent variables. Other statisticians treat them as if they were interval data. It is usually possible to assume that the underlying scale is continuous, and that because of a lack of a sophisticated measuring instrument, the investigator is not measuring with an interval scale.

The question really is, How far off is the ordinal scale from an interval scale? If it is close, then using an interval scale makes sense; otherwise, not. Further discussion on this point and a method for converting ordinal or rank variables to interval variables is given in Abelson and Tukey (1959 and 1963).

Although assumptions are sometimes stretched so that ordinal variables can be treated as if they were interval, this stretching should *never* be done with nominal data, because complete nonsense is likely to occur.

5.4 APPROPRIATE MULTIVARIATE ANALYSES UNDER STEVENS'S CLASSIFICATION

To decide on appropriate analyses, we must classify variables as follows:

1. Independent or dependent
2. Nominal, ordinal, interval, or ratio

The classification of independent or dependent may differ from analysis to analysis, but the classification into Stevens's system should remain constant throughout the analysis phase of the study. Once these classifications are determined, it is possible to refer to Table 5.2 and decide what analysis should be considered.

In Table 5.2 nominal and ordinal variables have been combined because this book does not cover analyses appropriate only to nominal or ordinal data separately. An inexpensive summary of measures and tests for these types of variables is given in Reynolds (1977) and Hildebrand, Laing, and Rosenthal

TABLE 5.2. Suggested Data Analysis Under Stevens's Classification

Dependent Variable(s)	Independent Variable(s)			
	Nominal or Ordinal		Interval or Ratio	
	1 Variable	>1 Variable	1 Variable	>1 Variable
No dependent variables	χ^2 goodness of fit	Measures of association Log-linear model χ^2 test of independence	Univariate statistics (e.g., one-sample t tests) Descriptive measures (5) Outliers (3) Tests for normality (4) Transformations (4) Independence (4)	Correlation matrix (7) Principal components (13) Factor analysis (14) Cluster analysis (16)
Nominal or Ordinal				
1 variable	χ^2 test Fisher's exact test	Log-linear Logistic regression (12)	Discriminant function (11) Logistic regression (12) Univariate statistics (e.g., two-sample t tests)	Discriminant function (11) Logistic regression (12)
>1 variable	Log-linear model	Log-linear model	Discriminant function (11)	Discriminant function (11)
Interval or Ratio				
1 variable	t test Analysis of variance	Analysis of variance Multiple-classification analysis	Linear regression (6) Nonlinear regression (9) Correlation (6)	Multiple regression (7, 8, 10) Nonlinear regression (9)
>1 variable	Multivariate analysis of variance Analysis of variance on principal components Hotelling's T^2 Profile analysis (16)	Multivariate analysis of variance Analysis of variance on principal components	Canonical correlation (15)	Canonical correlation (15) Path analysis (10) Structural models (LISREL)

(1977). Interval and ratio variables have also been combined because the same analyses are used for both types of variables. There are many measures of association and many statistical methods not listed in the table. For further information on choosing analyses appropriate to various data types, see Gage (1963) or Andrews et al. (1981).

The first row of Table 5.2 includes analyses that can be done if there are no dependent variables. Note that if there is only one variable, it can be considered either dependent or independent. A single independent variable that is either interval or ratio can be screened by methods given in Chapters 3 and 4, and descriptive statistics can be obtained from many statistical programs. If there are several interval or ratio independent variables, then several techniques are listed in the table.

In Table 5.2 the numbers in the parentheses following some techniques refer to the chapters of this book where those techniques are described. For example, to determine the advantages of doing a principal components analysis and how to obtain results and interpret them, you would consult Chapter 11. A very brief description of this technique is also given in Chapter 1. If no number in parentheses is given, then that technique is not discussed in this book.

For interval or ratio dependent variables and nominal or ordinal independent variables, analysis of variance is the appropriate technique. Analysis of variance is not discussed in this book; for discussions of this topic, see Winer (1971), Dunn and Clark (1974), or Box, Hunter, and Hunter (1978). Multivariate analysis of variance and Hotelling's T^2 are discussed in Afifi and Azen (1979) and Morrison (1976). Discussion of multiple-classification analysis can be found in Andrews, Morgan, and Sonquist (1969). Structure models are discussed in Duncan (1975), and log-linear models are presented in Upton (1978).

Table 5.2 provides a general guide for what analyses should be *considered*. We do not mean that other analyses couldn't be done but simply that the usual analyses are the ones that are listed. For example, methods of performing discriminant function analyses have been studied for noninterval variables, but this technique was originally derived with interval or ratio data; and the available programs in BMDP, SAS, and SPSS–X are written for interval or ratio data.

Judgment will be called for when the investigator has, for example, five independent variables, three of which are interval, while one is ordinal and one is nominal, with one dependent variable that is interval. Most investigators would use multiple regression, as indicated in Table 5.2. They might pretend that the one ordinal variable is interval and use dummy variables for the nominal variable (see Chapter 10). Another possibility is to categorize all the independent variables and to perform an analysis of variance on the data. That is, analyses that require less assumptions in terms of types of variables can always be done. Sometimes, both analyses are done and the results are compared. Because the packaged programs are so simple to run, multiple analyses are a realistic option.

In the examples given in Chapters 6 through 16, the data used will often not be ideal. In some chapters a data set has been created that fits all the usual assumptions to explain the technique, but then a nonideal, real life data set is also run and analyzed. It should be noted that when inappropriate variables are used in a statistical analysis, the association between the statistical results and the real life situation is weakened. However, the statistical models do not have to fit perfectly in order for the investigator to obtain useful information from them.

SUMMARY

The Stevens system of classification of variables can be a helpful tool in deciding on the choice of descriptive measures as well as in sophisticated data analyses. In this chapter we presented a table to assist the investigator in each of these two areas. A beginning data analyst may benefit from practicing the advice given in this chapter and from consulting more experienced researchers.

The recommended analyses are intended as general guidelines and are by no means exclusive. It is a good idea to try more than one way of analyzing the data whenever possible. Also, special situations may require specific analyses, perhaps ones not covered thoroughly in this book. Some investigators may wish to consult the detailed recommendations given in Andrews et al. (1981).

BIBLIOGRAPHY

Abelson, R. P., and Tukey, J. W. 1959. Efficient conversion of nonmetric information into metric information. *Proceedings of the Social Statistics Section, American Statistical Association*: 226–230.

*———. 1963. Efficient utilization of non-numerical information in quantitative analysis: General theory and the case of simple order. *Annals of Mathematical Statistics* 34:1347–1369.

Afifi, A. A., and Azen, S. P. 1979. *Statistical analysis: A computer oriented approach*. 2nd ed., New York: Academic Press.

Andrews, F. M.; Klem, L.; Davidson, T. N.; O'Malley, P. M.; and Rodgers, W. L. 1981. *A guide for selecting statistical techniques for analyzing social sciences data*, 2nd ed. Ann Arbor: Institute for Social Research, University of Michigan.

Andrews, R.; Morgan, J.; and Sonquist, J. 1969. *Multiple classification analysis*. Ann Arbor: Institute for Social Research, University of Michigan.

Box, G. E. P.; Hunter, W. G.; and Hunter, J. S. 1978. *Statistics for experimenters*. New York: Wiley.

*Duncan, O. D. 1975. *Introduction to structural equation models*. New York: Academic Press.

Dunn, O. J., and Clark, V. A. 1974. *Applied statistics: Analysis of variance and regression*. New York: Wiley.

Gage, N. L., ed. 1963. *Handbook of research on teaching*. Chicago: Rand McNally.

Hildebrand, D. K.; Laing, J. D.; and Rosenthal, H. 1977. *Analysis of ordinal data*. Beverly Hills: Sage.

*Joreskog, K. G., and Sorbom, D. 1978. *LISREL IV: Analysis of linear structural relationships by the method of maximum likelihood*. Chicago: National Resources.

*Morrison, D. F. 1975. *Multivariate statistical methods*. New York: McGraw-Hill.

Reynolds, H. T. 1977. *Analysis of nominal data*. Beverly Hills: Sage.

Stevens, S. S. 1951. *Handbook of experimental psychology*. New York: Wiley.

Upton, G. J. G. 1978. *The analysis of cross-tabulated data*. New York: Wiley.

Winer, B. J. 1971. *Statistical principles in experimental design*. 2nd ed. New York: McGraw-Hill.

PROBLEMS

5.1 Compute an appropriate measure of the center of the distribution for the following variables from the depression data set: MARITAL, INCOME, AGE, and HEALTH.

5.2 An investigator is attempting to determine the health effects on families of living in crowded urban apartments. Several characteristics of the apartments have been measured, including square feet of living area per person, cleanliness, and age of the apartment. Several illness characteristics for the families have been measured also, such as number of infectious diseases and number of bed days per month for each child, and overall health rating for the mother. Suggest an analysis to use with these data.

5.3   A coach has made numerous measurements on successful basketball players, such as height, weight, and strength. He also knows which position each player is successful at. He would like to obtain a function from these data that would predict which position a new player would be best at. Suggest an analysis to use with these data.

5.4   A college admissions committee wishes to predict which prospective student will successfully graduate. To do so, the committee intends to obtain the college grade point averages for a sample of college seniors and compare these with their high school grade point averages and Scholastic Aptitude Test scores. Which analysis should the committee use?

5.5   Data on men and women who have died have been obtained from health maintenance organization records. These data include age at death, height, weight, and several physiological and life-style measurements such as blood pressure, smoking status, dietary intake, and usual amount of exercise. The immediate and underlying causes of death are also available. From these data we would like to find out which variables predict death due to various underlying causes. (This procedure is known as *risk factor analysis*.) Suggest possible analyses.

5.6   Large amounts of data are available from the United Nations and other international organizations on each country and sovereign state of the world, including health, education, and commercial data. An economist would like to invent a descriptive system for the degree of development of each country on the basis of these data. Suggest possible analyses.

5.7   For the data described in Problem 5.6 we wish to put together similar countries into groups. Suggest possible analyses.

5.8   For the data described in Problem 5.6 we wish to relate health data such as infant mortality (the proportion of children dying before the age of one year) and life expectancy (the expected age at death of a person born today if the death rates remain unchanged) to other data such as gross national product per capita, percentage of people older than 15 who can read and write (literacy), average daily caloric intake per capita, average energy consumption per year per capita, and number of persons per practicing physician. Suggest possible analyses. What other variables would you include in your analysis?

Two

APPLIED REGRESSION ANALYSIS

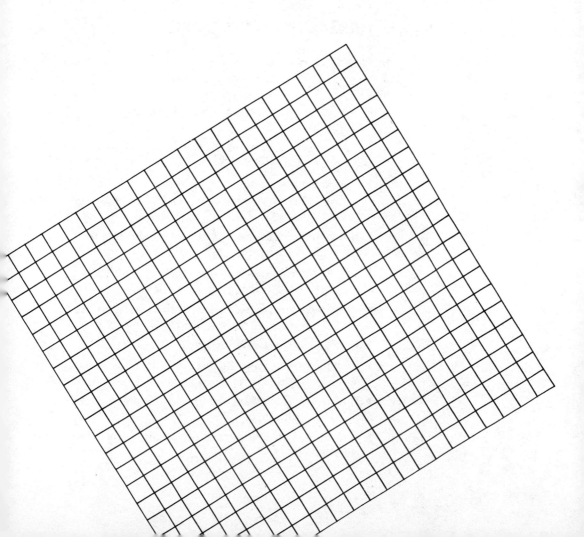

6 SIMPLE LINEAR REGRESSION AND CORRELATION

6.1 WHAT WILL YOU LEARN FROM THIS CHAPTER?

From this chapter you will learn how to examine the relationship between two variables based on a sample. In particular, you will learn:

▶ Why and where regression and correlation are used (6.2, 6.3).

▶ The correct meaning of the words *regression* and *correlation* (6.4, 6.5).

▶ How to obtain the appropriate regression equation and statistics that explain the relationship between the variables (6.4, 6.5).

▶ How to predict the value of one variable from the value of another (6.4, 6.5).

▶ How to test appropriate hypotheses regarding the strength of the relationship (6.6, 6.7).

▶ How to perform a residual analysis in order to validate or improve the derived equation (6.8).

▶ How and when to use weighted regression (6.9).

▶ How to correctly use the technique of regression in special contexts such as calibration (6.10).

▶ How to use regression analysis to analyze and explain paired data (6.10).

▶ How and when to use transformations to improve the equation (6.11).

▶ Where to go from this chapter (6.11).

▶ Which are the appropriate computational methods or programs to obtain the numerical results (6.12, 6.13).

If you are reading about simple linear regression for the first time, skip Sections 6.9, 6.10, and 6.11 in your first reading. If this chapter is a review for you, you can skim most of it, but read the above-mentioned sections in detail.

6.2 WHEN ARE REGRESSION AND CORRELATION USED?

The methods described in this chapter are appropriate for studying the relationship between two variables X and Y. By convention, X is called the *independent variable* and is plotted on the horizontal axis. The variable Y is called the *dependent variable* and is plotted on the vertical axis.

The data for regression analysis can arise in two forms:

1. *Fixed-X case*: The values of X are selected by the researchers or forced on them by the nature of the situation. For example, in the problem of predicting the sales for a company, the total sales are given for each year. Year is the fixed X variable, and its values are imposed on the investigator by nature. In an experiment to determine the growth of a plant as a function of temperature, a researcher could randomly assign plants to three different preset temperatures that are maintained in three greenhouses. The three temperature values then become the fixed values for X.

2. *Variable-X case*: The values of X and Y are both random variables. In this situation, cases are selected randomly from the population, and both X and Y are measured. All survey data are of this type, whereby individuals are chosen and various characteristics are measured on each.

Regression and correlation analysis can be used for either of two main purposes:

1. *Descriptive*: The kind of relationship and its strength are examined. This examination can be done graphically or by the use of descriptive equations. Tests of hypotheses and confidence intervals can serve to draw inferences regarding the relationship.

2. *Predictive*: The equation relating Y and X can be used to predict the value of Y for a given value of X. Prediction intervals can also be used to indicate a likely range of the predicted value of Y.

6.3 DATA EXAMPLE

In this section we present an example that we use in the remainder of the chapter to illustrate the methods of regression and correlation.

Data were obtained from a sample of factory workers in a particular industry. (Since these are proprietary data, no further identification will be made of their source. The data set will be referred to as the factory workers data.) In these factories management was concerned about possible effects of the working environment on respiratory function.

One of the major early indicators of reduced respiratory function is FEV1, or forced expiration volume in the first second (amount of air exhaled in 1 s). It is known that taller males tend to have a higher FEV1, and we wished to determine the relationship between height and FEV1. A particular age group was chosen (20–24 years old) since it is also known that FEV1 decreases with age in adults. We also excluded females since different relationships exist for women and most of the workers were males. We thus obtained a sample of 87 cases. These data belong in the variable-X case, where X is height (in inches) and Y is FEV1 (in liters). Here we may be concerned with describing the relationship between FEV1 and height, a descriptive purpose. We may also use the resulting equation to determine expected or normal FEV1 for a given height, a predictive use.

In Figure 6.1 a scatter diagram of the data is reproduced as printed by the computer program BMDP6D. In this graph the horizontal axis is divided into 70 units, and the vertical axis into 45 units. The actual values of the observations are rounded off to fit into this grid. For that reason more observations appear to coincide than is actually the case. The digits on the graph represent

FIGURE 6.1. Regression Equation of FEV1 Versus Height for Factory Workers, Age 20–24 Years Old

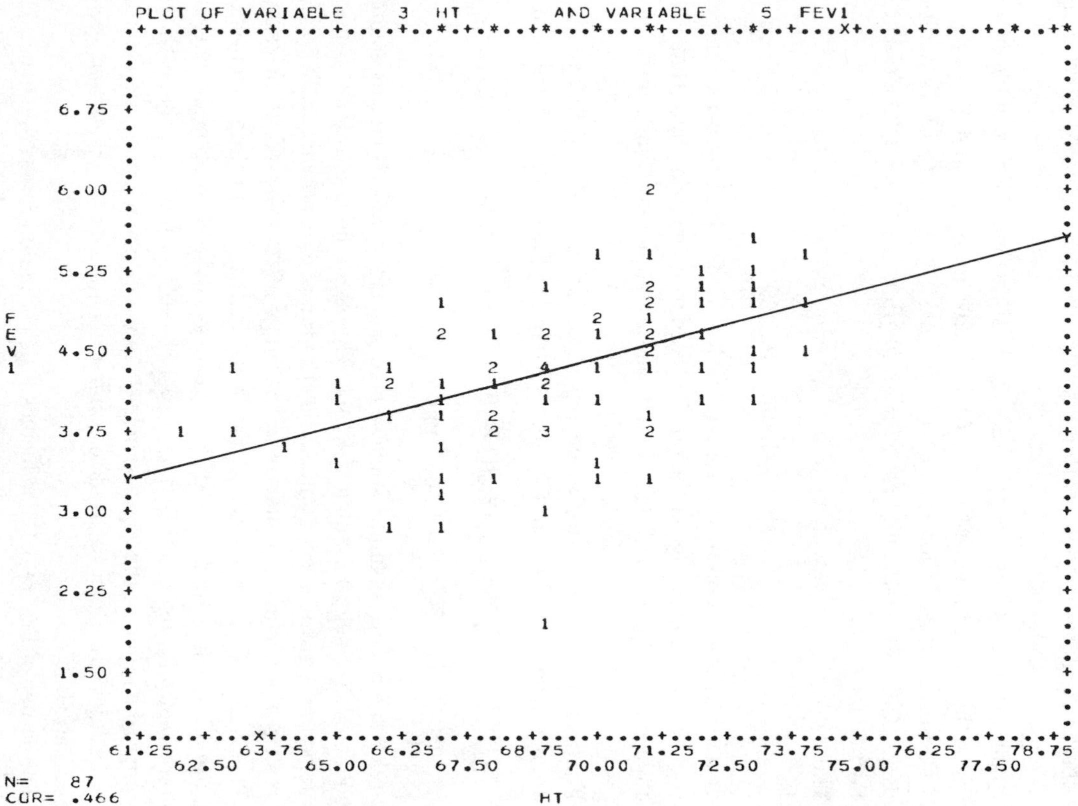

PLOT OF VARIABLE   3 HT   AND VARIABLE   5 FEV1

```
6.75 +
6.00 +                                    2
5.25 +                    1        1      1
                          1   1 1  1      1 1
4.50 +              1    2 2 1 2 2 2 1
                  2  1 2   2 1 1  1    1
                1  2   1 4 1 2  3          1
3.75 +        1 2  1 1 1 1 2 2  1 1
            1 1 1  1 1 1 1  1
          1 1 1  1       1
3.00 +  Y                              1
2.25 +
1.50 +
      +....+....+....+....+....+....+....+....+....+....+....+....+....+....+
      61.25 62.50 63.75 65.00 66.25 67.50 68.75 70.00 71.25 72.50 73.75 75.00 76.25 77.50 78.75
FEV1                                   HT

N=  87
COR= .466

            MEAN      ST.DEV.     REGRESSION LINE      RES.MS.
X         69.264     2.6390      X= 1.7276*Y+ 61.830    5.5174
Y          4.3031     .71153     Y=  .12559*X-4.3956     .40109
```

the number of points at that grid location. (If more than nine points occur at one location, some programs will use letters of the alphabet to indicate numbers.) A more detailed scatter diagram could be done by a computer-driven plotter or by hand.

Under the scatter plot are printed various statistics. These include $N = 87$ persons, the mean for each variable ($\overline{X} = 69.264$ in. height and $\overline{Y} = 4.303$ L), and the standard deviation ($S_X = 2.639$ in. and $S_Y = 0.7115$ L). In addition, COR = 0.466 is the estimated *correlation coefficient* (which will be discussed in Section 6.5).

The *regression line*, which will be defined in Section 6.4, is

$$Y = -4.396 + 0.126X$$

A visual representation of the regression line is obtained by connecting the two Y symbols appearing on the outside border of the grid. Note that FEV1 is an increasing function of height, and the rate of increase (the coefficient of X) is 0.126 L/in.

For predictive purposes we would expect a male factory worker, age 20–24 years old, whose height is 5 ft 10 in., or 70 in., to have an FEV1 value of

$$\text{FEV1} = -4.394 + (0.126)(70) = 4.3957$$

or $\qquad\qquad$ FEV1 = 4.40 (rounded off)

To take an extreme example, suppose a person is 2 ft tall. Then this equation would predict a negative value of FEV1. This example illustrates the danger of using the regression equation outside the range of concern. A safe policy is to restrict the use of the equation to the range of X observed in the sample.

The *residual mean square* (RES. MS.) in the bottom line of output in Figure 6.1 is an estimate of the variance around the regression line. The square root of this quantity [$S = (0.40109)^{1/2} = 0.63$] is the estimated standard deviation of the regression line. As will be shown in Section 6.4, this standard deviation can be used to compute confidence intervals and prediction intervals.

The output of Figure 6.1 also includes the estimated regression equation of height on FEV1 (height as the dependent variable and FEV1 as the independent variable). Although the regression of X and Y is not meaningful in this example, it may be useful in other applications such as calibration (discussed in Section 6.10).

6.4 DESCRIPTION OF METHODS OF REGRESSION: FIXED-X CASE

In this section we present the background assumptions, models, and formulas necessary for understanding simple linear regression. The theoretical background is simpler for the fixed-X case than for the variable-X case, so we will begin with it.

Assumptions and Background

For each value of X we conceptualize a distribution of values of Y. This distribution is described partly by its mean and variance at each fixed X value:

$$\text{mean of } Y \text{ values at a given } X = \alpha + \beta X$$

and $\qquad$ variance of Y values at a given $X = \sigma^2$

The basic idea of *simple linear regression* is that the means of Y lie on a straight line when plotted against X. Secondly, the variance of Y at a given X is assumed to be the same for all values of X. The latter assumption is called *homoscedasticity*, or homogeneity of variance. Figure 6.2 illustrates the distribution of Y at three values of X. Note from Figure 6.2 that σ^2 is not the variance of all the Y's from their mean but is, instead, the variance of Y at a given X. It is clear that the means lie on a straight line in the range of concern and that the variance or the degree of variation is the same at the different values of X. Outside the range of concern it is immaterial to our analysis what the curve looks like, and in most practical situations linearity will hold over a limited range of X. This figure illustrates that extrapolation of a linear relationship beyond the range of concern can be dangerous.

The expression $\alpha + \beta X$, relating the mean of Y to X, is called the *population least squares regression equation*. Figure 6.3 illustrates the meaning of the parameters α and β. The parameter α is the *intercept* of this line. That is, it is the mean of Y when $X = 0$. The *slope* β is the amount of change in the mean of Y when the value of X is increased by one unit. A negative value of β signifies that the mean of Y decreases as X increases.

FIGURE 6.2. Simple Linear Regression Model for Fixed X's

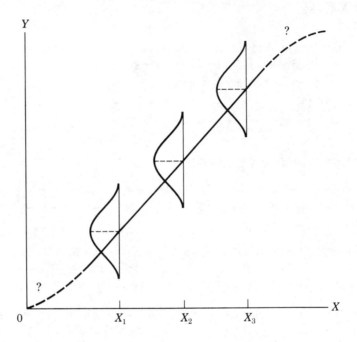

FIGURE 6.3. Theoretical Regression Line Illustrating α and β Geometrically

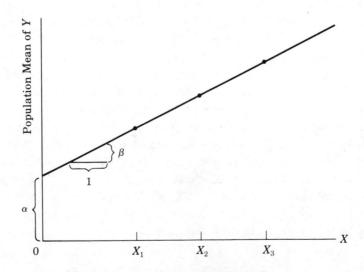

Least Squares Method

The parameters α and β are estimated from a sample collected according to the fixed-X model. The *sample estimates* of α and β are denoted by A and B, respectively, and the resulting regression line is called the *sample least squares regression equation*.

To illustrate the method, we consider a hypothetical sample of four points, where X is fixed at $X_1 = 5$, $X_2 = 5$, $X_3 = 10$, and $X_4 = 10$. The sample values of Y are $Y_1 = 14$, $Y_2 = 17$, $Y_3 = 27$, and $Y_4 = 22$. These points are plotted in Figure 6.4.

The output from the computer would include the following information:

	MEAN	ST.DEV.	REGRESSION LINE	RES.MS.
X	7.5	2.8868		
Y	20.0	5.7155	Y = 6.5 + 1.8X	8.5

The *least squares method* finds the line that minimizes the sum of squared vertical deviations from each point in the sample to the point *on* the line corresponding to the X value. It can be shown mathematically that the least squares line is

$$\hat{Y} = A + BX$$

where

$$B = \frac{\Sigma\,(X - \overline{X})(Y - \overline{Y})}{\Sigma\,(X - \overline{X})^2}$$

and

$$A = \overline{Y} - B\overline{X}$$

Here $\overline{X}$ and $\overline{Y}$ denote the sample means of X and Y, and $\hat{Y}$ denotes the predicted value of Y for a given X.

The *deviation* (or *residual*) for the first point is computed as follows:

$$Y(1) - \hat{Y}(1) = 14 - [6.5 + 1.8(5)] = -1.5$$

Similarly, the other residuals are $+1.5$, -2.5, and $+2.5$, respectively. The sum of the squares of these deviations is 17.0. No other line can be fitted to produce a smaller sum of squared deviations than 17.0.

The estimate of σ^2 is called the *residual mean square* (RES. MS.) and is computed as

$$S^2 = \text{RES. MS.} = \frac{\Sigma\,(Y - \hat{Y})^2}{N - 2}$$

FIGURE 6.4.   Simple Data Example Illustrating Computations of Output Given in Figure 6.1

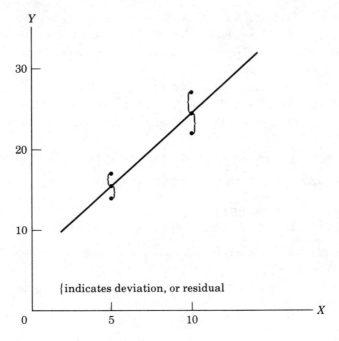

The number $N - 2$, called the *residual degrees of freedom*, is the sample size minus the number of parameters in the line (in this case, α and β). Using $N - 2$ as a divisor in computing S^2 produces an *unbiased estimate* of σ^2. In the example,

$$S^2 = \text{RES. MS.} = \frac{17}{4 - 2} = 8.5$$

The square root of the residual mean square is called the *standard deviation of the estimate* and is denoted by S.

Packaged regression programs will also produce the *standard errors* of A and B. These statistics are computed as

$$\text{SE}(A) = S\left[\frac{1}{N} + \frac{\overline{X}^2}{\Sigma\,(X - \overline{X})^2}\right]^{1/2}$$

and

$$\text{SE}(B) = \frac{S}{[\Sigma\,(X - \overline{X})^2]^{1/2}}$$

Confidence and Prediction Intervals

For each value of X under consideration a population of Y values is assumed to exist. Confidence intervals and tests of hypotheses concerning the intercept, slope, and line may be made with assurance when three assumptions hold: (1) The Y values are assumed to be normally distributed; (2) their means lie on a straight line; and (3) their variances are all equal.

For example, confidence for the slope B can be computed by using the standard error of B. The *confidence interval* (CI) for B is

$$CI = B \pm t \cdot SE (B)$$

where t is the $100(1 - \alpha/2)$ percentile of the t distribution with $N - 2$ df (*degrees of freedom*); see Appendix Table A.2. Similarly, the confidence interval for A is

$$CI = A \pm t \cdot SE (A)$$

where the same degrees of freedom are used for t.

The value $\hat{Y}$ computed for a particular X can be interpreted in two ways:

1. $\hat{Y}$ is the *point estimate of the mean* of Y at that value of X.

2. $\hat{Y}$ is the *estimate of the Y value* for any individual with the given value of X.

The investigator may supplement these point estimates with interval estimates. The *confidence interval* (CI) *for the mean of Y* at a given value of X, say X^*, is

$$CI = \hat{Y} \pm t \cdot S \left[\frac{1}{N} + \frac{(X^* - \overline{X})^2}{\sum (X - \overline{X})^2} \right]^{1/2}$$

where t is the $100(1 - \alpha/2)$ percentile of the t distribution with $N - 2$ df (see Appendix Table A.2).

For an individual Y value the confidence interval is called the *prediction interval* (PI). The prediction interval (PI) for an individual Y at X^* is computed as

$$PI = Y \pm tS \left[1 + \frac{1}{N} + \frac{(X^* - \overline{X})^2}{\sum (X - \overline{X})^2} \right]^{1/2}$$

where t is the same as for the confidence interval for the mean of Y.

In summary, for the fixed-X case we presented the model for simple regression analysis and methods for estimating the parameters of the model. Later in the chapter we will return to this model and present special cases and other uses. Next, we present the variable-X model.

6.5 DESCRIPTION OF METHODS OF REGRESSION AND CORRELATION: VARIABLE-X CASE

In this section we present, for the variable-X case, material similar to that given in the previous section.

For this model both X and Y are random variables measured on cases that are randomly selected from a population. One example is the factory workers data set, where FEV1 was predicted from height. The fixed-X regression model applies in this case when we treat the X values as if they were preselected. (This technique is justifiable theoretically by *conditioning* on the X values that happened to be obtained in the sample.) Therefore all the previous discussion and formulas are precisely the same for this case as for the fixed-X case. In addition, since both X and Y are considered random variables, other parameters can be useful for describing the model. These include the means and variances for X and Y over the entire population (μ_X, μ_Y, σ_X^2, and σ_Y^2). The sample estimates for these parameters are usually included in computer output. For example, in the analysis of the factory workers data these estimates were printed at the bottom of Figure 6.1.

As a measure of how the variables X and Y vary together, a parameter called the *population covariance* is often estimated. The population covariance of X and Y is defined as the average of the product $(X - \mu_X)(Y - \mu_Y)$ over the entire population. This parameter is denoted by σ_{XY}. If X and Y tend to increase together, σ_{XY} will be positive. If, on the other hand, one tends to increase as the other decreases, σ_{XY} will be negative.

So that the magnitude of σ_{XY} is standardized, its value is divided by the product $\sigma_X \sigma_Y$. The resulting parameter, denoted by ρ, is called the *product moment correlation coefficient*, or simply the *correlation coefficient*. The value of

$$\rho = \frac{\sigma_{XY}}{\sigma_X \sigma_Y}$$

lies between -1 and $+1$, inclusive. The sample estimate for ρ is the *sample correlation coefficient* r, or

$$r = \frac{S_{XY}}{S_X S_Y}$$

where

$$S_{XY} = \frac{\Sigma (X - \overline{X})(Y - \overline{Y})}{N - 1}$$

The sample statistic r also lies between -1 and $+1$, inclusive. Further interpretation of the correlation coefficient is given in Thorndike (1978).

Tests of hypotheses and confidence intervals for the variable-X case require that X and Y be jointly normally distributed. Formally, this requirement is that X and Y follow a bivariate normal distribution. Examples of the appearance of bivariate normal distributions are given in Section 6.7. If this condition is true, it can be shown that Y also satisfies the three conditions for the fixed-X case.

6.6 INTERPRETATION OF RESULTS: FIXED-X CASE

In this section we present methods for interpreting the results of a regression output.

First, the type of the sample must be determined. If it is a fixed-X sample, the statistics of interest are the intercept and slope of the line and the standard error of the estimate; point and interval estimates for α and β have already been discussed.

The investigator may also be interested in *testing hypotheses* concerning the parameters. A commonly used test is for the null hypothesis

$$H_0: \beta = \beta_0$$

The test statistic is

$$t = \frac{(B - \beta_0)[\Sigma (X - \overline{X})^2]^{1/2}}{S}$$

where S is the square root of the residual mean square and the computed value of t is compared with the tabled t percentiles (Appendix Table A.2) with $N - 2$ degrees of freedom to obtain the P value. Many computer programs will print the standard error of B. Then the t statistic is simply

$$t = \frac{B - \beta_0}{SE(B)}$$

A common value of β_0 is $\beta_0 = 0$, indicating independence of X and Y; i.e., the mean value of Y does not change as X changes.

Tests concerning α can also be performed for the null hypothesis H_0: $\alpha = \alpha_0$, using

$$t = \frac{A - \alpha_0}{S\{(1/N) + [\overline{X}^2/\Sigma\,(X - \overline{X})^2]\}^{1/2}}$$

Values of this statistic can also be compared with the tabled t percentiles (Appendix Table A.2) with $N - 2$ degrees of freedom to obtain the t value. If the standard error of A is printed by the program, the test statistic can be computed simply as

$$t = \frac{A - \alpha_0}{SE(A)}$$

For example, to test whether the line passes through the origin, the investigator would test the hypothesis of $\alpha_0 = 0$.

It should be noted that rejecting the null hypothesis H_0: $\beta = 0$ is no indication in and of itself of the magnitude of the slope. An observed $B = 0.1$, for instance, might be found significantly different from zero, while a slope of $B = 1.0$ might be considered inconsequential in a particular application. The importance and strength of a relationship between Y and X is a separate question from the question of whether certain parameters are significantly different from zero. The test of the hypothesis $\beta = 0$ is a preliminary step to determine whether the magnitude of B should be further examined. If the null hypothesis is rejected, then the magnitude of the effect of X on Y should be investigated.

One way of investigating the *magnitude of the effect* of a typical X value on Y is to multiply B by $\overline{X}$ and to contrast this result with $\overline{Y}$. If $B\overline{X}$ is small relative to $\overline{Y}$, then the magnitude of the effect of B in predicting Y is small. Another interpretation of B can be obtained by first deciding on two typical values of X, say X_1 and X_2, and then calculating the difference $B(X_2 - X_1)$. This difference measures the change in Y when X goes from X_1 to X_2.

To infer *causality*, we must justify that all other factors possibly affecting

Y have been controlled in the study. One way of accomplishing this control is to design an experiment in which such intervening factors are held fixed while only the variable X is set at various levels. Standard statistical wisdom also requires randomization in the assignment to the various X levels in the hope of controlling for any other factors not accounted for (see Box 1966).

6.7 INTERPRETATION OF RESULTS: VARIABLE-X CASE

In this section we present methods for interpreting the results of a regression and correlation output. In particular, we will look at the ellipse of concentration and the coefficient of correlation.

For the variable-X model the regression line and its interpretation remain valid. Strictly speaking, however, causality cannot be inferred from this model. Here we are concerned with the bivariate distribution of X and Y. We can safely estimate the means, variances, covariance, and correlation of X and Y, i.e., the distribution of pairs of values of X and Y measured on the same individual. (Although these parameter estimates are printed by the computer, they are meaningless in the fixed-X model.) For the variable-X model the interpretations of the means and variances of X and Y are the usual measures of location (center of the distribution) and variability. We will now concentrate on how the *correlation coefficient* should be interpreted.

Ellipse of Concentration

The *bivariate distribution* of X and Y is best interpreted by a look at the scatter diagram from a random sample. If the sample comes from a bivariate normal distribution, the data will tend to cluster around the means of X and Y and will approximate an ellipse called the *ellipse of concentration*. Note in Figure 6.1 that the points representing the data could be enclosed by an ellipse of concentration.

An ellipse can be characterized by the following:

1. The center.
2. The major axis, i.e., the line going from the center to the farthest point on the ellipse.
3. The minor axis, i.e., the line going from the center

to the nearest point on the ellipse (the minor axis is always perpendicular to the major axis).

4. The ratio of the length of the minor axis to the length of the major axis. If this ratio is small, the ellipse is thin and elongated; otherwise, the ellipse is fat and round-shaped.

Interpreting the Correlation Coefficient

For ellipses of concentration the center is at the point defined by the means of X and Y. The directions and lengths of the major and minor axes are determined by the two variances and the correlation coefficient. For fixed values of the variances the ratio of the length of the minor axis to that of the major axis, and hence the shape of the ellipse, is determined by the correlation coefficient ρ.

In Figure 6.5 we represent ellipses of concentration for various bivariate normal distributions in which the means of X and Y are both zero and the variances are both one (standardized X's and Y's). The case $\rho = 0$, Figure 6.5a, represents independence of X and Y. That is, the value of one variable has no effect on the value of the other, and the ellipse is a perfect circle. Higher values of ρ correspond to more elongated ellipses, as indicated in Figures 6.5b through 6.5f.

We see that for very high values of ρ one variable conveys a lot of information about the other. That is, if we are given a value of X, we can guess the corresponding Y value quite accurately. We can do so because the range of the possible values of Y for a given X is determined by the width of the ellipse at that value of X. This width is small for a large value of ρ. For negative values of ρ similar ellipses could be drawn where the major axis (long axis) has a negative slope, i.e., in the northwest/southeast direction.

Another interpretation of ρ stems from the concept of the *conditional distribution*. For a specific value of X the distribution of the Y value is called the conditional distribution of Y given that value of X. The word *given* is translated symbolically by a vertical line, so Y given X is written as $Y|X$. The variance of the conditional distribution of Y, variance $(Y|X)$, can be expressed as

$$\text{variance } (Y|X) = \sigma_Y^2 \, (1 - \rho^2)$$

or

$$\sigma_{Y|X}^2 = \sigma_Y^2 \, (1 - \rho^2)$$

FIGURE 6.5. Hypothetical Ellipses of Concentration for Various ρ Values

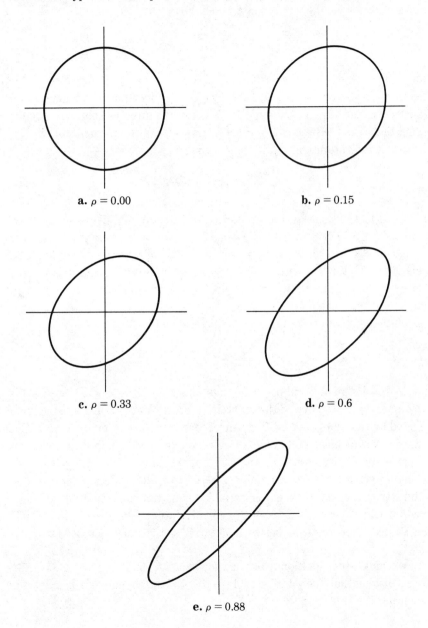

a. $\rho = 0.00$ **b.** $\rho = 0.15$

c. $\rho = 0.33$ **d.** $\rho = 0.6$

e. $\rho = 0.88$

Note that $\sigma_{Y|X} = \sigma$ as given in Section 6.4.

This equation can be written in another form, as

$$\rho^2 = \frac{\sigma_Y^2 - \sigma_{Y|X}^2}{\sigma_Y^2}$$

The term $\sigma_{Y|X}^2$ measures the variance of Y when X has a specific fixed value. Therefore this equation states that ρ^2 is the proportion of variance of Y reduced because of knowledge of X. This result is often loosely expressed by saying that ρ^2 is the proportion of the variance of Y "explained" by X.

A better interpretation of ρ is to note that

$$\frac{\sigma_{Y|X}}{\sigma_Y} = (1 - \rho^2)^{1/2}$$

This value is a measure of the proportion of the standard deviation of Y not explained by X. For example, if $\rho = \pm 0.8$, then 64% of the variance of Y is explained by X. However, $(1 - 0.8^2)^{1/2} = 0.6$, saying that 60% of the standard deviation of Y is not explained by X. Since the standard deviation is a better measure of variability than the variance, it is seen that when $\rho = 0.8$, more than half of the variability of Y is still not explained by X. If instead of using ρ^2 from the population we use r^2 from a sample, then

$$S_{Y|X}^2 = S^2 = \left(\frac{N-1}{N-2}\right) S_Y^2 (1 - r^2)$$

and the results must be adjusted for sample size.

An important property of the correlation coefficient is that its value is not affected by the units of X or Y or any linear transformation of X or Y. For instance, X was measured in inches in the example shown in Figure 6.1, but the correlation between height and FEV1 is the same if we change the units of height to centimeters or the units of FEV1 to milliliters. In general, adding (subtracting) a constant to either variable or multiplying either variable by a constant will not alter the value of the correlation. Since $\hat{Y} = A + BX$, it follows that the correlation between Y and $\hat{Y}$ is the same as that between Y and X.

If we make the additional assumption that X and Y have a bivariate normal distribution, then it is possible to test the null hypothesis $H_0: \rho = 0$ by computing the test statistic

$$t = \frac{r(N-2)^{1/2}}{(1 - r^2)^{1/2}}$$

FIGURE 6.6. Example of Nonlinear Regression

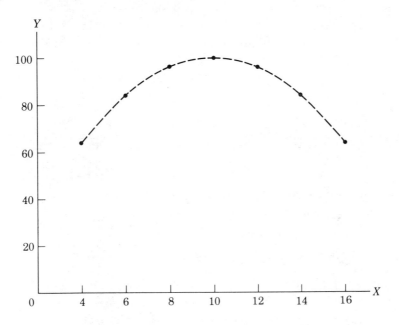

with $N - 2$ degrees of freedom. For the factory workers example of Figure 6.1, $r = 0.466$. To test the hypothesis H_0: $\rho = 0$ versus the alternative H_1: $\rho > 0$, we compute

$$t = \frac{0.466(87 - 2)^{1/2}}{(1 - 0.466^2)^{1/2}} = 4.86$$

with 85 df. This statistic results in $P < 0.001$, and the observed r is significantly greater than zero.

Tests of null hypotheses other than $\rho = 0$ and confidence intervals for ρ can be found in many textbooks (see Brownlee 1965; Dunn and Clark 1974; Afifi and Azen 1979). As before, a test of $\rho = 0$ should be made before attempting to interpret the magnitude of the sample correlation coefficient. Note that the test of $\rho = 0$ is equivalent to the test of $\beta = 0$ given earlier.

All of the above interpretations were made with the assumption that the data follow a bivariate normal distribution, which implies that the mean of Y is related to X in a linear fashion. If the regression of Y on X is nonlinear, it is conceivable that the sample correlation coefficient is near zero when Y is, in fact, strongly related to X. For example, in Figure 6.6 we can quite accurately predict Y from X [in fact, the points fit a curve $Y = 100 - (X - 10)^2$ exactly]. However, the sample correlation coefficient is $r = 0.0$. (An appropriate regression equation can be fitted by the techniques of polynomial regression; see Section 7.8. Also, in Section 6.11 we discuss the role of transformations in reducing nonlinearities.)

6.8 ADDITIONAL STATISTICS FOUND IN COMPUTER OUTPUTS

Most packaged regression programs include other statistics in the printed output that are useful for linear regression. In this section we discuss those most commonly found in the output.

Standardized Regression Coefficient

The *standardized regression coefficient* is the slope in the regression equation if X and Y are standardized. Standardization of X and Y is achieved by subtracting the respective means from each set of observations and dividing the differences by the respective standard deviations. The resulting set of standardized sample values will have a mean of zero and a standard deviation of one for both X and Y. After standardization the intercept in the regression equation will be zero, and for simple linear regression (one X variable) the standardized slope will be equal to the correlation coefficient r. In multiple regression, where several X variables are used, the standardized regression coefficients help quantify the relative contribution of each X variable (see Chapter 7).

Analysis of Variance Table

The test for H_0: $\beta = 0$ was discussed in Section 6.6 using the t statistic. This test allows one-sided or two-sided alternatives. When the two-sided alternative is chosen, it is possible to represent the test in the form of an *analysis of variance (ANOVA) table*. A typical ANOVA table is represented in Table 6.1. If X were useless in predicting Y, our best guess of the Y value would be $\overline{Y}$ regardless of the value of X. To measure how different our fitted line $\hat{Y}$ is from $\overline{Y}$, we calculate the sums of squares for regression as

TABLE 6.1. ANOVA Table for Simple Linear Regression

Source of Variation	Sums of Squares	df	Mean Square	F
Regression	$\Sigma(\hat{Y} - \overline{Y})^2$	1	$SS_{reg}/1$	MS_{reg}/MS_{res}
Residual	$\Sigma(Y - \hat{Y})^2$	$N - 2$	$SS_{res}/(N - 2)$	
Total	$\Sigma(Y - \overline{Y})^2$	$N - 1$		

TABLE 6.2. ANOVA Example from Figure 6.1

Source of Variation	Sums of Squares	df	Mean Square	F
Regression	9.4469	1	9.4469	23.55
Residual	34.0927	85	0.40109	
Total	43.5396	86		

$\Sigma(\hat{Y} - \overline{Y})^2$, summed over each data point. (Note that $\overline{Y}$ is the average of all the $\hat{Y}$ values.) The residual mean square is a measure of how poorly or how well the regression line fits the actual data points. A large residual mean square indicates a poor fit. The F ratio is, in fact, the squared value of the t statistic described in Section 6.6 for testing $H_0: \beta = 0$.

Table 6.2 shows the ANOVA table for the factory workers data example. Note that the F ratio of 23.55 is the square of the t value of 4.86 computed previously. Also, the residual mean square (0.401) is the same as that given in Figure 6.1.

Residual Analysis

Several of the regression programs offer *options to print lists and plots of the residuals*, i.e., the values $Y - \hat{Y}$ for the data points. Table 6.3 shows the data for the hypothetical example represented in Figure 6.4. Also shown are the predicted values $\hat{Y}$ and the residuals. Note that as expected, the mean of the $\hat{Y}$ values is equal to $\overline{Y}$. Also, it will always be the case that the mean of the residuals is zero.

Residuals do not all have the same variance and are therefore not directly

TABLE 6.3. Hypothetical Data Example from Figure 6.4

i	X	Y	$\hat{Y} = 6.5 + 1.8X$	$Y - \hat{Y}$ = Residual
1	5	14	15.5	−1.5
2	5	17	15.5	1.5
3	10	27	24.5	2.5
4	10	22	24.5	−2.5
Mean	7.5	20	20	0

comparable. To remedy this situation, packaged programs include output of *standardized residuals*, i.e., each residual divided by its standard deviation. Another type of residual that is also available is the *studentized residual*. These residuals are computed in the same manner as standardized residuals are computed, with the additional feature that the ith studentized residual is computed from a regression line fitted to all but the ith observation. Studentized residuals have the additional advantage of minimizing the effect of an extreme outlier. In the remainder of this book we will discuss the use of residuals for various purposes. The recommended analyses apply to raw residuals as well as to their standardized and studentized forms.

The list of residuals is useful in drawing attention to particularly large or small residuals in order to examine them as possible *outliers* or blunders. Detection of outliers was discussed in Chapter 3, and the same techniques can be applied to these residuals. That is, histograms of the residual values can be plotted and examined for unusual values.

A more recent concept related to residual analysis is that of the *influence of individual observations*. Although examination of residual plots may uncover potential outliers, the effect of certain observations on the fitted regression equation may not be obvious from the plots. The magnitude of this influence can be examined by comparing the regression equation with another equation derived from the data after the one observation has been deleted. Cook (1977 and 1979) proposed a distance (sometimes called the *Cook's distance*) to measure the influence of each observation. An optional output in some programs enables the user to obtain a list of these distances. A large distance indicates a strong influence of that observation on the overall equation. The investigator may wish to fit a regression equation after removing the most influential observation to examine the stability of the estimated parameters (see Cook and Weisberg 1980 for additional discussion).

Some regression programs (BMDP1R and SPSS–X REGRESSION, e.g.) will provide *normal probability plots* of residuals to enable the user to decide whether the data approximate a normal distribution (SAS UNIVARIATE can be used to produce such a plot from the residuals). If the residuals are not normally distributed, then the distribution of Y at each value of X is not normal.

Another violation is *lack of independence among residuals*. If the observations are recorded in a certain order—e.g. successively in time—then the Durbin-Watson statistic calculated by some packaged programs can be used to test whether the residuals are "serially" correlated. Its value should be close to 2 if the residuals are uncorrelated. Other regression programs print the serial correlation, which is simply the correlation coefficient between pairs of successive residuals. For a sufficiently large N the levels of significance taken from the usual tests for $\rho = 0$ apply approximately to the serial correlation.

Most regression programs offer the option of plotting the residuals versus the X values or versus the predicted $\hat{Y}$ values. Again, potential outliers and blunders can be identified by examining these plots. Since the residuals have a zero mean, it is useful to draw a horizontal line through the zero point on the vertical axis. This line will aid in checking whether the residuals are symmetric around their mean (which is expected if the residuals are normally distributed). Unusual clusters of points can alert the investigator to possible anomalies in the data.

Plots are also useful in checking the *linearity of the regression* of Y on X. These plots can be done directly from scatter diagrams or from residual plots. Figure 6.7 illustrates four possible situations.

In Figure 6.7a the idealized bivariate normal distribution (variable-X model) is illustrated, using contours similar to those for Figure 6.5. In this case the residuals plotted against X would also approximate an ellipse.

In Figure 6.7b the ellipse is replaced by a fan-shaped figure. This shape suggests that as X increases, the standard deviation of Y also increases (not necessarily linearly). This result is a strong indication that the assumption of homoscedasticity is violated. In this case the use of weighted least squares is recommended (see Section 6.9).

In Figure 6.7c the ellipse is replaced by a kidney-shaped form, indicating that the regression of Y is not linear in X. One possibility for solving this

FIGURE 6.7. Hypothetical Scatter Plots and Corresponding Residual Plots

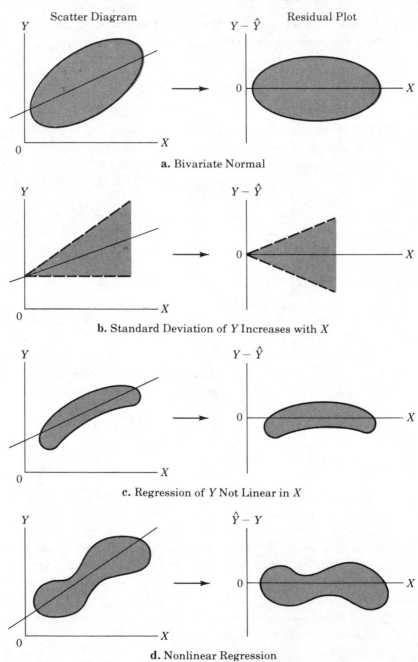

a. Bivariate Normal

b. Standard Deviation of Y Increases with X

c. Regression of Y Not Linear in X

d. Nonlinear Regression

problem is to fit a quadratic equation, as discussed in Chapter 7. Another possibility is to transform X into $\log X$ or some other function of X and then fit a straight line to the transformed X values and Y. This case will be discussed in Section 6.9.

In Figure 6.7d another case is illustrated. The regression equation is nonlinear, and transformations do not lead to a straight line. For this case the techniques of nonlinear regression, discussed in Chapter 9, may be used.

6.9 OTHER OPTIONS IN COMPUTER PROGRAMS

In this section we discuss two options available from computer programs: regression through the origin and weighted regression.

Regression Through the Origin

Sometimes an investigator is convinced that the *regression line* should pass *through the origin*. In this case the appropriate model is simply the mean of

$$Y = \beta X$$

That is, the intercept is forced to be zero. The programs usually give the option of using this model and estimate β as

$$B = \frac{\Sigma\, XY}{\Sigma\, X^2}$$

To test $H_0: \beta = \beta_0$, the test statistic is

$$t = \frac{B - \beta_0}{S^2/(\Sigma\, X^2)^{1/2}}$$

where

$$S^2 = \frac{\Sigma\,(Y - BX)^{1/2}}{N - 1}$$

and t has $N - 1$ degrees of freedom.

Weighted Least Squares Regression

The investigator may also request a *weighted least squares regression line*. In weighted least squares each observation is given an individual weight reflecting its importance or degree of variability. There are three common situations in which a weighted regression line is appropriate:

1. The variance of the distribution at a given X is a function of the X value. An example of this situation was shown in Figure 6.7b.

2. Each Y observation is, in fact, the mean of several determinations, and that number varies from one value of X to another.

3. The investigator wishes to assign different levels of importance to different points. For example, data from different countries could be weighted either by the size of the population or by the perceived accuracy of the data.

Formulas for weighted least squares are discussed in Dunn and Clark (1974) and Brownlee (1965). In case 1 the weights are the inverse of the variances of the point. In case 2 the weights are the number of determinations at each X point. In case 3 the investigator must make up numerical weights to reflect the perception of importance or accuracy of the data points.

In weighted least squares regression the estimates of α and β and other statistics are adjusted to reflect these special characteristics of the observations. In most situations the weights will not affect the results appreciably unless they are quite different from each other. Since it is considerably more work to compute a weighted least squares regression equation, it is recommended that one of the computer programs listed in Section 6.13 be used, rather than hand calculations.

6.10 SPECIAL APPLICATIONS OF REGRESSION

In this section we discuss some important applications of regression analysis that require some caution when the investigator is using regression techniques.

Calibration

A common situation in laboratories or industry is one in which an instrument that is designed to measure a certain characteristic needs calibration. For example, the concentration of a certain chemical could be measured by an instrument. For calibration of the instrument several compounds with known concentrations, denoted by X, could be used, and the measurements, denoted by Y, could be determined by the instrument for each compound. As a second example, a costly or destructive method of measurement that is very accurate, denoted by X, could be compared with another method of measurement, denoted by Y, that is a less costly or nondestructive method. In either situation more than one determination of the level of Y could be made for each value of X.

The object of the analysis is to derive, from the calibration data, a formula or a graph that can be used to predict the unknown value of X from the measured Y value in future determinations. We will present two methods (indirect and direct) that have been developed and a technique for choosing between them.

1. *Indirect method*: Since X is assumed known with little error and Y is the random variable, this classical method begins with finding the usual regression of Y on X,

$$\hat{Y} = A + BX$$

Then for a determination Y^* the investigator obtains the indirect estimate of X, $\hat{X}_{in}$, as

$$\hat{X}_{in} = \frac{Y^* - A}{B}$$

This method is illustrated in Figure 6.8a.

2. *Direct method*: Here we pretend that X is the dependent variable and Y is the independent variable and fit the regression of X on Y as illustrated in Figure 6.8b. We denote the estimate of X by $\hat{X}_{dr}$; thus

$$\hat{X}_{dr} = C + DY^*$$

To compare the two methods, we compute the quantities $\hat{X}_{in}$ and $\hat{X}_{dr}$ for all of the data in the sample. Then we compute the correlation between $\hat{X}_{in}$ and X, denoted by $r(in)$, and the correlation between $\hat{X}_{dr}$ and X, denoted by $r(dr)$. We then use the method that results in the higher correlation. Further discus-

FIGURE 6.8.   Illustration of Indirect and Direct Calibration Methods

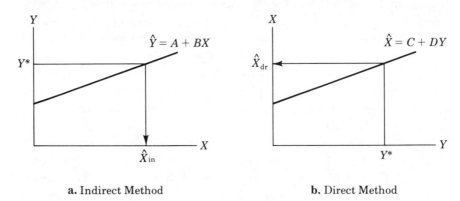

a. Indirect Method   **b.** Direct Method

sion of this problem is found in Krutchkoff (1967), Berkson (1969), Shukla (1972), and Lwin and Maritz (1982).

It is advisable, before using either of these methods, that the investigator test whether the slope B is significantly different from zero. This test can be done by the usual t test. If the hypothesis $H_0: \beta = 0$ is not rejected, this result is an indication that the instrument should not be used, and the investigator should not use either equation.

Forecasting

Forecasting is the technique of analyzing historical data in order to provide estimates of the future values of certain variables of interest. If data taken over time (X) approximate a straight line, forecasters might assume that this relationship would continue in the future and would use simple linear regression to fit this relationship. This line is extended to some point in the future. The difficulty in using this method lies in the fact that we are never sure that the linear trend will continue in the future. At any rate, it is advisable not to make long-range forecasts by using this method. Further discussion and additional methods are found in Montgomery and Johnson (1976) and Levenbach and Cleary (1981).

Paired Data

Another situation where regression analysis may be appropriate is the *paired-sample case*. A paired sample consists of observations taken at two time points (pre- and post-) on the same individual or consists of matched pairs, where one member of the pair is subject to one treatment and the other is a control or is subject to another treatment.

Many investigators apply only the paired t test to such data and as a result may miss an important relationship that could exist in the data. The investigator should, at least, plot a scatter diagram of the data and look for possible linear and nonlinear relationships. If a *linear relationship* seems appropriate, then the techniques of regression and correlation described in this chapter are useful in describing this relationship. If a *nonlinear relationship* exists, the methods described in Chapter 9 could be used.

Focus on Residuals

In certain cases performing the regression analysis may only be a preliminary step to obtaining the residuals, which are themselves the quantities of interest to be saved for further analysis. For example, in the factory workers data example a regression equation was obtained for FEV1 on height for a restricted age group. It may be useful to analyze the residuals, rather than the original FEV1, for possible effects of different locations or types of jobs in the factory. Other uses of the residuals will be described in Chapter 7.

6.11 ROBUSTNESS AND TRANSFORMATIONS FOR REGRESSION ANALYSIS

In this section we define the concept of robustness in statistical analysis, and we discuss the role of transformations in regression and correlation.

Robustness and Assumptions

Regression and correlation analysis make certain assumptions about the population from which the data were obtained. A *robust analysis* is one that is useful even though all the assumptions are not met. For the purpose of fitting a straight line, we assume that the Y values are normally distributed, the population regression equation is linear in the range of concern, and the variance of Y is the same for all values of X. Linearity can be checked graphically, and transformations can help straighten out a nonlinear regression line.

The *assumption of homogeneity of variance* is not crucial for the resulting least squares line. In fact, the least squares estimates of α and β are unbiased whether or not the assumption is valid. However, if glaring irregularities of variance exist, weighted least squares can improve the fit. In this case the weights are chosen to be proportional to the inverse of the variance. For example, if the variance is a linear function of X, then the weight is $1/X$.

The *assumption of normality* of the Y values of each value of X is made only when tests of hypotheses are performed or confidence intervals are calculated. It is generally agreed in the statistical literature that slight departures from this assumption do not appreciably alter our inferences if the sample size is sufficiently large.

The *lack of randomness* in the sample can seriously invalidate our inferences. Confidence intervals are often optimistically narrow because the sample is not truly a random one from the whole population to which we wish to generalize.

In all of the preceding analyses *linearity* of the relationship between X and Y was assumed. Thus careful examination of the scatter diagram should be the first step in any regression analysis. It is advisable to explore various transformations of Y and/or X if nonlinearity of the original measurements is apparent.

Transformations

The subject of *transformations* has been discussed in detail in the literature; for examples, see Hald (1952) and Mosteller and Tukey (1977). In this subsection we present some typical graphs of the relationship between Y and X and some practical transformations.

In Chapter 4 we discussed the effects of transformations on the frequency distribution. There it was shown that taking the logarithm or square root of a number condensed the magnitude of larger numbers and stretched the magnitude of values less than one. Conversely, raising a number to a power greater than one stretches the large values and condenses the values less

FIGURE 6.9. Choice of Transformation: Typical Curves and Appropriate Transformation

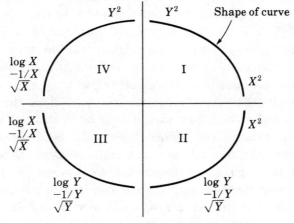

a. Curves Not Linear in X

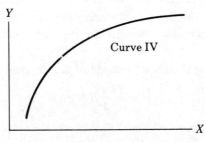

b. Detail of Fourth Quadrant

than one. These properties are useful in selecting the appropriate transformation to straighten out a nonlinear graph of one variable as a function of another.

Mosteller and Tukey (1977) present typical regression curves that are not linear in X as belonging to one of the quadrants of Figure 6.9a. A very common case is illustrated in Figure 6.9b, which is represented by the fourth quadrant of the circle in Figure 6.9a. For example, the curve in Figure 6.9b might be made linear by transforming X to log X, to $-1/X$, or to $X^{1/2}$. Another possibility would be to transform Y to Y^2. The other three cases are also indicated in Figure 6.9a. The remaining quadrants are interpreted in a similar fashion.

Other transformations could also be attempted, such as powers other than those indicated. It may also be useful to first add or subtract a constant from all values of X or Y and then take a power or logarithms. For example, sometimes taking log X does not straighten out the curve sufficiently. Subtracting a constant C (which must be smaller than the smallest X value) and then taking the logarithm has a greater effect.

The availability of packaged programs greatly facilitates the choice of an appropriate transformation. New variables can be created that are functions of the original variables, and scatter diagrams can be obtained of the new transformed variables. Visual inspection will often indicate the best transformation. Also, the magnitude of the correlation coefficient r will indicate the best linear fit since it is a measure of linear association. Attention should also be paid to transformations that are commonly used in the field of application and that have a particular scientific basis or physical rationale.

Once the transformation is selected, all subsequent estimates and tests are performed in terms of the transformed values. Since the variable to be predicted is usually the dependent variable Y, transforming Y can complicate the interpretation of the resulting regression equation more than if X is transformed.

For example, if log X is used instead of X, the resulting equation is

$$Y = A + B \log_{10} X$$

This equation presents no problems in interpreting the predicted values of Y, and most investigators accept the transformation of $\log_{10} X$ as reasonable in certain situations.

However, if log Y is used instead of Y, the resulting equation is

$$\log_{10} Y = A + BX$$

Then the predicted value of Y, say Y^*, must be detransformed, that is,

$$Y^* = 10^{A+BX}$$

Thus slight biases in fitting log Y could be detransformed into large biases in predicting Y. For this reason most investigators look for transformations of X first.

6.12 COMPUTATIONAL AIDS FOR HAND CALCULATIONS

When no transformations are necessary and the sample size is small, computations for simple linear regression and correlation are manageable on a nonsophisticated hand calculator. The necessary formulas require only the ability to obtain sums, sums of squares, and sums of products.

If computation is done by hand, then formulas equivalent to the ones given earlier are more manageable. For example, the sample variance of X was defined as

$$S_X^2 = \frac{\Sigma (X - \overline{X})^2}{N - 1}$$

For hand calculation the following equivalent formula,

$$S_X^2 = \frac{N \Sigma X^2 - (\Sigma X)^2}{N(N - 1)}$$

is more manageable, since we can obtain ΣX and ΣX^2 directly and then substitute them into the variance formula.

Similarly, we use

$$S_Y^2 = \frac{N \Sigma Y^2 - (\Sigma Y)^2}{N(N - 1)}$$

and

$$S_{XY} = \frac{N \Sigma XY - (\Sigma X)(\Sigma Y)}{N(N - 1)}$$

Then

$$B = \frac{S_{XY}}{S_X^2}$$

$$r = \frac{S_{XY}}{S_x S_Y}$$

$$A = \frac{\Sigma Y - B \Sigma X}{N}$$

and the residual mean square is

$$S^2 = \left(\frac{N - 1}{N - 2}\right) S_Y^2 (1 - r^2)$$

The above formulas should be used for hand calculators and small sample sizes but not for programming computers. The original formulas are more appropriate for use with computers.

6.13 DISCUSSION OF COMPUTER PROGRAMS

The programs 1R, REG, and REGRESSION listed in Table 6.4 can be used to obtain simple regression equations, as well as multiple regression equations when the investigator has more than one independent variable. Thus parts of the program outputs that are not essential for simple linear regression are not presented in that table. Other regression programs also exist in the BMDP, SAS, or SPSS-X packages that can be used to do simple linear regression as well as other types of regression.

If you just want a scatter plot and the regression equation, they can be obtained from simpler descriptive programs. For example, in SAS the PLOT procedure can be used to obtain the scatter diagram. In this procedure numerous options are available to enable the user to control the scale and the appearance of the axes and points. The UNIVARIATE and CORR procedures enable the user to obtain descriptive statistics including means, standard deviations, and the correlation between X and Y. Since $B = rS_Y/S_X$ and $A = \overline{Y} - B\overline{X}$, the regression equation can be easily obtained from these two procedures.

The PLOT procedure is very straightforward to use. Suppose you wish to plot FVC and FEV1 against height for males and females. This information is available in the factory workers data set. The factory worker data are numeric and are punched on cards as follows:

Column	Data
1–18	ID and job description
19–20	Age
21	Blank
22	Sex
23–24	Height in inches
25–28	FVC in liters × 100
29–32	FEV1 in liters × 100

TABLE 6.4. Summary of Computer Output from BMDP, SAS, and SPSS–X for Simple Linear Regression

Output	BMDP*	SAS	SPSS–X
Means and variances/standard deviations	1R	REG	REGRESSION
Regression line	1R	REG	REGRESSION
Test $\beta = 0$	1R	REG	REGRESSION
Standard error of B	1R	REG	REGRESSION
Test $\alpha = 0$	9R	REG	REGRESSION
Standard error of A	9R	REG	REGRESSION
ANOVA	1R	REG	REGRESSION
List residuals	1R	REG	REGRESSION
Plot residuals	1R	REG	REGRESSION
Scatter plots	1R, 6D	PLOT	SCATTERGRAM
Normal probability plots of residuals	1R	UNIVARIATE	REGRESSION
Serial correlation residuals	1R	REG	
Durbin–Watson D	9R	REG	REGRESSION
Cook's distance	9R	REG	REGRESSION
Regression through zero	1R	REG	REGRESSION
Standard error of slope through zero	1R	REG	REGRESSION
Weighted regression	1R	REG	REGRESSION + WEIGHT
Standard error of weight B	1R	REG	REGRESSION + WEIGHT
Standardized B	1R	REG	REGRESSION
Standard error of standardized B		REG	REGRESSION
Correlation coefficient	1R	REG, CORR	REGRESSION

*BMDP2R also performs many of these functions. However, it is more appropriate for multiple stepwise regression (see Chapter 8). Likewise, STEPWISE and GLM for SAS will perform most of the functions of REG.

The following control statements will result in four scatter diagrams: FEV1 by height for males, FVC by height for males, FEV1 by height for females, and FVC by height for females:

```
DATA    LUNGFN;
INPUT   AGE 19-20  SEX 22 HEIGHT  23-24  FVC 25-28 FEV1 29-32;
CARDS;
   (data cards)
PROC    SORT   DATA = LUNGFN;
        BY SEX;
PROC    PLOT;
        PLOT FEV1*HEIGHT   FVC*HEIGHT;
        BY SEX;
```

Note that the data cards themselves should *not* end with semicolons, except for the last one. (The same type of information is needed, as was illustrated in Chapter 3, for an SPSS–X program.)

The first SAS statement says that you want to read data and put it into a SAS data set called LUNGFN. All SAS statements end with a semicolon. The second and third statements are the data definition statements. The second statement tells where on the cards the data are located, and the third statement says the data are on cards. Note how straightforward the INPUT card is in SAS for locating and naming variables.

It is necessary to use the PROC SORT to plot the data only because the data must be ordered by sex before the separate plots can be obtained.

In SPSS–X the SCATTERGRAM program can be used to get the desired scatter diagram, the correlation coefficient, and the regression equation. The BMDP6D program, whose output was shown in Figure 6.1, directly provides the scatter plot, location of the regression line on the plot (connect the Y's on the border of the grid); and the essential statistics (sample size, means, standard deviations, correlation, and slope and intercept).

The advantage of using 1R, REG, or REGRESSION is that numerous other options listed in Table 6.4 are available. Because these programs can be used when you have several independent variables, it is sensible to first become familiar with how to run them and how to look at their output in a simple regression case. In particular, the examination of residuals and tests of hypotheses may be useful. The ease of use of these packaged programs makes it possible to try different transformations, weights, etc. to see their effect on real data. In Section 7.10 an explanation of the control statements for BMDP1R is presented along with discussion of the output.

SUMMARY

In this chapter we gave a conventional presentation of simple linear regression, similar to the presentations found in many statistical textbooks. Thus the reader familiar with the subject should be able to make the transition to the mode we follow in the remainder of the book. Because only two interval or ratio variables were used (one dependent and one independent), it was relatively simple to present algebraic formulas and methods for hand calculation.

We made a clear distinction between random- and fixed-X variable regression. Whereas most of the statistics computed apply to both cases, certain statistics apply only to the random-X case. Computer programs are written to provide more output than is sometimes needed or appropriate. You should be aware of which model is being assumed so that you can make the proper interpretation of the results.

If packaged programs are used, it may be sufficient to run one of the simple plotting programs in order to obtain both the plot and the desired statistics. It is good practice to plot many variables against each other in the preliminary stages of data analysis. This practice allows you to examine the data in a straightforward fashion. However, it is also advisable to use the more sophisticated programs as warm-ups for the more complex data analyses discussed in later chapters.

The concepts of simple linear regression and correlation presented in this chapter can be extended in several directions. Chapter 7 treats the fitting of linear regression equations to more than one independent variable. Chapter 8 gives methods for selecting independent variables. Chapter 9 discusses fitting nonlinear regression equations to one or more independent variables. Additional topics in regression analysis are covered in Chapter 10. These topics include missing values, dummy variables, segmented regression, ridge regression, and path analysis.

BIBLIOGRAPHY

General References

Acton, F. S. 1959. *The analysis of straight-line data*, New York: Wiley.

Afifi, A. A., and Azen, S. P., 1979. *Statistical analysis: A computer oriented approach*. 2nd ed. New York: Academic Press.

Allen, D. M., and Cady, F. B., 1982. *Analyzing experimental data by regression.* Belmont, Calif.: Lifetime Learning.

Box, G. E. P. 1966. Use and abuse of regression. *Technometrics* 8: 625–629.

Brownlee, K. A. 1965. *Statistical theory and methodology in science and engineering.* 2nd ed. New York: Wiley.

Dixon, W. J., and Massey, F. J. 1983. *Introduction to statistical analysis.* 4th ed. New York: McGraw-Hill.

Dunn, O. J., and Clark, V. A. 1974. *Applied statistics: Analysis of variance and regression.* New York: Wiley.

Hald, A. 1952. *Statistical theory with engineering applications.* New York: Wiley.

Lewis-Beck, M. S. 1980. *Applied regression: An introduction.* Beverly Hills: Sage.

Mosteller, F., and Tukey, J. W. 1977. *Data analysis and regression.* Reading, Mass.: Addison-Wesley.

Thorndike, R. M. 1978. *Correlational procedures for research.* New York: Gardner Press.

Williams, E. J. 1959. *Regression analysis.* New York: Wiley.

Calibration

Berkson, J. 1969. Estimation of a linear function for a calibration line. *Technometrics* 11:649–660.

*Hunter, W. G., and Lamboy, W. F. 1981. A Bayesian analysis of the linear calibration problem. *Technometrics* 23:323–343.

Krutchkoff, R. G. 1967. Classical and inverse regression methods of calibration. *Technometrics* 9:425–440.

Lwin, T., and Maritz, J. S. 1982. An analysis of the linear-calibration controversy from the perspective of compound estimation. *Technometrics* 24:235–242.

Naszodi, L. J. 1978. Elimination of bias in the course of calibration. *Technometrics* 20:201–206.

Shukla, G. K. 1972. On the problem of calibration. *Technometrics* 14:547–554.

Outlier Detection

Belsley, D. A.; Kuh, E.; and Welsch, R. E. 1980. *Regression diagnostics: Identifying influential data and sources of colinearity.* New York: Wiley.

Cook, R. D. 1977. Detection of influential observations in linear regression. *Technometrics* 19:15–18.

*———. 1979. Influential observations in linear regression. *Journal of the American Statistical Association* 74:169–174.

*Cook, R. D., and Weisberg, S. 1980. Characterization of an empirical influence function for detecting influential cases in regression. *Technometrics* 22:495–508.

Transformations

Bartlett, M. S. 1947. The use of transformations. *Biometrics* 3:39–52.

Bennett, C. A., and Franklin, N. L. 1954. *Statistical analysis in chemistry and the chemical industry*. New York: Wiley.

*Bickel, P. J., and Doksum, K. A. 1981. An analysis of transformations revisited. *Journal of the American Statistical Association* 76:296–311.

*Box, G. E. P., and Cox, D. R. 1964. An analysis of transformations. *Journal of the Royal Statistical Society* Series B, 26:211–252.

Draper, N. R., and Hunter, W. G. 1969. Transformations: Some examples revisited. *Technometrics* 11:23–40.

Hald, A. 1952. *Statistical theory with engineering applications*. New York: Wiley.

Kowalski, C. J. 1970. The performance of some rough tests for bivariate normality before and after coordinate transformations to normality. *Technometrics*. 12:517–544.

Mosteller, F., and Tukey, J. W. 1977. *Data analysis and regression*. Reading, Mass.: Addison-Wesley.

*Tukey, J. W. 1957. On the comparative anatomy of transformations. *Annals of Mathematical Statistics* 28:602–632.

Time Series and Forecasting

*Box, G. E. P., and Jenkins, G. M. 1976. *Time series analysis: Forecasting and control*. Rev. ed. San Francisco: Holden-Day.

Levenbach, H., and Cleary, J. P. 1981. *The beginning forecaster: The forecasting process through data analysis*. Belmont, Calif.: Lifetime Learning.

———. 1982. *The professional forecaster: The forecasting process through data analysis*, Belmont, Calif.: Lifetime Learning.

McCleary, R., and Hay, R. A. 1980. *Applied time series analysis for the social sciences*. Beverly Hills: Sage.

Montgomery, D. C., and Johnson, L. A. 1976. *Forecasting and time series analysis*. New York: McGraw-Hill.

PROBLEMS

6.1   In Table 8.1 financial performance data of 30 chemical companies are presented. Use growth in earnings per share, labeled EPS5, as the dependent variable and growth in sales, labeled SALESGR5, as the independent variable. (A description of these variables is given in Section 8.3.) Plot the data, compute a regression line, and test that $\beta = 0$ and $\alpha = 0$. Are earnings affected by sales growth for these chemical companies? Which company's earnings were highest, considering its growth in sales?

6.2   From the family lung function data set in Appendix B, perform a regression analysis of weight on height for fathers. Repeat for mothers. Determine the correlation coefficient and the regression equation for fathers and mothers. Test that the coefficients are significantly different from zero for both sexes. Also, find the standardized regression equation and report it. Would you suggest removing the woman who weighs 267 lb from the data set? Discuss why the correlation for fathers appears higher than that for mothers.

6.3   In Problem 4.6 the New York Stock Exchange Composite Index for August 9 through September 17, 1982, was presented. Data for the daily volume of transactions, in millions of shares, for all of the stocks traded on the exchange and included in the Composite Index are given below, together with the Composite Index values, for the same period as that in Problem 4.6. Describe how volume appears to be affected by the price index, using regression analysis. Describe whether or not the residuals from your regression analysis are serially correlated. Plot the index versus time and volume versus time, and describe the relationships you see in these plots.

Day	Month	Index	Volume
9	Aug.	59.3	63
10	Aug.	59.1	63
11	Aug.	60.0	59
12	Aug.	58.8	59
13	Aug.	59.5	53
16	Aug.	59.8	66
17	Aug.	62.4	106
18	Aug.	62.3	150
19	Aug.	62.6	93
20	Aug.	64.6	113
23	Aug.	66.4	129
24	Aug.	66.1	143
25	Aug.	67.4	123
26	Aug.	68.0	160
27	Aug.	67.2	87
30	Aug.	67.5	70
31	Aug.	68.5	100
1	Sept.	67.9	98
2	Sept.	69.0	87
3	Sept.	70.3	150
6	Sept.	–	–
7	Sept.	69.6	81
8	Sept.	70.0	91
9	Sept.	70.0	87
10	Sept.	69.4	82
13	Sept.	70.0	71
14	Sept.	70.6	98
15	Sept.	71.2	83
16	Sept.	71.0	93
17	Sept.	70.4	77

6.4   For the data in Problem 6.3, pretend that the index increases linearly in time and use linear regression to obtain an equation to forecast the index value as a function of time. Using "volume" as a weight variable, obtain a weighted least squares forecasting equation. Does weighted least squares help the fit? Obtain a recent value of the index (from a newspaper). Does either forecasting equation predict the true value correctly (or at least somewhere near it)? Explain.

6.5   Repeat Problem 6.2, using log(weight) and log(height) in place of the original variables. Using graphical and numerical devices, decide whether the transformations help.

6.6   Examine the plot you produced in Problem 6.1 and choose some transformation for X and/or Y and repeat the analysis described there. Compare the correlation coefficients for the original and transformed variables, and decide whether the transformation helped. If so, which transformation was helpful?

7 | **MULTIPLE REGRESSION AND CORRELATION**

7.1 WHAT WILL YOU LEARN FROM THIS CHAPTER?

From this chapter you will learn how to examine the relationship between one dependent or response variable Y and a set of independent predictor variables X_1 to X_P. In particular you will learn:

- ▶ When multiple regression is used (7.2, 7.3).
- ▶ The correct meaning of the phrases multiple linear regression, multiple correlation, and partial correlation (7.4, 7.5).
- ▶ How to obtain appropriate equations and statistics to describe the interrelationships (7.4, 7.5, 7.6, 7.7).
- ▶ How to predict the value of one variable from a set of values of other variables (7.4).
- ▶ How to read and interpret a correlation or covariance matrix (7.5, 7.9).
- ▶ How to perform a residual analysis in order to improve the derived equations (7.8).
- ▶ How and when to use transformations to improve the equations (7.8).
- ▶ About the role of interactions in regression analysis (7.8).

▶ How to compute and compare regression equations for subgroups of data (7.9).

▶ Which are the appropriate computer programs (7.10).

The methods described in this chapter for examining the interrelationships among a set of variables are fundamental to an understanding of all of the multivariate techniques discussed in the remainder of the book. On a first reading Sections 7.8 and 7.9 may be skipped.

7.2 WHEN ARE MULTIPLE REGRESSION AND CORRELATION USED?

The methods described in this chapter are appropriate for studying the relationship between several X variables and one Y variable. By convention, the X variables are called *independent variables*, although they do not have to be statistically independent and are permitted to be intercorrelated. The Y variable is called the *dependent variable*.

As in Chapter 6, the data for multiple regression analysis can come from one of two situations:

1. *Fixed-X case*: The levels of the various X's are selected by the researcher or dictated by the nature of the situation. For example, in a chemical process an investigator can set the temperature, the pressure, and the length of time that a vat of material is agitated and then measure the concentration of a certain chemical. A regression analysis can be performed to *describe* or *explain* the relationship between the independent variables and the dependent variable (the concentration of a certain chemical) or to *predict* the dependent variable.

2. *Variable-X case*: A random sample of individuals is taken from a population, and the X variables and the Y variable are measured on each individual. For example, a sample of adults might be taken from Los Angeles and information obtained on their age, income, and education in order to see whether these variables predict their attitudes toward air pollution.

Regression and correlation analyses can be performed in the variable-X case. Both *descriptive* and *predictive information* can be obtained from the results.

Multiple regression is one of the most commonly used statistical methods, and an understanding of its use will help you comprehend other multivariate techniques.

7.3 DATA EXAMPLE

In Chapter 6 the data from a sample of factory workers, age 20–24, were analyzed. These data fit the variable-X case. Height was used as the X variable in order to predict FEV1, and the following equation was obtained:

$$FEV1 = -4.39 + .126(\text{height, in inches})$$

Several such equations could have been obtained, one for each age interval. However, in analyzing such data, investigators usually want to obtain a single equation that is useful with a wider age range. Because FEV1 decreases with age in adults, we need to fit a multiple regression with both age and height as independent variables and FEV1 as the dependent variable. We expect the slope coefficient for age to be negative and the slope coefficient for height to be positive. Also, since prolonged smoking is known to decrease FEV1, separate regression equations should be fitted for smokers and nonsmokers.

A geometric representation of the simple regression of FEV1 on age and height, respectively, is shown in Figures 7.1a and 7.1b. The multiple regression equation is represented by a plane, as shown in Figure 7.1c. Note that the plane slopes upward as a function of height and downward as a function of age. A hypothetical individual whose FEV1 is large relative to his age and height appears above both simple regression lines as well as above the multiple regression plane.

For illustrative purposes Figure 7.2 shows how a plane can be constructed. Constructing such a regression plane involves the following steps:

1. Draw lines on the X_1, Y wall, setting $X_2 = 0$.
2. Draw lines on the X_2, Y wall, setting $X_1 = 0$.

3. Drive nails in the walls at the lines drawn in steps 1 and 2.

4. Connect the pairs of nails by strings and tighten the strings.

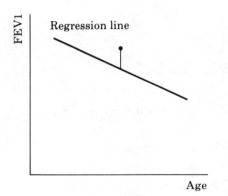

a. Simple Regression of FEV1 on Age

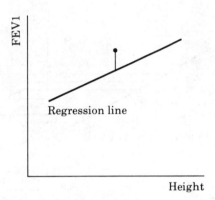

b. Simple Regression of FEV1 on Height

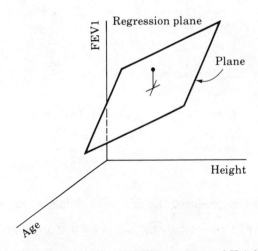

c. Multiple Regression of FEV1 on Age and Height

FIGURE 7.1. Hypothetical Representation of Simple and Multiple Regression Equations of FEV1 on Age and Height

The resulting strings in step 4 form a plane. This plane is the regression plane of Y on X_1 and X_2. The mean of Y at a given X_1 and X_2 is the point on the plane vertically above the point X_1, X_2.

Data for 448 smokers were analyzed by BMDP1R, and the following descriptive statistics were printed:

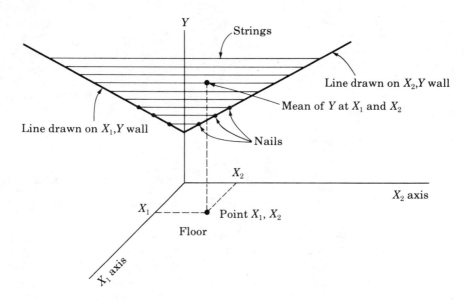

FIGURE 7.2. Visualizing the Construction of a Plane

VARIABLE	MEAN	STD DEVIATION
AGE	39.43	11.87
HEIGHT	68.76	2.63
FEV1	3.62	0.82

The program also produced the following estimated regression equation:

$$\hat{\text{FEV1}} = -1.40 - 0.036(\text{age}) + 0.093(\text{height})$$

As expected, there is a positive slope associated with height and a negative slope associated with age. For predictive purposes, we would expect a 22-year-old male whose height is 70 in., for example, to have an FEV1 value of $-1.40 - 0.036(22) + 0.093(70) = 4.32$ L.

Note that a value of $-4.39 + 0.126(70) = 4.4$ L will be obtained for a person in the age group 20–24 for the same height when we are using the *first* equation given in this section. In the single-predictor equation the coefficient for height is 0.126. This value is the rate of change of FEV1 for young males as a function of height when no other variables are taken into account. With two predictors the coefficient for height is 0.093, which is interpreted as the rate of change of FEV1 as a function of height *after adjusting for age*. The latter slope is also called the *partial regression coefficient* of FEV1 on height after adjusting for age. Even when both equations are derived from the same sample, the simple and partial regression coefficients may not be comparable.

The output of this program includes other items that will be discussed later in this chapter.

7.4 DESCRIPTION OF TECHNIQUES: FIXED X CASE

In this section we present the background assumptions, model, and formulas necessary for an understanding of multiple linear regression for the fixed-X case. Computations can quickly become tedious when there are several independent variables, so we assume that you will obtain output from a packaged computer program. Therefore we present a minimum of formulas and place the main emphasis on the techniques and interpretation of results. This section is slow reading and requires concentration.

Since there is more than one X variable, we use the notation $X_1, X_2, \ldots, X_P$ to represent P possible variables. In packaged programs these variables may appear in the output as X(1), X1, VAR 1, etc. For the fixed-X case values of the X variables are assumed to be fixed in advance in the sample. At each combination of levels of the X variables, we conceptualize a distribution of the values of Y. This distribution of Y values is assumed to have a mean value equal to $\alpha + \beta_1 X_1 + \beta_2 X_2 + \cdots + \beta_P X_P$ and a variance equal to σ^2 at given levels of $X_1, X_2, \ldots, X_P$.

When $P = 2$, the surface is a plane, as depicted in Figure 7.1c or 7.2. The parameter β_1 is the rate of change of the mean of Y as a function of X_1, where the value of X_2 is held fixed. Similarly, β_2 is the rate of change of the mean of Y as a function of X_2 when X_1 is fixed. Thus β_1 and β_2 are the slopes of the regression plane with regard to X_1 and X_2.

When $P > 2$, the regression plane generalizes to a so-called *hyperplane*, which cannot be represented geometrically on two-dimensional paper. Some people conceive the vertical axis as always representing the Y variable, and they think of all of the X variables as being represented by the horizontal plane. In this situation the hyperplane can still be imagined as in Figure 7.1c. The parameters $\beta_1, \beta_2, \ldots, \beta_P$ represent the slope of the regression hyperplane with respect to $X_1, X_2, \ldots, X_P$, respectively. The betas are called *partial regression coefficients*. For example, β_1 is the rate of change of the mean of Y as a function of X_1 when the levels of $X_2, \ldots, X_P$ are held fixed. In this sense it represents the change of the mean of Y as a function of X_1 after adjusting for $X_2, \ldots, X_P$.

Again we assume that the variance of Y is homogeneous over the range of concern of the X variables. Usually, for the fixed-X case P is rarely larger than three or four.

Least Squares Method

As for simple linear regression, the method of least squares is used to obtain estimates of the parameters. These estimates, denoted by $A, B_1, B_2, \ldots, B_P$, are printed in the output of any multiple regression program. The estimate A is usually labeled "intercept," and $B_1, B_2, \ldots, B_P$ are usually given in tabular form under the label "coefficient" or "regression coefficient" by variable name. The formulas for these estimates are mathematically derived by minimizing the sums of the squared vertical deviations. Formulas may be found in many of the books listed in the Bibliography at the end of this chapter.

The *predicted value* $\hat{Y}$ for a given set of values $X_1^*, X_2^*, \ldots, X_P^*$ is then calculated as

$$\hat{Y} = A + B_1X_1^* + B_2X_2^* + \cdots + B_PX_P^*$$

The estimate σ^2 is the *residual mean square*, obtained as

$$S^2 = \text{RES. MS.} = \frac{\Sigma(Y - \hat{Y})^2}{N - P - 1}$$

The square root of the residual mean square is called the *standard error of the estimate*. It represents the variation unexplained by the regression plane.

Packaged programs usually print the value of S^2 and the standard errors for the regression coefficients (and sometimes for the intercept). These quantities are useful in computing confidence intervals and prediction intervals around the computed $\hat{Y}$. These intervals can be computed from the output in the following manner.

Prediction and Confidence Intervals

To obtain *prediction* and *confidence intervals*, we need access to the estimated covariances or correlations among the regression coefficients. For example, BMDP1R prints these correlations as an optional output. When these estimates are available, we compute the estimated variance of $\hat{Y}$ as

$$\text{Var } \hat{Y} = \frac{S^2}{N} + [(X_1^* - \overline{X}_1)^2 \text{ Var } B_1 + (X_2^* - \overline{X}_2)^2 \text{ Var } B_2 + \cdots$$

$$+ (X_P^* - \overline{X}_P)^2 \text{ Var } B_P] + [2(X_1^* - \overline{X}_1)(X_2^* - \overline{X}_2) \text{ Cov}(B_1, B_2)$$

$$+ 2(X_1^* - \overline{X}_1)(X_3^* - \overline{X}_3) \text{ Cov}(B_1, B_3) + \cdots]$$

The variances (Var) of the various B_i are computed as the squares of the standard errors of the B_i, i going from 1 to P. The covariances (Cov) are computed from the standard errors and the correlations among the regression coefficients. For example,

$$\text{Cov } (B_1, B_2) = (\text{standard error } B_1) (\text{standard error } B_2) [\text{Corr}(B_1, B_2)]$$

If $\hat{Y}$ is interpreted as an estimate of the mean of Y at $X_1^*, X_2^*, \ldots, X_P^*$, then the confidence interval for this mean is computed as

$$\text{CI(mean } Y \text{ at } X_1^*, X_2^*, \ldots, X_P^*) = \hat{Y} \pm t \sqrt{\text{Var } \hat{Y}}$$

where t is the $100(1 - \alpha/2)$ percentile of the t distribution with $N - P - 1$ degrees of freedom.

When $\hat{Y}$ is interpreted as an estimate of the Y value for an individual, then the variance of $\hat{Y}$ is increased by S^2, similar to what is done in simple linear regression. Then the prediction interval is

$$\text{PI(individual } Y \text{ at } X_1^*, X_2^*, \ldots, X_P^*) = \hat{Y} \pm t \sqrt{S^2 + \text{Var } \hat{Y}}$$

where t is the same as it is for the confidence interval above. Note that these intervals require the additional assumption that for any set of levels of X the values of Y are normally distributed.

In summary, for the fixed-X case we presented the model for multiple linear regression analysis and discussed estimates of the parameters. Next, we present the variable-X model.

7.5 DESCRIPTION OF TECHNIQUES: VARIABLE-X CASE

For this model the X's and the Y variable are random variables measured on cases that are randomly selected from a population. An example is the factory worker data, where FEV1, height, age, and smoking status were measured on a sample of workers. As in Chapter 6, the previous discussion and formulas given for the fixed-X case apply to the variable-X case. (This result is justifiable theoretically by conditioning on the X variable values that happen to be obtained in the sample.) Furthermore, since the X variables are also random variables, the joint distribution of Y, X_1, $X_2, \ldots$, X_P is of interest.

When there is only one X variable, it was shown in Chapter 6 that the bivariate distribution of X and Y can be characterized by its region of concentration. These regions are ellipses if the data come from a bivariate normal distribution. Two such ellipses and their corresponding regression lines are illustrated in Figures 7.3a and 7.3b.

When two X variables exist, the regions of concentration become three-dimensional forms. These forms take the shape of ellipsoids if the joint distribution is multivariate normal. In Figure 7.3c such an ellipsoid with the corresponding regression plane of Y on X_1 and X_2 is illustrated. The regression plane passes through the point representing the population means of the three variables and intersects the vertical axis at a distance α from the origin. Its position is determined by the slope coefficients β_1 and β_2. When $P > 2$, we can imagine the horizontal plane representing all the X variables and the vertical plane representing Y. The ellipsoid of concentration becomes the so-called *hyperellipsoid*.

Estimation of Parameters

For the variable-X case several additional parameters are needed to characterize the joint distribution. These include the population means, $\mu_1, \mu_2, \ldots,$ μ_P and μ_Y, the population standard deviations, $\sigma_1, \sigma_2, \ldots, \sigma_P$ and σ_Y, and the covariances of the X and Y variables.

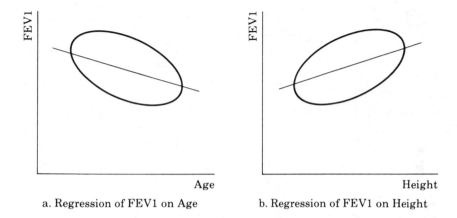

a. Regression of FEV1 on Age   b. Regression of FEV1 on Height

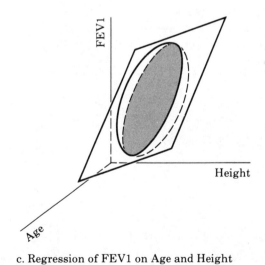

c. Regression of FEV1 on Age and Height

FIGURE 7.3. Hypothetical Regions of Concentration and Corresponding Regression Lines and Planes for the Population Variable-X Model

The variances and covariances are usually arranged in the form of a square array called the *covariance matrix*. For example, if $P = 3$, the covariance matrix is a four-by-four array of the form

		1	2	3	Y
		\multicolumn{3}{c}{X}	
	1	σ_1^2	σ_{12}	σ_{13}	σ_{1Y}
X	2	σ_{12}	σ_2^2	σ_{23}	σ_{2Y}
	3	σ_{13}	σ_{23}	σ_3^2	σ_{3Y}
	Y	σ_{1Y}	σ_{2Y}	σ_{3Y}	σ_Y^2

A dashed line is included to separate the dependent variable Y from the independent variables. The estimates of the means, variances, and covariances are available in the output of most regression programs. The estimated variances and covariances are denoted by S instead of σ.

In addition to the estimated covariance matrix, an estimated *correlation matrix* is also available, which is given in the same format:

		1	2	3	Y
		\multicolumn{3}{c}{X}	
	1	1	r_{12}	r_{13}	r_{1Y}
X	2	r_{12}	1	r_{23}	r_{2Y}
	3	r_{13}	r_{23}	1	r_{3Y}
	Y	r_{1Y}	r_{2Y}	r_{3Y}	1

Since the correlation of a variable with itself must be equal to 1, the diagonal elements of the correlation matrix are equal to 1. The off-diagonal elements are the simple correlation coefficients described in Section 6.5. As before, their numerical values always lie between +1 and −1.

As an example, the following covariance matrix is obtained from data from the 448 factory workers who are smokers. For the sake of illustration we include a third X variable, FVC, defined as the forced vital capacity, which is the total volume of air exhaled by the worker in one breath:

		X			Y
		Age	Height	FVC	FEV1
X	Age	140.80	−3.62	−5.41	−5.35
	Height	−3.62	6.94	1.14	0.78
	FVC	−5.41	1.14	0.85	0.65
Y	FEV1	−5.35	0.78	0.65	0.67

Note that this matrix is a symmetric matrix. That is, if a mirror were placed along the diagonal, then the elements in the lower triangle would be mirror images of those in the upper triangle. For example, the covariance of age and height is −3.62. Also, the covariance between height and age is −3.62.

Symmetry holds also for the correlation matrix that follows:

		X			Y
		Age	Height	FVC	FEV1
X	Age	1	−0.12	−0.49	−0.55
	Height	−0.12	1	0.47	0.36
	FVC	−0.49	0.47	1	0.87
Y	FEV1	−0.55	0.36	0.87	1

All of the correlations are computed from the appropriate elements of the covariance matrix. For example, the correlation between age and height is

$$r_{12} = \frac{S_{12}}{\sqrt{S_1^2 S_2^2}} = \frac{S_{12}}{S_1 S_2}$$

or

$$-0.12 = \frac{-3.62}{\sqrt{140.8}\sqrt{6.94}}$$

Note that the correlations are computed between Y and each X as well as among the X variables. We will return to interpreting these simple correlations in Section 7.7.

Multiple Correlation

So far, we have discussed the concept of correlation between two variables. It is also possible to describe the strength of the linear relationship between Y and a set of X variables by using the *multiple correlation coefficient*. In the population we will denote this multiple correlation by $\mathcal{R}$. It represents the simple correlation between Y and the corresponding point on the regression plane for all possible combinations of the X variables. Each individual in the population has a Y value and a corresponding point on the plane computed as

$$Y' = \alpha + \beta_1 X_1 + \beta_2 X_2 + \cdots + \beta_P X_P$$

Correlation $\mathcal{R}$ is the population simple correlation between all such Y and Y' values. The numerical value of $\mathcal{R}$ cannot be negative. The maximum possible value for $\mathcal{R}$ is 1.0, indicating a perfect fit of the plane to the points in the population.

Another interpretation of the $\mathcal{R}$ coefficient for multivariate normal distributions involves the concept of the *conditional distribution*. This distribution describes all of the Y values of individuals whose X values are specified at certain levels. The variance of the conditional distribution is the variance of the Y values about the regression plane in a vertical direction. For multivariate normal distributions this variance is the same at all combinations of levels of the X variables and is denoted by σ^2. The following fundamental expression relates to σ^2 and σ_Y^2:

$$\sigma^2 = \sigma_Y^2 (1 - \mathcal{R}^2)$$

Rearrangement of this expression shows that

$$\mathcal{R}^2 = \frac{\sigma_Y^2 - \sigma^2}{\sigma_Y^2} = 1 - \frac{\sigma^2}{\sigma_Y^2}$$

When the variance about the plane, or σ^2, is small relative to σ_Y^2, then the squared multiple correlation $\mathcal{R}^2$ is close to 1. When the variance σ^2 about the plane is almost as large as the variance σ_Y^2 of Y, then $\mathcal{R}^2$ is close to zero. In this case the regression plane does not fit the Y values much better than μ_Y. Thus the multiple correlation squared suggests the proportion of the variation accounted for by the regression plane. As in the case of simple linear regression, another interpretation of $\mathcal{R}$ is that $\sqrt{(1 - \mathcal{R}^2)} \times 100$ is the percentage of σ_Y *not* explained by X_1 to X_P.

Note that σ_Y^2 and σ^2 can be estimated from a computer output as follows:

$$S_Y^2 = \frac{\Sigma (Y - \overline{Y})^2}{N - 1} = \frac{\text{total sum of squares}}{N - 1}$$

and

$$S^2 = \frac{\Sigma (Y - \hat{Y})^2}{N - P - 1} = \frac{\text{residual sum of squares}}{N - P - 1}$$

Partial Correlation

Another correlation coefficient is useful in measuring the degree of dependence between two variables after adjusting for the linear effect of one or more of the other X variables. For example, suppose an educator is interested in the correlation between the total scores of two tests, T_1 and T_2, given to twelfth graders. Since both scores are probably related to the student's IQ, it would be reasonable to first remove the linear effect of IQ from both T_1 and T_2 and then find the correlation between the adjusted scores. The resulting correlation coefficient is called a *partial correlation coefficient* between T_1 and T_2 given IQ.

For this example we first derive the regression equations of T_1 on IQ and T_2 on IQ. These equations are displayed in Figures 7.4a and 7.4b. Consider an individual whose IQ is IQ* and whose actual scores are T_1^* and T_2^*. The test scores defined by the population regression lines are $\tilde{T}_1$ and $\tilde{T}_2$, respectively. The adjusted test scores are the residuals $T_1^* - \tilde{T}_1$ and $T_2^* - \tilde{T}_2$. These adjusted scores are calculated for each individual in the population, and the simple correlation between them is computed. The resulting value is defined to be the population partial correlation coefficient between T_1 and T_2 with the linear effects of IQ removed.

Formulas exist for computing partial correlations directly from the population simple correlations or other parameters without obtaining the residuals. In the above case suppose that the simple correlation for T_1 and T_2 is denoted by ρ_{12}, for T_1 and IQ by ρ_{1q}, and T_2 and IQ by ρ_{2q}. Then the partial correlation of T_1 and T_2 given IQ is derived as

$$\rho_{12.q} = \frac{\rho_{12} - \rho_{1q}\rho_{2q}}{\sqrt{(1 - \rho_{1q}^2)(1 - \rho_{2q}^2)}}$$

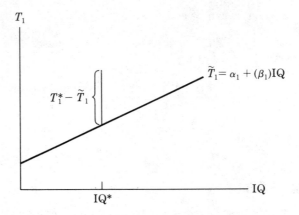

a. Regression Line for T_1

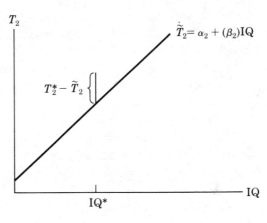

b. Regression Line for T_2

FIGURE 7.4. Hypothetical Population Regressions of T_1 and T_2 Scores, Illustrating the Computation of a Partial Correlation Coefficient

In general, for any three variables denoted by i, j, and k,

$$\rho_{ij.k} = \frac{\rho_{ij} - \rho_{ik}\rho_{jk}}{\sqrt{(1 - \rho_{ik}^2)(1 - \rho_{jk}^2)}}$$

It is also possible to compute partial correlations between any two variables after removing the linear effect of two or more other variables. In this case the residuals are deviations from the regression planes on the variables whose effects are removed. The partial correlation is the simple correlation between the residuals.

Packaged computer programs, discussed further in Section 7.10, calculate estimates of multiple and partial correlations. We note that the formula given above can be used with sample simple correlations to obtain sample partial correlations.

This section has covered the basic concepts in multiple regression and correlation. Additional items found in output will be presented in Section 7.8 and in Chapters 8, 9, and 10.

7.6 HOW TO INTERPRET THE RESULTS: FIXED-X CASE

In this section we discuss interpretation of the estimated parameters and present some tests of hypotheses for the fixed-X case.

As mentioned previously, the regression analysis might be performed for prediction: deriving an equation to predict Y from the X variables. This situation was discussed in Section 7.4. The analysis can also be performed for description: an understanding of the relationship between the Y and X variables.

In the fixed-X case the number of X variables is generaly small. The values of the slopes describe how Y changes with changes in the levels of X variables. In most situations a plane does not describe the response surface for the whole possible range of X values. Therefore it is important to define the region of concern for which the plane applies. The magnitudes of the β's and their estimates depend on this choice. In the interpretation of the relative magnitudes of the estimated B's, it is useful to compute $B_1\overline{X}_1$, $B_2\overline{X}_2$,..., $B_P\overline{X}_P$ and compare the resulting values. A large (relative) value of the magnitude of $B_i\overline{X}_i$ indicates a relatively important contribution of the variable X_i. Here the $\overline{X}$'s represent typical values within the region of concern.

If we restrict the range of concern here, the *additive model*, $\alpha + \beta_1X_1 + \beta_2X_2 + \cdots + \beta_PX_P$, is often an appropriate description of the underlying relationship. It is also sometimes useful to incorporate interactions or nonlinearities. This step can partly be achieved by transformations and will be discussed in Section 7.8. Examination of residuals can help you assess the adequacy of the model, and this analysis is discussed in Section 7.8.

Analysis of Variance

For a test of whether the regression plane is at all helpful in predicting the values of Y, the ANOVA table printed in most regression outputs can be used. The null hypothesis being tested is

$$H_0 : \beta_1 = \beta_2 = \cdots = \beta_P = 0$$

That is, the mean of Y is as accurate in predicting Y as the regression plane. The ANOVA table for multiple regression has the form given in Table 7.1, which is similar to that of Table 6.1.

When the fitted plane differs from a horizontal plane in a significant fashion, then the term $\Sigma (\hat{Y} - \overline{Y})^2$ will be large relative to the residuals from the plane $\Sigma (Y - \hat{Y})^2$. This result is the case in the factory workers example, where $F = 144.6$ with 2 and 445 degrees of freedom, as indicated in Table 7.2. Note that only data on smokers are used in this table; regression analysis for the rest of the workers is given later in Section 7.9. The computed value of $F = 144.6$ with $P = 2$ degrees of freedom in the numerator and $N - P - 1 = 445$ degrees of freedom in the denominator is compared with the printed F value in Appendix Table A.4 with v_1 (numerator) degrees of freedom across the top of the table and v_2 (denominator) degrees of freedom in the first column of the table for an appropriate α. For example, for $\alpha = 0.05$, $v_1 = 2$, and $v_2 = 120$ ($v_2 = 445$ is not tabled, so we take the next lower level), the tabled $F(1 - \alpha) = F(0.95) = 3.07$, from the body of the table. Since the computed F, 144.6, is much larger than the tabled F, 3.07, the null hypothesis is rejected.

The P value can be determined more precisely. In fact, it is printed in the output as $P = 0.00000$. This value should be interpreted and reported as $P < 0.00001$. Thus the variables "age" and "height" together help significantly in predicting an individual's FEV1.

If this blanket hypothesis is rejected, then the degree to which the regression equation fits the data can be assessed by examining a quantity called the *coefficient of determination*, defined as

$$\text{coefficient of determination} = \frac{\text{sum of squares regression}}{\text{sum of squares total}}$$

If the regression sum of squares is not in the program output, it can be obtained by subtraction, as follows:

TABLE 7.1. ANOVA Table for Multiple Regression

Source of Variation	Sums of Squares	df	Mean Square	F
Regression	$\Sigma\,(\hat{Y} - \overline{Y})^2$	P	SS_{reg}/P	MS_{reg}/MS_{res}
Residual	$\Sigma\,(Y - \hat{Y})^2$	$N - P - 1$	$SS_{res}/(N - P - 1)$	
Total	$\Sigma\,(Y - \overline{Y})^2$	$N - 1$		

TABLE 7.2. ANOVA Example from Factory Workers Data (Smokers Only)

Source of Variation	Sums of Squares	df	Mean Square	F
Regression	117.492	2	58.746	144.6
Residual	180.746	445	0.406	
Total	298.238	447		

regression sum of squares = total sum of squares − residual sum of squares

For the factory workers example,

$$\text{coefficient of determination} = \frac{117.492}{298.238} = 0.394$$

This value is an indication of the reduction in the variance of Y achieved by using X_1 and X_2 as predictors. In this case the variance around the regression plane is 39.4% less than the variance of the original Y values. Numerically, the coefficient of determination is equal to the square of the multiple correlation coefficient, and therefore it is called RSQUARE in the output of some computer programs. Although the multiple correlation coefficient is not meaningful in the fixed-X case, the interpretation of its square is valid.

Other Tests of Hypotheses

Tests of hypotheses can be used to assess whether variables are contributing significantly to the regression equation. It is possible to test these variables either singularly or in groups. For any specific variable X_i we can test the null hypothesis

$$H_0: \beta_i = 0$$

by computing

$$t = \frac{B_i - 0}{\text{SE}(B_i)}$$

and performing a one- or two-sided t test with $N - P - 1$ degrees of freedom. These t statistics are often printed in the output of packaged programs. Other programs print the corresponding F statistics ($t^2 = F$), which can test the same null hypothesis against a two-sided alternative.

When this test is performed for each X variable, the joint significance level cannot be determined. A method designed to overcome this uncertainty makes use of the so-called *Bonferroni inequality*. In this method, to compute the appropriate joint P value for any test statistic, we multiply the single P value obtained from the printout by the number of X variables we wish to test. The joint P value is then compared with the desired overall level.

For example, to test that age alone has an effect, we use

$$H_0: \beta_1 = 0$$

and from the computer output

$$t = \frac{-0.36 - 0}{0.0026} = -13.8$$

with 445 degrees of freedom. This result is also highly significant, indicating that the equation using age and height is significantly better than an equation using height alone.

Most computer outputs will display the P level of the t statistic for each regression coefficient. To get joint P values according to the Bonferroni inequality, we multiply each individual P value by the number of X variables before we compare it with the overall significance level α. For example, if $P = 0.015$ and there are two X's, then $2(0.015) = 0.03$, which could be compared with an α level of 0.05.

7.7 HOW TO INTERPRET THE RESULTS: VARIABLE-X CASE

In the variable-X case all of the Y and X variables are considered to be random variables. The means and standard deviations printed by packaged programs are used to estimate the corresponding population parameters. Frequently, the covariance and/or the correlation matrices are printed. In the remainder of this section we discuss the interpretation of the various correlation and regression coefficients found in the output.

Correlation Matrix

Most people find the correlation matrix more useful than the covariance matrix since all of the correlations are limited to the range −1 to +1. Note that because of the symmetry of these matrices, some programs will print only the terms on one side of the diagonal. The diagonal terms of a correlation matrix will always be 1 and thus sometimes are not printed.

Initial screening of the correlation matrix is helpful in obtaining a preliminary impression of the interrelationships among the X variables and between Y and each of the X variables. One way of accomplishing this screening is to first determine a cutoff point on the magnitude of the correlation, for example, 0.3 or 0.4. Then each correlation greater in magnitude than this number is underlined or highlighted. The resulting pattern can give a visual impression of the underlying interrelationships; i.e., highlighted correlations are indications of possible relationships. Correlations near +1 or −1 among the X variables indicate that the two variables are nearly perfect functions of each other, and the investigator should consider dropping one of them since they convey nearly the same information. Also, if the physical situation leads the investigator to expect larger correlations than those found in the printed matrix, then the presence of outliers in the data or of nonlinearities among the variables may be causing the discrepancies.

If tests of significance are desired, the following test can be performed. To test the hypothesis

$$H_0: \rho = 0$$

use the statistic given by

$$t = \frac{r\sqrt{N-2}}{\sqrt{1-r^2}}$$

where ρ is a particular population simple correlation and r is the estimated simple correlation. This statistic can be compared with the t table value with df $= N - 2$ to obtain one-sided or two-sided P values. Again, $t^2 = F$ with 1 and $N - 2$ degrees of freedom can be used for two-sided alternatives.

Some programs, such as the SPSS–X procedure PEARSON CORR, will print the P value for each correlation in the matrix. Since several tests are being performed in this case, it may be advisable to use the Bonferroni correction to the P value; i.e., each P is multiplied by the number of correlations (say M). This number of correlations can be determined as follows:

$$M = \frac{(\text{no. of rows})(\text{no. of rows} - 1)}{2}$$

After this adjustment to the P values is made, they can be compared with the nominal significance level α.

Standardized Coefficients

The interpretations of the regression plane, the associated regression coefficients, and the standard error around the plane are the same as those in the fixed-X case. All the tests presented in Section 7.6 apply here as well.

Another way to interpret the regression coefficients is to examine *standardized coefficients*. These are printed in the output of many regression programs and can be computed easily as

$$\text{standardized } B_i = B_i \left(\frac{\text{standard deviation of } X_i}{\text{standard deviation of } Y} \right)$$

These coefficients are the ones that would be obtained if the Y and X variables were standardized prior to performing the regression analysis. The standardized coefficients of the various X variables can be directly compared in order to determine the relative contribution of each to the regression plane. The larger the magnitude of the standardized B_i, the more X_i contributes to the prediction of Y. Comparing the unstandardized B directly does not achieve this result because of the different units and degrees of variability of the X variables. The regression equation itself should be reported for future use in terms of the unstandardized coefficients so that prediction can be made directly from the raw X values. Standardized regression coefficients are also useful in path analysis (see Section 10.6).

Multiple and Partial Correlations

As mentioned earlier, the multiple correlation coefficient is a measure of the strength of the linear relationship between Y and the set of variables X_1, $X_2, \ldots, X_P$. The multiple correlation has another useful property: It is the highest possible simple correlation between Y and any linear combination of X_1 to X_P. This property explains why R (the computed correlation) is never negative. In this sense the least squares regression plane maximizes the correlation between the set of X variables and the dependent variable Y. It therefore presents a numerical measure of how well the regression plane fits the Y values. When the multiple correlation R is close to zero, the plane barely predicts Y better than simply using $\overline{Y}$ to predict Y. A value of R close to 1 indicates a very good fit.

As in the case of simple linear regression, discussed in Chapter 6, R^2 is an estimate of the proportional reduction in the variance of Y achieved by fitting the plane. Again, the proportion of the standard deviation of Y around the plane is estimated by $\sqrt{1 - R^2}$. If a test of significance is desired, the hypothesis

$$H_0: \mathcal{R} = 0$$

can be tested by the statistic

$$F = \frac{R^2/P}{(1 - R^2)/(N - P - 1)}$$

which is compared with a tabled F with P and $N - P - 1$ degrees of freedom. Since this hypothesis is equivalent to

$$H_0: \beta_1 = \beta_2 = \cdots = \beta_P = 0$$

the F statistic is equivalent to the one calculated in Table 7.1.

As mentioned earlier, the partial correlation coefficient is a measure of the linear relationship between two variables after adjusting for the linear effect of a group of other variables. For example, suppose a regression of Y on X_1 and X_2 is fitted. The square of the partial correlation of Y on X_3 after adjusting X_1 and X_2 is the proportion of the variance of Y reduced by using X_3 as an additional X variable. The hypothesis

$$H_0: \text{partial } \rho = 0$$

can be tested by using a test statistic similar to the one for simple r, namely,

$$t = \frac{(\text{partial } r)\sqrt{N - Q - 2}}{\sqrt{1 - (\text{partial } r^2)}}$$

where Q is the number of variables adjusted for. The value of t is compared with the tabled t value with $N - Q - 2$ degrees of freedom. The square of the t statistic is an F statistic with df $= 1$ and $N - Q - 2$; the F statistic may be used to test the same hypothesis against a two-sided alternative.

In the special case where Y is regressed on X_1 to X_P, a test that the partial correlation between Y and any of the X variables, say X_i, after adjusting for the remaining X variables zero is, is equivalent to testing that the regression coefficient β_i is zero.

In Chapter 8, where variable selection is discussed, partial correlations will play an important role.

7.8 RESIDUAL ANALYSIS AND TRANSFORMATIONS

In this section we discuss residual analysis, transformations, polynomial regression, and the incorporation of interaction terms into the equation.

Residuals

The use of residuals was discussed in Section 6.8 in the case of simple linear regression. Residual analysis is even more important in multiple regression because the scatter diagram of Y and all of the X variables cannot be adequately portrayed on two-dimensional paper. Scatter plots of the residual $Y - \hat{Y}$ against $\hat{Y}$ and against each of the X_i variables can be obtained from various packaged regression programs. Initial scanning of these plots can help the investigator identify individual potential outliers or blunders as well as groups of suspect observations. A listing of the residuals together with an identifying variable allows the investigator to determine which cases contain the suspected outliers. The types of residuals and methods for their analysis described in Section 6.8 also apply here.

As in the case of simple linear regression, Cook's distance statistic is useful in determining the effect of an observation on the regression plane. This distance function determines the degree of influence of each observation on the set of estimated B_i. For a given observation Cook's distance is a measure of the distance between the values of B_i with all cases present and the values of B_i with that observation removed. If Cook's distance is greater than the 50th percentile of the F distribution with df = $P + 1$ and $N - P - 1$, then that case should be carefully examined and possibly removed (see Cook 1977). That percentile is in the range 0.7–0.9 for $P = 1, 2, \ldots, 10$.

The similarity of the spread of the residuals around the horizontal line through zero allows the investigator to assess the validity of the homogeneity-of-variance assumption. If, for example, a *fan-shaped pattern* of residuals (Figure 6.7b) emerges as a function of one of the X variables and not of the others, transforming this variable might help. On the other hand, a fan-shaped appearance of the residuals with each of the X_i may suggest taking the logarithm or square root of the Y values and performing the regression on the transformed values. As in the case of simple linear regression, transformations of Y should be avoided, when possible, because they tend to obscure the interpretation of the regression equation (see Section 6.11).

Transformations

In examining the adequacy of the multiple linear regression model, the investigator may wonder whether the transformations of some or all of the X variables might improve the fit. (See Section 6.11 for review.) For this purpose the residual plots against the X variables may suggest certain transformations. For example, if the residuals against X_i take the humped-shaped form shown in Figure 6.7c, it may be useful to transform X_i. Reference to Figure 6.9a suggests that log X_i may be the appropriate transformation. In this situation log X_i can be used in place of or together with X_i in the same equation. When both X_i and log X_i are used as predictors, this method of analysis is an instance of attempting to improve the fit by including an additional variable. Other commonly used additional variables are the square (X_i^2) and square root $(X_i^{1/2})$ of appropriate X_i variables; other candidates may be new variables altogether. Plots against candidate variables should be obtained in order to check whether the residuals are related to them.

Caution should be taken at this stage not to be deluged by a multitude of additional variables. For example, it may be preferable in certain cases to use log X_i instead of X_i and X_i^2; the interpretation of the equation becomes more difficult as the number of predictors increases. The subject of variable selection will be discussed in Chapter 8.

Polynomial Regression

A subject related to variable transformations is the so-called *polynomial regression*. For example, suppose that the investigator starts with a single variable X and that the scatter diagram of Y on X indicates a curvilinear function, as shown in Figure 7.5a. An appropriate regression equation in this case has the form

$$\hat{Y} = A + B_1X + B_2X^2$$

This equation is, in effect, a multiple regression equation with $X_1 = X$ and $X_2 = X^2$. Thus the multiple regression programs can be used to obtain the estimated curve. Similarly, for Figure 7.5b the regression curve has the form

$$\hat{Y} = A + B_1X + B_2X^2 + B_3X^3$$

which is a multiple regression equation with $X_1 = X$, $X_2 = X^2$, and $X_3 = X^3$. Both of these equations are examples of polynomial regression equations with degrees two and three, respectively. Some computer packages have special programs that perform polynomial regression without the need for the user to make transformations (see Section 7.10).

Interactions

Another issue in model fitting is to determine whether the X variables interact. If the effects of two variables X_i and X_j are *not* interactive, then they appear as $B_iX_i + B_jX_j$ in the regression equation. In this case the effects of the two variables are said to be *additive*. Another way of expressing this concept is to say that there is *no interaction* between X_i and X_j. If the additive terms for these variables do not completely specify their effects on the dependent variable Y, then interaction of X_i and X_j is said to be present. This phenomenon can be observed in many situations. For example, in chemical processes the additive effects of two agents are often not an accurate reflection of their combined effect since synergies or catalytic effects are often present. Similarly, in studies of economic growth the interactions between the two major factors labor and capital are important in predicting outcome. It is therefore often advisable to include additional variables in the regression equation to represent interactions. A commonly used practice is to add the product X_iX_j to the set of X variables to represent the

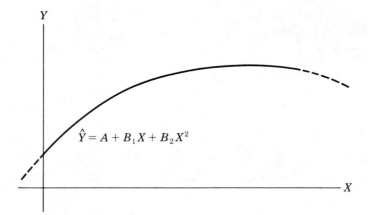

a. Single Inflection Point

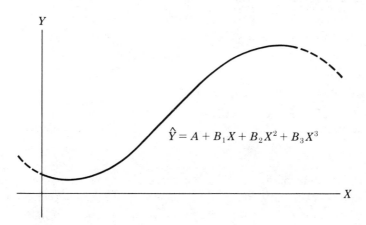

b. Two Inflection Points

FIGURE 7.5. Hypothetical Polynomial Regression Curves

interaction between X_i and X_j. The use of so-called dummy variables is a
method of incorporating interactions and will be discussed in Chapter 10.

As a check for the presence of interactions, tables can be constructed from
a list of residuals, as follows: The ranges of X_i and X_j are divided into
intervals, as shown in the examples in Tables 7.3 and 7.4. The residuals for
each cell are found from the output and simply classified as positive or

TABLE 7.3. Examination of Percentage of Positive Residuals for Detecting Interactions: No Interaction

	X_j		
X_j	Low	Medium	High
Low	50%	52%	48%
Medium	52%	48%	50%
High	49%	50%	51%

TABLE 7.4. Examination of Percentage of Positive Residuals for Detecting Interactions: Interactions Present

	X_j		
X_j	Low	Medium	High
Low	20%	40%	45%
Medium	40%	50%	60%
High	55%	60%	80%

negative. The percentage of *positive residuals* in each cell is then recorded. If these percentages are around 50%, as shown in Table 7.3, then no interaction term is required. If the percentages vary greatly from 50% in the cells, as in Table 7.4, then an interaction term could most likely improve the fit. This search for interaction effects should be made after consideration of transformations and selection of additional variables.

7.9 OTHER OPTIONS IN COMPUTER PROGRAMS

Because regression analysis is a very commonly used technique, packaged programs offer a bewildering number of options to the user. But do not be tempted into using options that you do not understand well. As a guide in selecting options, in this section we briefly discuss some of the more popular ones. Other options are often, but not always, described in user guides.

The options for the use of *weights for the observations* and for *regression through the origin* were discussed in Section 6.9 for simple linear regression. These options are also available for multiple regression, and the discussion in Section 6.9 applies here as well. Be aware that regression through the origin means that the mean of Y is assumed to be zero when *all* X_i equal zero. This assumption is often unrealistic.

In Section 7.5 we described the matrix of correlations among the estimated slopes, and we indicated how this matrix can be used to find the covariances among the slopes. Covariances can be obtained directly as optional output in some programs.

Multicollinearity

In practice, a problem called *multicollinearity* occurs when some of the X variables are highly intercorrelated. A considerable amount of effort is devoted in the statistical literature to dealing with this problem (see Wonnacott and Wonnacott 1979; Belsley, Kuh, and Welsch 1980). The main point is that when multicollinearity is present, the computed estimates of the regression coefficients are unstable and their interpretation becomes tenuous.

One simple way to check for multicollinearity is to examine the correlations among the X variables. If, for example, X_1 and X_2 are highly correlated (say greater than 0.95), then it may be simplest to use only one of them since one variable conveys essentially all of the information contained in the other.

When more subtle patterns of correlations exist, the *tolerance option* available in many programs is useful. The tolerance associated with any variable X_i is defined as 1 minus the squared multiple correlation between that X_i and the remaining X variables. When tolerance is small, say less than 0.01, then the investigator should discard the variable with the smallest tolerance. The inverse of the tolerance, called the *variance inflation factor* (VIF), is also printed for each variable by some programs. Some programs offer more complicated options for investigating multicollinearity, but the use of tolerance as indicated is usually sufficient for handling the problem. If moderate multicollinearity is present, the investigator may consider using ridge regression, as discussed in Chapter 10.

Comparing Regression Planes

Sometimes, it is desirable to examine the regression equations for *subgroups* of the population. For example, different regression equations for subgroups subjected to different treatments can be derived. In the program it is often convenient to designate the subgroups as various levels of a *grouping varia-ble*. In the factory worker example, for instance, we presented the regression equation for 448 smokers:

$$FEV1 = -1.397 - 0.036(age) + 0.093(height)$$

An equation was derived also for 231 nonsmokers, as

$$FEV1 = -0.906 - 0.030(age) + 0.084(height)$$

In Table 7.5 the statistical results for both groups, as well as for smokers and nonsmokers combined, are presented. From the table we see that 448/679, or 66%, of the males smoke. This result is much higher than the 42% male smokers found in the depression study described in Chapter 3. The mean age for smokers among the factory workers is 3 years older than for nonsmokers, but the workers' heights are very similar.

In this type of analysis it is useful for an investigator to present the means and standard deviations along with the regression coefficients so that a reader of the report can try typical values of the independent variables in the regression equation in order to assess the numerical effects of the independent variables. Presentation of these statistics also enables readers to assess the characteristics of the sample under study. For example, suppose two 70-in.-tall males are 20 and 60 years old, respectively. From the two equations presented above we would predict FEV1 to be as follows:

	Age (Years)	
Subgroup	20	60
Smoker	4.39	2.95
Nonsmoker	4.37	3.17

Thus for the 20-year-old male the predicted FEV1 is almost the same whether or not he smokes, while for a 60-year-old male the FEV1 is predicted to be somewhat less if he smokes than it would be if he did not.

TABLE 7.5. Statistical Output from Factory Worker Example for Males Aged 20–64, Overall and by Smoking Status

	Mean	Standard Deviation	Regression Coefficient	Standard Error	Standardized Regression Coefficient
Overall (N = 679, df = 676)					
Intercept			−1.163		0
Age	38.45	12.09	−0.034	0.002	−0.525
Height	68.68	2.67	0.090	0.009	0.305
FEV1	3.68	0.78			
$R = 0.63,\ R^2 = 0.40,\ S = 0.61$					
Smokers (N_S = 448, df = 445)					
Intercept			−1.397		0
Age	39.43	11.87	−0.036	0.003	−0.517
Height	68.76	2.63	0.093	0.012	0.301
FEV1	3.62	0.82			
$R = 0.63,\ R^2 = 0.39,\ S = 0.64$					
Nonsmokers (N_{ns} = 231, df = 228)					
Intercept			−0.906		0
Age	36.57	12.33	−0.030	0.003	−0.527
Height	68.54	2.75	0.084	0.013	0.330
FEV1	3.79	0.70			
$R = 0.65,\ R^2 = 0.42,\ S = 0.54$					

The standardized regression coefficient is also useful to the reader in assessing the relative effects of the two variables. Note that even though the unstandardized regression coefficients for age are smaller in magnitude than those for height, the magnitudes of the standardized regression coefficients for age are larger than those for height. Thus the relative contribution of age is greater than that of height.

If the regression coefficients are divided by their standard errors, then highly significant t values are obtained. The levels of significance are sometimes included in the table, or asterisks are placed beside the regression coefficients to indicate their level of significance. The asterisks are then usually explained in footnotes to the table.

The standard errors could also be used to test equality of the individual coefficients for the two groups. For example, to compare the regression coefficients for height for the smokers and the nonsmokers and to test the null hypothesis of equal coefficients, we compute

$$Z = \frac{B_s - B_{ns}}{[\text{SE}^2(B_s) + \text{SE}^2(B_{ns})]^{1/2}}$$

or

$$Z = \frac{0.093 - 0.084}{(0.013^2 + 0.012^2)^{1/2}} = \frac{0.009}{(0.000313)^{1/2}} = 0.51$$

The computed value of Z can be compared with the percentiles of the normal distribution in Appendix Table A.1 to obtain an approximate P value for large samples. In this case the standard error for age was reported with too few significant digits, so its use would be questionable.

The standard deviations of the regression planes are useful to readers in making comparisons with their own results and in obtaining confidence intervals at the point where $X_i^* = \overline{X}$ (see Section 7.4).

The grouping option, as mentioned earlier, is useful in comparing relevant subgroups such as smokers and nonsmokers. An F test is available as optional output from BMDP1R and the SAS GLM procedure to check whether the regression equations derived from the different subgroups are significantly different from each other. This test is an example of the general F test that is often used in statistics and is given in its usual form in Section 8.5.

For our factory workers example the null hypothesis H_0 is that a single population plane for both smokers and nonsmokers combined is the true plane. The alternative hypothesis H_1 is that separate planes should be fitted for smokers and nonsmokers. The residual sum of squares obtained from $\Sigma (Y - \hat{Y})^2$ where the two separate planes are fitted will always be less than or equal to the residual sum of squares from the single plane for the overall sample. These residual sums of squares are printed in the analysis of variance table that accompanies the regression programs. Table 7.2 is an example of such a table for smokers.

The general F test compares the residual sums of squares when a single plane is fitted to what is obtained when two planes are fitted. If these two residuals are not very different, then the null hypothesis cannot be rejected. If fitting two separate planes results in a much smaller residual sum of squares, then the null hypothesis is rejected. The smaller residual sum of squares for the two planes indicates that the regression coefficients differ beyond chance between the two groups. This difference could be a difference in either the intercepts or the slopes or both. In this test normality of errors is assumed.

The F statistic for the test is

$$F = \frac{[\text{SS}_{\text{res}}(H_0) - \text{SS}_{\text{res}}(H_1)]/[\text{df}(H_0) - \text{df}(H_1)]}{\text{SS}_{\text{res}}(H_1)/\text{df}(H_1)}$$

In the factory workers example we have

$$F = \frac{[249.008 - (180.746 + 66.197)]/[676 - (445 + 228)]}{(180.746 + 66.197)/(445 + 228)}$$

where 249.008 is the residual sum of squares for the overall plane with 676 degrees of freedom, 180.746 is the residual sum of squares for the smokers with 445 degrees of freedom, and 66.197 is the residual sum of squares for the nonsmokers with 228 degrees of freedom. Thus

$$F = \frac{(249.008 - 246.943)/3}{246.943/673} = \frac{0.688}{0.367} = 1.88$$

with 3 and 673 degrees of freedom. The printed P value from the program was 0.13, indicating that the two regression equations are not significantly different from each other.

Other topics of regression analysis and corresponding options will be discussed in the next three chapters. In particular, selecting the best predictors from a set of variables is given in Chapter 8.

7.10 DISCUSSION OF COMPUTER PROGRAMS

The amount of computation necessary for a multiple regression or correlation analysis increases rapidly with the number of variables. With only two X variables and a small number of observations, it may be feasible to perform the computations on a good hand calculator. Detailed formulas for this case are given in Brownlee (1965) and Dunn and Clark (1974).

With more than two variables or with larger sample sizes, the investigator is well advised to use one of the many packaged programs available for this purpose. For example, the regression analysis for the factory workers data was performed by using the BMDP1R program. The control statements for this run are given below to indicate the data definition instructions and the instructions that tell the computer which features of the analysis to use. These statements are explained in succeeding paragraphs. All BMDP programs use a similar set of data definition instructions, which are explained in Chapters 3 through 6 of the manual.

```
/PROBLEM TITLE IS 'REGRESSION ANALYSIS OF FACTORY WORKER
DATA'.
/INPUT          UNIT=9.
                FORMAT IS '(18X,F2.0,1X,F1.0,F2.0,2F4.2,15X,F1.0)'.
                CASES ARE 679.
                VARIABLES ARE 6.
/VARIABLE       NAMES ARE AGE,SEX,HT,FVC,FEV1,SMOKER.
                GROUPING IS SMOKER.
```

The BMDP statements are written in sentences that are grouped into paragraphs. Sentences always end with a period. Paragraphs are headed by a / followed by the paragraph name. PROBLEM, INPUT, and VARIABLE paragraphs are the same for all BMDP programs.

The PROBLEM paragraph allows the user to label the output. The INPUT paragraph provides four essential pieces of information: where the data are located, the format of the data, the number of cases, and the number of variables per case. The sentences for describing where the data are located are quite specific to the computer used. For example, in programs run on a large university computing system, file names and location are included in the job control language cards that proceed the BMDP instructions. The statement UNIT = 9 is included in the INPUT paragraph to indicate that the data are on tape. In programs run on a smaller interactive system, the data file name is specified directly in the input paragraph, such as FILE = 'FACTORY'.

The format used for the factory workers data is a standard F format. The notation 18X tells the program to skip columns 1–18 and to start reading in column 19. The notation F2.0 specifies that the number is a whole number and occupies two columns (columns 19 and 20). The notation F4.2 signifies that the data have at most four digits and that the decimal point comes before the last two digits. The 2 in front of F4.2 (2F4.2) signifies that two numbers each with the F4.2 format occur.

The names used in the VARIABLE paragraph must be in the same order as the order used for the variables in the format sentence. The names must have eight or less characters and must be enclosed in apostrophes if they include blanks or symbols or do not begin with a letter. A grouping sentence is included to tell the computer to separate the cases into smokers and nonsmokers. Regression planes were obtained for all cases combined, for smokers, and for nonsmokers.

Five additional control statements are needed to complete the instructions for the computer. These statements for the factory workers example are given next and are explained in the paragraphs that follow.

```
/GROUP          CODES(6) ARE 1,2,
                NAMES(6) ARE YES,NO,
/REGRESSION     DEPENDENT=FEV1,
                INDEPENDENT=HT,AGE,
/PLOT           RESIDUAL,
                VARIABLES ARE HT,AGE,
                NORMAL,
/END
```

The paragraph GROUP is used to show how the program is to classify the cases into groups. The sentence CODES(6) ARE 1,2 says that the sixth variable (smoking status) takes on two values, 1 or 2. The names for the sixth variable are YES or NO, as signified in the next sentence.

The REGRESSION paragraph tells the computer which variables are the independent variables and which one is the dependent variable.

The PRINT paragraph lists which output, in addition to the standard or default output, the user wants printed. In this example just the standard output was needed, so this paragraph was omitted.

The PLOT paragraph specifies what scatter diagrams are to be printed. In this example, the residual and the residual squared versus the dependent variable are requested by using the RESIDUAL sentence. Then the VARIABLES ARE HT,AGE sentence is included to obtain the predicted and observed dependent variables plotted first against height. The residual is also plotted against height. To interpret this residual, we draw by hand a horizontal line through 0.0. If statistical tests or confidence limits are going to be reported, these residuals are assumed to be normally distributed. They should always be examined for outliers. Similar output follows for age. The normality of the residuals is examined by looking at the normal probability plot of the residuals obtained by including the NORMAL sentence.

The END paragraph tells the program that the investigator wishes to terminate the instructions for this problem. *Note*: Do not use a period after the word END.

We now have shown all the control statements for the factory workers regression analysis. The results shown in Table 7.5 were abstracted from the printout. In Table 7.6 we reproduce a part of the actual output for the total sample of smokers and nonsmokers. This output corresponds to the upper third of Table 7.5, where some rounding off was done. In addition to the items discussed in Table 7.5, Table 7.6 identifies the dependent variable, indicates the tolerance level used, and gives a printout of the ANOVA table. This ANOVA table presents a test of the hypothesis that β_1 and β_2 are both zero. The P value is very low, indicating that the regression equation is useful in predicting FEV1.

Figure 7.6 shows a plot of the residuals from the regression plane plotted against height. The horizontal line at 0.0 has been drawn in. The average of all the residuals will always be zero. Note that there are no extreme outliers, although several residuals may be suspect. Also, there is no clumping of residuals or fan-shaped pattern.

The normality of the residuals can be assessed by examining Figure 7.7, which gives the normal probability plot of the residuals from the regression plane of FEV1. The observed residuals are plotted on the horizontal axis. On the vertical axis the normal values of Z are plotted, as described in Section 4.4. If the residuals are normally distributed, then the normal probability plot of the residuals should approximate a straight line. The extreme upper and lower points should be ignored, since they represent very few points. In this example the residuals appear to be either normally distributed or close enough to it so that the investigator can safely obtain confidence intervals or tests of hypotheses.

Some of the most used items of output are shown in Table 7.7 for the BMDP, SAS, and SPSS-X programs. These programs also include other output, which will be discussed in Chapters 8 and 10.

Regression analysis is one of the most widely used statistical techniques, so the programs have included a wide array of possible options, which can make it difficult to interpret the output. We recommend that you carefully review objectives in running a program before submitting program statements, to determine just what is needed. A second run can always be used to obtain additional output. The output should be read, and the results that are significant to you should be either highlighted, encircled, or written out in English on the printed output while it is still fresh. Many researchers find it useful to

TABLE 7.6. BMDP1R Output for All Factory Workers

```
        FEV1 ON HT.AGE  20-64

REGRESSION TITLE. . . . . . . . . . . . . . . . . . .FEV1 ON HT.AGE 20-64
DEPENDENT VARIABLE. . . . . . . . . . . . . . .    5 FEV1
TOLERANCE . . . . . . . . . . . . . .    0.0100
ALL DATA CONSIDERED AS A SINGLE GROUP

MULTIPLE R           0.6338      STD. ERROR OF EST.      0.6069
MULTIPLE R-SQUARE    0.4017

ANALYSIS OF VARIANCE
                 SUM OF SQUARES   DF    MEAN SQUARE    F RATIO    P(TAIL)
  REGRESSION        167.150        2      83.575       226.887    0.00000
  RESIDUAL          249.008      676       0.368

                                              STD. REG
  VARIABLE      COEFFICIENT   STD. ERROR     COEFF        T      P(2 TAIL)  TOLERANCE

INTERCEPT       -1.16281
AGE       1     -0.03403       0.002       -0.525     -17.562      0.0      0.989706
HT        3      0.08950       0.009        0.305      10.211      0.0      0.989705
```

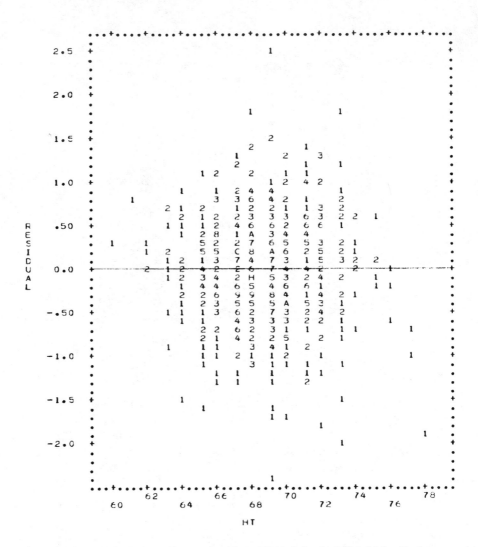

FIGURE 7.6. Residuals from Regression Plane Plotted Against Height for Factory Workers; Data Set for Smokers and Nonsmokers

write a paragraph describing the output on the first page of the computer printout, which is equivalent to a paragraph in the results section of an article or a report. It is very easy to forget why a program was run and what the output signifies after a few months. Therefore such notations on the output can be useful reminders.

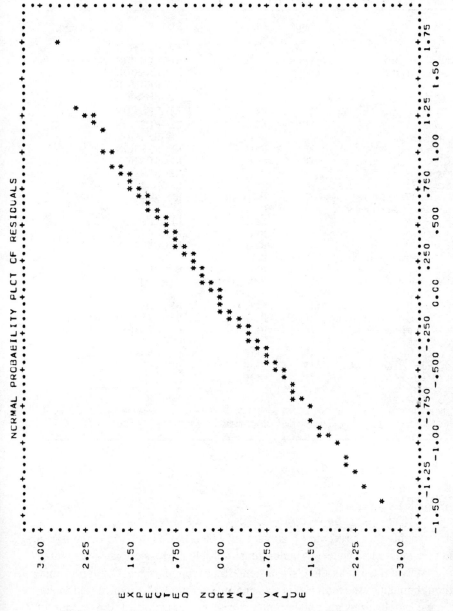

FIGURE 7.7. Normal Probability Plot of Residuals from Regression Plane for Factory Workers; Data Set for Smokers and Nonsmokers

TABLE 7.7. Summary of Computer Output from BMDP, SAS, and SPSS–X for Multiple Linear Regression

Output	BMDP*	SAS	SPSS–X
Correlation matrix	1R	REG	REGRESSION
Covariance matrix	1R	REG	REGRESSION
Regression plane	1R	REG	REGRESSION
Standard error B_i	1R	REG	REGRESSION
Test $\beta_i = 0$	1R	REG	REGRESSION
Standardized B	1R	REG	REGRESSION
Standard error of A	9R	REG	REGRESSION
Test $\alpha = 0$	9R	REG	REGRESSION
ANOVA	1R	REG	REGRESSION
List residuals	1R	REG	REGRESSION
Plot residuals	1R	REG	REGRESSION
Normal probability plots	1R	UNIVARIATE	REGRESSION
Serial correlation residuals	1R	REG	
Durbin-Watson	9R	REG	REGRESSION
Cook's distance	9R	REG	REGRESSION
Tolerance	1R	REG	REGRESSION
Other multicollinearity tests		REG	
Test $\mathcal{R} = 0$	1R	REG	REGRESSION
Partial correlations	6R	PRINCOMP + PARTIAL	PARTIAL CORR
Correlation of B's	1R	REG	REGRESSION
Weighted regression coefficient	1R	REG	REGRESSION + WEIGHT
Test weighted regression coefficient	1R	REG	REGRESSION + WEIGHT
Regression through zero	1R	REG	REGRESSION
Regression for subgroups	1R	REG	REGRESSION
Test equality of regression for subgroups	1R	GLM	

* If the output is available in 1R, then only 1R is listed. Much of the same output is available in 2R and 9R. Also, other SAS regression programs provide similar output.

SUMMARY

In this chapter we presented a subject in which the use of the computer is almost a must. The concepts underlying multiple regression were presented along with a fairly detailed discussion of packaged regression programs. Although much of the material in this chapter is found in standard textbooks, we have emphasized an understanding and presentation of the output of packaged programs. This philosophy is one we will follow in the rest of the book. In fact, in future chapters the emphasis on interpretation of output will be even more pronounced.

Certain topics included in this chapter but not covered in usual courses include how to handle interactions among the X variables and how to compare regression equations from different subgroups. We also included discussion of recent developments, such as the examination of residuals and influential observations. A review of the literature on these and other topics in regression analysis is found in Hocking (1983).

Several texts contain excellent presentations of the subject of regression analysis. A partial list includes Neter and Wasserman (1974), Chatterjee and Price (1977), Wonnacott and Wonnacott (1979), Gunst and Mason (1980), and Draper and Smith (1981).

BIBLIOGRAPHY

Afifi, A. A., and Azen, S. P. 1979. *Statistical analysis: A computer oriented approach*. 2nd ed. New York: Academic Press.

Anscombe, F. J., and Tukey, J. W. 1963. The examination and analysis of residuals. *Technometrics* 5:141–160.

Belsley, D. A.; Kuh, E.; and Welsch, R. E. 1980. *Regression diagnostics: Identifying influential data and sources of collinearity*. New York: Wiley.

Brownlee, K. A. 1965. *Statistical theory and methodology in science and engineering*. 2nd ed. New York: Wiley.

Chatterjee, S., and Price, B. 1977. *Regress analysis by example*. New York: Wiley.

Cook, R. D. 1977. Detection of influential observations in linear regression. *Technometrics* 19:15–18.

*———. 1979. Influential observations in linear regression. *Journal of the American Statistical Association* 74:169–174.

*Cook, R. D., and Weisberg, S. 1980. Characterization of an empirical influence function for detecting influential cases in regression. *Technometrics* 22:495–508.

Daniel, C., and Wood, F. S. 1980. *Fitting equations to data*. New York: Wiley.

*Devlin, S. J.; Gnanadesekan, R.; and Kettenring, J. R. 1981. Robust estimation of dispersion matrices and principal components. *Journal of the American Statistical Association* 76:354–362.

Draper, N. R., and Smith, H. 1981. *Applied regression analysis*. 2nd ed. New York: Wiley.

Dunn, O. J., and Clark, V. A. 1974. *Applied statistics: Analysis of variance and regression*. New York: Wiley.

*Graybill, F. A. 1976. *Theory and application of the linear model*. N. Scituate, Mass.: Duxbury Press.

Gunst, R. F., and Mason, R. L. 1980. *Regression analysis and its application*. New York: Dekker.

Hald, A. 1952. *Statistical theory with engineering applications*. New York: Wiley.

Hanushek, E. A., and Jackson, J. E. 1977. *Statistical methods for social scientists*. New York: Academic Press.

Hocking, R. R. 1983. Developments in linear regression methodology. *Technometrics* 25 : 219–229.

Neter, J., and Wasserman, W. 1974. *Applied linear statistical models*. Homewood, Ill.: Irwin.

Theil, H. 1971. *Principles of econometrics*. New York: Wiley.

Thorndike, R. M. 1978. *Correlational procedures for research*. New York: Gardner Press.

Williams, E. J. 1959. *Regression analysis*. New York: Wiley.

Wonnacott, R. J., and Wonnacott, T. H. 1979. *Econometrics*. 2nd ed. New York: Wiley.

Younger, M. S. 1979. *A handbook for linear regression*. N. Scituate, Mass.: Duxbury Press.

PROBLEMS

7.1 Using the chemical companies data given in Table 8.1, predict the price/earnings (P/E) ratio from the debt-to-equity (D/E) ratio, the annual dividends divided by the latest 12-months' earnings-per-share (PAYOUTR1), and the percentage net profit margin (NPM1). Obtain the correlation matrix, and check the intercorrelations of the variables. Summarize the results, including appropriate tests of hypotheses.

7.2 Fit regression planes for FEV1 on height and age for mothers alone, fathers alone, and overall from the lung function data given in Appendix B. Summarize the results in tabular form. Test whether the two regression planes for fathers and mothers are significantly different.

7.3 Compare the regression equation for fathers obtained in Problem 7.2 with the equation for nonsmokers given in Section 7.8.

7.4 In Problem 7.2, compute 95% confidence and prediction intervals for a 40-year-old father who is 68 in. tall. Do the same for a mother with the same age and height. Comment.

7.5 From the depression data set described in Table 3.2, predict the reported level of depression as given by CESD, using INCOME and AGE as independent variables, for females and males separately. Analyze the residuals and decide whether or not it is reasonable to assume that they follow a normal distribution.

7.6 The control cards for the BMDP6D program run on the factory workers data set used in Chapter 6 are as follows:

```
/PROBLEM     TITLE='FEV1 ON HT FOR MALE WORKERS AGE 20-24'.
/INPUT       UNIT=9.
             FORMAT IS '(18X,F2.0,1X,F1.0,F2.0,2F4.2)'.
             CASES ARE 679.
             VARIABLES ARE 5.
/VARIABLE    NAMES ARE AGE,SEX,HT,FVC,FEV1.
             MAXIMUM IS (4)7.2,(5)7.2.
             MINIMUM IS (4)2,(5)1.
/TRANSFORM   IF (SEX NE 1 OR AGE GE 25 OR AGE LT 20) THEN USE =0.
/PLOT        YVAR IS FEV1,FVC.
             XVAR IS HT.
             CROSS.
             STATISTICS.
/END
```

Only the first plot with FEV1 was given in Chapter 6. In this data set the variable SEX is coded 1 if male and 2 if female. Read the TRANSFORM and the VARIABLE paragraphs instructions in the manual to see how the cases are restricted to the desired ones. Also, check why the above instructions are used in the PLOT paragraph.

7.7 Create a hypothetical data set, which you will use for exercises in this chapter and some of the following chapters. Begin by generating 100 independent cases for each of 10 variables according to the standard normal distribution (means = 0, variances = one). Call these variables X1, X2,..., X9, Y. This data set can be obtained by using BMDP1D, starting without input variables and generating them by means of transformation statements. The program specifications follow, interspersed with explanations. These specifications were prepared for a microcomputer and therefore are in lowercase letters.

```
/problem     title = 'generation of multivariate normal data'.
/input       format = free.
             cases = 100.
             variables = 0.
/variable    names are x1,x2,x3,x4,x5,x6,x7,x8,x9,y.
             add = 10.
/transform   x1=rndg(36541).
             x2=rndg(43893).
             x3=rndg(45671).
             x4=rndg(65431).
             x5=rndg(98753).
             x6=rndg(78965).
             x7=rndg(67893).
             x8=rndg(34521).
             x9=rndg(98431).
             y=rndg(67895).
```

We now have 10 independent, random, normal numbers for each of 100 cases. The population mean of each variable is zero, and the population standard deviation is 1. Further transformations are done in order to make the variables intercorrelated. The transformations are accomplished by making some of the variables functions of other variables, as follows:

```
          x1=5*x1.
          x2=3*x2.
          x3=x1 + x2 + 4*x3.
          x4=x4.
          x5=4*x5.
          x6=x5 - x4 + 6*x6.
          x7=2*x7.
          x8=x7 + 2*x8.
          x9=4*x9.
          y=5 + x1 + 2*x2 + x3 + 10*y.
/print    data.
/end
```

We now have created a random sample of 100 cases on 10 variables: X1, X2, X3, X4, X5, X6, X7, X8, X9, Y. The population distribution is multivariate normal. It can be shown that the population means and variances are as follows:

Population	X1	X2	X3	X4	X5	X6	X7	X8	X9	Y
Mean	0	0	0	0	0	0	0	0	0	5
Variance	25	9	50	1	16	53	4	8	16	297

The population correlation matrix is as follows:

	X1	X2	X3	X4	X5	X6	X7	X8	X9	Y
X1	1	0	0.71	0	0	0	0	0	0	0.58
X2	0	1	0.42	0	0	0	0	0	0	0.52
X3	0.71	0.42	1	0	0	0	0	0	0	0.76
X4	0	0	0	1	0	-0.14	0	0	0	0
X5	0	0	0	0	1	0.55	0	0	0	0
X6	0	0	0	-0.14	0.55	1	0	0	0	0
X7	0	0	0	0	0	0	1	0.71	0	0
X8	0	0	0	0	0	0	0.71	1	0	0
X9	0	0	0	0	0	0	0	0	1	0
Y	0.58	0.52	0.76	0	0	0	0	0	0	1

The population squared multiple correlation coefficient between Y and X1 to X9 is 0.6633, between Y and X1, X2, X3 is 0.6633, and between Y and X4 to X9 is zero. Also, the population regression line of Y on X1 to X9 has $\alpha = 5$, $\beta_1 = 1$, $\beta_2 = 2$, $\beta_3 = 1$, $\beta_4 = \beta_5 = \cdots = \beta_9 = 0$. The following table lists the first five cases from a sample obtained from the above program specification with its particular random number initializations:

X1	X2	X3	X4	X5
−9.280	0.356	−10.986	1.816	−1.127
−13.124	0.108	−10.260	0.412	−3.560
−4.583	6.216	6.400	−0.0383	−5.559
10.387	−0.442	10.709	−1.146	−0.381
−4.261	−1.341	− 1.642	−1.198	1.584

X6	X7	X8	X9	Y
−2.455	1.335	0.972	−2.224	−8.316
−2.942	0.360	−1.683	−2.693	−14.214
0.627	3.192	2.244	0.341	27.687
8.420	2.350	2.578	2.455	16.473
−0.271	−0.169	−2.733	−4.454	11.423

Now, using the data you generated, obtain the sample statistics from a computer packaged program, and compare them with the parameters given above. Using the Bonferroni inequality, test the simple correlations, and determine which are significantly different from zero. Comment.

7.8 Repeat Problem 7.7, using SAS.

7.9 Continuation of Problem 7.7: Calculate the population partial correlation coefficient between X2 and X3 after removing the linear effect of X1. Is it larger or smaller than ρ_{23}? Explain. Also, obtain the corresponding sample partial correlation. Test whether it is equal to zero.

7.10 Continuation of Problem 7.7: Using a multiple regression program, perform an analysis, with the dependent variable = Y and the independent variables = X1 to X9, on the 100 generated cases. Summarize the results and state whether they came out the way you expected them to, considering how the data were generated. Perform appropriate tests of hypotheses. Comment.

8 | VARIABLE SELECTION IN REGRESSION ANALYSIS

8.1 WHAT WILL YOU LEARN FROM THIS CHAPTER?

From this chapter you will learn how to select independent variables from the entire set of possible variables. This selection may be necessary for either descriptive or predictive purposes. In particular, you will learn:

▶ When variable selection methods are used (8.2, 8.3).

▶ How to use criteria for variable selection (8.4).

▶ About a general method for testing the significance of the contribution of a subgroup of independent variables (8.5).

▶ How to use forward and backward variable selection (8.6).

▶ How to use stepwise selection (8.6).

▶ About stopping rules for deciding how many and which variables to include (8.6).

▶ How to use all possible subsets and the best subsets of variables (8.7).

▶ How to force variables that you know are important into the equation (8.6, 8.7).

▶ Which options are available in the computer programs (8.8).

▶ About stagewise regression (8.9).

8.2 WHEN ARE VARIABLE SELECTION METHODS USED?

Variable selection methods are used mainly in exploratory situations where many independent variables have been measured and a final model explaining the situation has not been reached. Thus, for example, if an investigator has done an experiment where the levels of only a few variables have been preset, then the methods used in this chapter are not necessary and the techniques described in Chapter 7 should be used. In contrast, variable selection methods are useful in, say, a survey situation in which numerous characteristics and responses of the individuals are recorded in order to determine their medical expenses over the past year. The investigator may have prior justification for using certain variables but be open to suggestion for the remaining variables. The researcher may have one of two goals:

1. To use the resulting regression equation to identify variables that best explain the amount of expenditure by an individual. The equation is then obtained for *descriptive or explanatory purposes*.

2. To obtain an equation that predicts the future expenditure of a given individual with as little error as possible. The purpose is then a *predictive purpose*.

The variable selection methods discussed in this chapter can sometimes serve one purpose better than the other, as will be discussed after the methods are presented.

8.3 DATA EXAMPLE

Table 8.1 presents various characteristics reported by the 30 largest chemical companies; the data are taken from a January 1981 issue of *Forbes*. This data set will henceforth be called the chemical companies data.

The variables listed in Table 8.1 are defined as follows (see Brigham 1978):

▶ *P/E*: Price-to-earnings ratio, which is the price of one share of common stock divided by the earnings per share for the past year. This ratio shows the

TABLE 8.1. Chemical Companies Financial Performance

Company	P/E	ROR5	D/E	SALESGR5	EPS5	NPM1	PAYOUTR1
Diamond Shamrock	9	13.0%	0.7	20.2%	15.5%	7.2%	0.43
Dow Chemical	8	13.0%	0.7	17.2%	12.7%	7.3%	0.38
Stauffer Chemical	8	13.0%	0.4	14.5%	15.1%	7.9%	0.41
E. I. du Pont	9	12.2%	0.2	12.9%	11.1%	5.4%	0.57
Union Carbide	5	10.0%	0.4	13.6%	8.0%	6.7%	0.32
Pennwalt	6	9.8%	0.5	12.1%	14.5%	3.8%	0.51
W. R. Grace	10	9.9%	0.5	10.2%	7.0%	4.8%	0.38
Hercules	9	10.3%	0.3	11.4%	8.7%	4.5%	0.48
Monsanto	11	9.5%	0.4	13.5%	5.9%	3.5%	0.57
American Cyanamid	9	9.9%	0.4	12.1%	4.2%	4.6%	0.49
Celanese	7	7.9%	0.4	10.8%	16.0%	3.4%	0.49
Allied Chemical	7	7.3%	0.6	15.4%	4.9%	5.1%	0.27
Rohm & Haas	7	7.8%	0.4	11.0%	3.0%	5.6%	0.32
Reichhold Chemicals	10	6.5%	0.4	18.7%	- 3.1%	1.3%	0.38
Lubrizol	13	24.9%	0.0	16.2%	16.9%	12.5%	0.32
Nalco Chemical	14	24.6%	0.0	16.1%	16.9%	11.2%	0.47
Sun Chemical	5	14.9%	1.1	13.7%	48.9%	5.8%	0.10
Cabot	6	13.8%	0.6	20.9%	36.0%	10.9%	0.16
International Minerals & Chemical	10	13.5%	0.5	14.3%	16.0%	8.4%	0.40
Dexter	12	14.9%	0.3	29.1%	22.8%	4.9%	0.36
Freeport Minerals	14	15.4%	0.3	15.2%	15.1%	21.9%	0.23
Air Products & Chemicals	13	11.6%	0.4	18.7%	22.1%	8.1%	0.20
Mallinckrodt	12	14.2%	0.2	16.7%	18.7%	8.2%	0.37
Thiokol	12	13.8%	0.1	12.6%	18.0%	5.6%	0.34
Witco Chemical	7	12.0%	0.5	15.0%	14.9%	3.6%	0.36
Ethyl	7	11.0%	0.3	12.8%	10.8%	5.0%	0.34
Ferro	6	13.8%	0.2	14.9%	9.6%	4.4%	0.31
Liquid Air of North America	12	11.5%	0.4	15.4%	11.7%	7.2%	0.51
Williams Companies	9	6.4%	0.7	16.1%	- 2.8%	6.8%	0.22
Akzona	14	3.8%	0.6	6.8%	-11.1%	0.9%	1.00
Mean	9.37	12.01%	0.42	14.94%	12.93%	6.55%	0.39
Standard deviation	2.80	4.50%	0.23	4.05%	11.15%	3.92%	0.16

Data abstracted from *Forbes*, 127, no. 1 (January 5, 1981).

dollar amount investors are willing to pay for the stock per dollar of current earnings of the company.

▶ *ROR5*: Percent rate of return on total capital (invested plus debt) averaged over the past five years.

▶ *D/E*: Debt-to-equity (invested capital) ratio for the past year. This ratio indicates the extent to which management is using borrowed funds to operate the company.

▶ *SALESGR5*: Percent annual compound growth rate of sales, computed from the most recent five years compared with the previous five years.

▶ *EPS5*: Percent annual compound growth in earnings per share, computed from the most recent five years compared with the preceding five years.

▶ *NPM1*: Percent net profit margin, which is the net profits divided by the net sales for the past year, expressed as a percentage.

▶ *PAYOUTR1*: Annual dividend divided by the latest 12-month earnings per share. This value represents the proportion of earnings paid out to shareholders rather than retained to operate and expand the company.

The P/E ratio is usually high for growth stocks and low for mature or troubled firms. Company managers generally want high P/E ratios, because high ratios make it possible to raise substantial amounts of capital for a small number of shares and make acquisitions of other companies easier. Also, investors consider P/E ratios of companies, both over time and in relation to similar companies, as a factor in valuation of stocks for possible purchase and/or sale. Therefore it is of interest to investigate which of the other variables reported in Table 8.1 influence the level of the P/E ratio. In this chapter we use the regression of the P/E ratio on the remaining variables to illustrate various variable selection procedures. Note that all the firms of the data set are from the chemical industry; such a regression equation could vary appreciably from one industry to another.

To get a preliminary impression of the data, examine the means and standard deviations shown at the bottom of Table 8.1. Also examine Table 8.2, which shows the simple correlation matrix. Note that P/E, the

TABLE 8.2. Correlation Matrix of Chemical Companies Data

	P/E	ROR5	D/E	SALESGR5	EPS5	NPM1	PAYOUTR1
P/E	1						
ROR5	0.32	1					
D/E	−0.47	−0.46	1				
SALESGR5	0.13	0.36	−0.02	1			
EPS5	−0.20	0.56	0.19	0.39	1		
NPM1	0.35	0.62	−0.22	0.25	0.15	1	
PAYOUTR1	0.33	−0.30	−0.16	−0.45	−0.57	−0.43	1

dependent variable, is most highly correlated with D/E. The association, however, is negative. Thus investors tend to pay less per dollar earned for companies with relatively heavy indebtedness. The sample correlations of P/E with ROR5, NPM1, and PAYOUTR1 range from 0.32 to 0.35 and thus are quite similar. Investors tend to pay more per dollar earned for stocks of companies with higher rates of return on total capital, higher net profit margins, and higher proportion of earnings paid out to them as dividends. Similar interpretations can be made for the other correlations.

If a single independent variable were to be selected for "explaining" P/E, the variable of choice would be D/E since it is the most highly correlated. But clearly there are other variables that represent differences in management style, dynamics of growth, and efficiency of operation that should also be considered.

It is possible to derive a regression equation using all the independent variables, as discussed in Chapter 7. However, some of the independent variables are strongly interrelated. For example, from Table 8.2 we see that growth of earnings (EPS5) and growth of sales (SALESGR5) are positively correlated, suggesting that both variables measure related aspects of dynamically growing companies. Further, growth of earnings (EPS5) and growth of sales (SALESGR5) are positively correlated with return on total capital (ROR5), suggesting that dynamically growing companies show higher returns than mature, stable companies. In contrast, growth of earnings (EPS5) and growth of sales (SALESGR5) are both negatively correlated with the proportion of earnings paid out to stockholders (PAYOUTR1), suggesting that earnings must be plowed back into operations to achieve growth. The profit margin (NPM1) shows the highest positive correlation with rate of return (ROR5), suggesting that efficient conversion of sales into earnings is consistent with high return on total capital employed.

Since the independent variables are interrelated, it may be better to use a subset of the independent variables to derive the regression equation. Most investigators prefer an equation with a small number of variables since such an equation will be easier to interpret. For future predictive purposes it is often possible to do at least as well with a subset as with the total set of independent variables. Methods and criteria for subset selection are given in the subsequent sections. It should be noted that these methods are fairly sensitive to gross outliers. The data sets should therefore be carefully edited prior to analysis.

8.4 CRITERIA FOR VARIABLE SELECTION

In many situations where regression analysis is useful, the investigator has strong justification for including certain variables in the equation. The justification may be to produce results comparable to previous studies or to conform to accepted theory. But often the investigator has no preconceived assessment of the importance of some or all of the independent variables. It is in the latter situation that variable selection procedures can be useful.

Any variable selection procedure requires a criterion for deciding how many and which variables to select. As discussed in Chapter 7, the least squares method of estimation minimizes the *residual sum of squares* (RSS) about the regression plane [$RSS = \Sigma\,(Y - \hat{Y})^2$]. Therefore an implicit criterion is the value of RSS. In deciding between alternative subsets of variables, the investigator would select the one producing the smaller RSS if this criterion were used in a mechanical fashion. Note, however, that

$$RSS = \Sigma\,(Y - \overline{Y})^2(1 - R^2)$$

where R is the multiple correlation coefficient. Therefore minimizing RSS is equivalent to maximizing the multiple correlation coefficient. If the criterion of maximizing R were used, the investigator would always select all of the independent variables, because the value of R will never decrease by including additional variables.

Since the multiple correlation coefficient, on the average, overestimates the population correlation coefficient, the investigator may be misled into including too many variables. For example, if the population multiple correlation coefficient is, in fact, equal to zero, the average of all possible values of R^2 from samples of size N from a multivariate normal population is $P/(N - 1)$, where P is the number of independent variables (see Kendall and Stuart 1967). An estimated multiple correlation coefficient that reduces the bias is the *adjusted multiple correlation coefficient*, denoted by $\overline{R}$. It is related to R by the following equation:

$$\overline{R}^2 = R^2 - \frac{P(1 - R^2)}{N - P - 1}$$

where P is the number of independent variables *in the equation* (see Theil 1971). Note that in this chapter the notation P will sometimes signify less than the total number of available independent variables.

The investigator may proceed now to select the independent variables that maximize $\overline{R}^2$. Note that excluding some independent variables may, in fact, result in a higher value of $\overline{R}^2$. As will be seen, this result occurs for the chemical companies data. Note also that if N is very large relative to P, then $\overline{R}^2$ will be approximately equal to R^2. Conversely, maximizing $\overline{R}^2$ can give different results from those obtained in maximizing R^2 when N is small.

Another method suggested by statisticians is to minimize the *residual mean square*, which is defined as

$$\text{RMS or RES. MS.} = \frac{\text{RSS}}{N - P - 1}$$

This quantity is related to the adjusted $\overline{R}^2$ as follows:

$$\overline{R}^2 = 1 - \frac{(\text{RMS})}{S_Y^2}$$

where

$$S_Y^2 = \frac{\Sigma (Y - \overline{Y})^2}{N - 1}$$

Since S_Y^2 does not involve any independent variables, minimizing RMS is equivalent to maximizing $\overline{R}^2$.

Another quantity used in variable selection and found in standard computer packages is the so-called C_p *criterion*. The theoretical derivation of this quantity is beyond the scope of this book (see Mallows 1973). However, C_p can be expressed as follows:

$$C_p = (N - P - 1) \left(\frac{\mathrm{RMS}}{\hat{\sigma}^2} - 1 \right) + (P + 1)$$

where RMS is the residual mean square based on the P selected variables and $\hat{\sigma}^2$ is the RMS derived from the total set of independent variables. The quantity C_p is the sum of two components, $(P + 1)$ and the remainder of the expression. While $(P + 1)$ increases as we choose more independent variables, the other part of C_p will tend to decrease. When all variables are chosen, $P + 1$ is at its maximum but the other part of C_p is zero since RMS $= \hat{\sigma}^2$. Many investigators recommend selecting those independent variables that minimize the value of C_p.

Another criterion, which is appropriate for the variable-X case, is the theoretical *average of the mean squared error* averaged over the values of the X variables. An estimate of this quantity is found in Bendel and Afifi (1977). The variables can be selected to minimize this average mean squared error, which is equivalent to minimizing the quantity

$$U_p = \frac{1 - R^2}{(N - P - 1)(N - P - 2)}$$

The above criteria are highly interrelated. No criterion will serve under all circumstances, and the criteria should not be used mechanically, particularly when the sample size is small. Practical methods for using these criteria for variable selection will be discussed in Sections 8.6 and 8.7. Hocking (1976) compared these and other criteria and recommends the following uses:

1. In a given sample the value of the unadjusted R^2 can be used as a measure of data description. Many investigators exclude variables that add only a very small amount to the R^2 value obtained from the remaining variables.

2. If the object of the regression is extrapolation or estimation of the regression parameters, the investigator is advised to select those variables that maximize the adjusted $\overline{R}^2$ (or, equivalently, minimize RMS).

3. For prediction, finding the variables that make the value of Cp approximately equal to $P + 1$ is a reasonable strategy.

Bendel and Afifi (1977) also examined these and other criteria in the context of stepwise regression (see Section 8.6). They found that minimizing Cp or Up often selects the best variables when N is large relative to P.

The discussion above was concerned with criteria for judging among alternative subsets of independent variables. How to select the candidate subsets is our next subject of concern. Before describing methods to implement the selection, though, we first discuss the use of the F test for determining the effectiveness of a subset of independent variables relative to the total set.

8.5 A GENERAL F TEST

Suppose we are convinced that the variables $X_1, X_2 \ldots , X_P$ should be used in the regression equation. Suppose also that measurements on Q additional variables, $X_{P+1}, X_{P+2}, \ldots, X_{P+Q}$, are available. Before deciding whether any of the additional variables should be included, we can test the hypothesis that, as a group, the Q variables do not improve the regression equation.

If the regression equation in the population has the form

$$Y = \alpha + \beta_1 X_1 + \beta_2 X_2 + \cdots + \beta_P X_P + \beta_{P+1} X_{p+1} + \cdots + \beta_{P+Q} X_{P+Q} + e$$

we test the hypothesis $H_0: \beta_{P+1} = \beta_{P+2} = \cdots = \beta_{P+Q} = 0$. To perform the test, we first obtain an equation that includes all the $P + Q$ variables, and we obtain the residual sum of squares (RSS_{P+Q}). Similarly, we obtain an equation that includes only the first P variables and the corresponding residual sum of squares (RSS_P). Then the test statistic is computed as

$$F = \frac{(\text{RSS}_P - \text{RSS}_{P+Q})/Q}{\text{RSS}_{P+Q}/(N - P - Q - 1)}$$

The numerator measures the improvement in the equation from using the additional Q variables. This quantity is never negative. The hypothesis is rejected if the computed F exceeds the tabled $F(1 - \alpha)$ with Q and $N - P - Q - 1$ degrees of freedom.

This very general test is sometimes referred to as the *generalized linear hypothesis test*. Essentially, this same test was used in Section 7.9 to test whether or not it is necessary to report the regression analyses by subgroups. The quantities P and Q can take on any integer values greater than or equal to one. For example, suppose that six variables are available. If we take P equal to 5 and Q equal to 1, then we are testing H_0: $\beta_6 = 0$ in the equation $Y = \alpha + \beta_1 X_1 + \beta_2 X_2 + \beta_3 X_3 + \beta_4 X_4 + \beta_5 X_5 + \beta_6 X_6 + e$. This test is the same as the test that was discussed in Section 7.6 for the significance of a single regression coefficient.

As another example, in the chemical companies data it was already observed that D/E is the best single predictor of the P/E ratio. A relevant hypothesis is whether the remaining five variables improve the prediction obtained by D/E alone. Two regressions were run (one with all six variables and one with just D/E), and the results were as follows:

$$RSS_6 = 103.86$$

$$RSS_1 = 176.08$$
$$P = 1 = Q = 5$$

Therefore the test statistic is

$$F = \frac{(176.08 - 103.86)/5}{103.86/(30 - 1 - 5 - 1)} = \frac{14.44}{4.52} = 3.20$$

with 5 and 23 degrees of freedom. Comparing this value with the value found in Appendix Table A.4, we find the P value to be less than 0.025. Thus at the 5% significance level we conclude that one or more of the other five variables significantly improves the prediction of the P/E ratio. Note, however, that the D/E ratio is the most highly correlated variable with the P/E ratio.

This selection process affects the inference so that the true P value of the test is unknown, but it is perhaps greater than 0.025. Strictly speaking, the general linear hypothesis test is valid only when the hypothesis is determined prior to examining the data.

The general linear hypothesis test is the basis for several selection procedures, as will be discussed in the next two sections.

8.6 STEPWISE REGRESSION

The variable selection problem can be described as considering certain subsets of independent variables and selecting that subset that either maximizes or minimizes an appropriate criterion. Two obvious subsets are the *best single variable* and the *complete set of independent variables*, as considered in Section 8.5. The problem lies in selecting an *intermediate subset* that may be better than both these extremes. In this section we discuss some methods for making this choice.

Forward Selection Method

Selecting the best single variable is a simple matter: We choose the variable with the highest absolute value of the simple correlation with Y. In the chemical companies data example D/E is the best single predictor of P/E because the correlation between D/E and P/E is -0.47, which has a larger magnitude than any other correlation with P/E (see Table 8.2).

To choose a second variable to combine with D/E, we could naively select NPM1 since it has the second highest absolute correlation with P/E (0.35). This choice may not be wise, however, since another variable together with D/E may give a higher multiple correlation than D/E combined with NPM1. One strategy therefore is to search for that variable that maximizes multiple R^2 when combined with D/E. Note that this procedure is equivalent to choosing the second variable to minimize the residual sum of squares given that D/E is kept in the equation. This procedure is also equivalent to choosing the second variable to maximize the magnitude of the partial correlation with P/E after removing the linear effect of D/E. The variable thus selected will also maximize the F statistic for testing that the above-mentioned partial correlation is zero (see Section 7.7). This F test is a special application of the general linear hypothesis test described in Section 8.5.

In the chemical companies data example the partial correlations between P/E and each of the other variables after removing the linear effect of D/E are as follows:

	Partial Correlations
ROR5	0.126
SALESGR5	0.138
EPS5	-0.114
NPM1	0.285
PAYOUTR1	0.286

Therefore, the method described above, called the *forward selection method*, would choose D/E as the first variable and PAYOUTR1 as the second variable. Note that it is almost a toss-up between PAYOUTR1 and NPM1 as the second variable. The computer programs implementing this method will ignore such reasoning and select PAYOUTR1 as the second variable because 0.286 is greater than 0.285. But to the investigator the choice may be more difficult since NPM1 and PAYOUTR1 measure quite different aspects of the companies.

Similarly, the third variable chosen by the forward selection procedure is the variable with the highest absolute partial correlation with P/E after removing the linear effects of D/E and PAYOUTR1. Again, this procedure is equivalent to maximizing multiple R and F and minimizing the residual mean square, given that D/E and PAYOUTR1 are kept in the equation.

The forward selection method proceeds in this manner, each time adding one variable to the variables previously selected, until a specified *stopping rule* is satisfied. The most commonly used stopping rule in packaged programs is based on the F test of the hypothesis that the partial correlation of the variable entered is equal to zero. One version of the stopping rule terminates entering variables when the computed value of F is less than a specified value. This cutoff value is often called the *minimum F-to-enter*.

Equivalently, the P value corresponding to the computed F statistic could be calculated and the forward selection stopped when this P value is greater than a specified level. Note that here also the P value is affected by the fact that the variables are selected from the data and therefore should not be used in the hypothesis-testing context.

Bendel and Afifi (1977) compared various levels of the minimum F-to-enter used in forward selection. A recommended value is the F percentile corresponding to a P value equal to 0.15. For example, if the sample size is large, the recommended minimum F-to-enter is the 85th percentile of the F distribution with 1 and ∞ degrees of freedom, or 2.07.

Any of the criteria discussed in Section 8.4 could also be used as the basis for a stopping rule. For instance, the multiple R^2 is always presented in the output of the commonly used programs whenever an additional variable is selected. The user can examine this series of values of R^2 and stop the process when the increase in R^2 is a very small amount. Alternatively, the series of adjusted $\overline{R}^2$ values can be examined. The process stops when the adjusted $\overline{R}^2$ is maximized.

TABLE 8.3. Forward Selection of Variables in Chemical Companies Data Example

Variables Added	Computed F-to-enter	Multiple R^2	Multiple $\overline{R}^2$
1. D/E	8.09	0.224	0.197
2. PAYOUTR1	2.41	0.288	0.235
3. NPM1	9.05	0.472	0.411
4. SALESGR5	3.33	0.534	0.459
5. EPS5	0.37	0.541	0.445
6. ROR5	0.08	0.542	0.423

As an example, data from the chemical companies analysis are given in Table 8.3. Using the P value of 0.15 (or the minimum F of 2.07) as a cutoff, we would enter only the first four variables since the computed F-to-enter for the fifth variable is only 0.37. In examining the multiple R^2, we note that going from five to six variables increased R^2 only by 0.001. It is therefore obvious that the sixth variable should not be entered. However, other readers may disagree about the practical value of a 0.007 increase in R^2 obtained by including the fifth variable. We would be inclined not to include it. Finally, in examination of the adjusted multiple $\overline{R}^2$ we note that it is also maximized by including only the first four variables.

This selection of variables by the forward selection method for the chemical companies data agrees well with our understanding gained from study of the correlations between variables. As we noted in Section 8.3, return on total capital (ROR5) is correlated with NPM1, SALESGR5, and EPS5 and so adds little to the regression when they are already included. Similarly, the correlation between SALESGR5 and EPS5, both measuring aspects of company growth, corroborates the result shown in Table 8.3 that EPS5 adds little to the regression after SALESGR5 is already included.

It is interesting to note that the computed F-to-enter follows no particular pattern as we include additional variables. Note also that even though R^2 is always increasing, the amount by which it increases varies as new variables are added. Finally, this example verifies that the adjusted multiple $\overline{R}^2$ increases to a maximum and then decreases.

The other criteria discussed in Section 8.4 are C_p and U_p. These criteria can also be used for stopping rules. Specifically, the combination of variables

minimizing their values can be chosen. A numerical example for this procedure will be given in Section 8.7.

Computer programs may also terminate the process of forward selection
when the tolerance level (see Section 7.9) is smaller than a specified minimum tolerance.

Backward Elimination Method

An alternative strategy for variable selection is the *backward elimination
method*. This technique begins with all of the variables in the equation and
proceeds by eliminating the least useful variables one at a time.

As an example, Table 8.4 lists the regression coefficient (both standardized and unstandardized), the P value, and the corresponding computed F
for testing that each coefficient is zero for the chemical company data. The F
statistic here is called the *computed F-to-remove*.

Since ROR5 has the smallest computed F-to-remove, it is a candidate for
removal. The user must specify the *maximum F*-to-remove; if the computed
F-to-remove is less than that maximum, ROR5 is removed. No recommended
value can be given for the maximum F-to-remove. We suggest, however,
that a reasonable choice is the 70th percentile of the F distribution (or,
equivalently, $P = 0.30$). For a large value of the residual degrees of freedom,
this procedure results in a maximum F-to-remove of 1.07. In Table 8.4, since
the computed F-to-remove for ROR5 is 0.08, this variable is removed first.
Note that ROR5 also has the smallest standardized regression coefficient and
the largest P value.

The backward elimination procedure proceeds by computing a new equation with the remaining five variables and examining the computed F-to-
remove for another likely candidate. The process continues until no variable
can be removed according to the stopping rule.

In comparing the forward selection and the backward elimination methods,
we note that one advantage of the former is that it involves a smaller amount
of computation than the latter. However, it may happen that two or more
variables can together be a good predictive set while each variable taken
alone is not very effective. In this case backward elimination would produce a
better equation than forward selection. Neither method is expected to produce the best possible equation for a given number of variables to be included
(other than one or the total set).

TABLE 8.4. Computed Coefficients and F and P Values for Chemical Companies Data

Variable	Coefficients		Computed F-to-Remove	P
	Unstandardized	Standardized		
Intercept	1.26	—	—	—
D/E	−2.51	−0.20	1.03	0.32
PAYOUTR1	9.76	0.57	8.48	0.01
NPM1	0.35	0.49	6.25	0.02
SALESGR5	0.20	0.29	3.13	0.09
EPS5	−0.04	−0.15	0.40	0.53
ROR5	0.05	0.08	0.08	0.78

Stepwise Procedure

One very commonly used technique that combines both of the above methods is called the *stepwise procedure*. In fact, the forward selection method is often called the *forward stepwise method*. In this case a step consists of adding a variable to the predictive equation. At step 0 the only "variable" used is the mean of the Y variable. At that step the program normally prints the computed F-to-enter for each variable. At step 1 the variable with the highest computed F-to-enter is entered; and so forth.

Similarly, the backward elimination method is often called the *backward stepwise method*. At step 0 the computed F-to-remove is calculated for each variable. In successive steps the variables are removed one at a time, as described above.

The standard stepwise regression programs do forward selection with the option of removing some variables already selected. Thus at step 0 only the Y mean is included. At step 1 the variable with the highest computed F-to-enter is selected. At step 2 a second variable is entered if any variable qualifies (i.e., if at least one computed F-to-enter is greater than the minimum F-to-enter). After the second variable is entered, the F-to-remove is computed for both variables. If either of them is lower than the maximum F-to-remove, that variable is removed. If not, a third variable is included if its computed F-to-enter is large enough. In successive steps this process is repeated. For a given equation, variables with small enough computed F-to-remove values are removed, and the variables with large enough computed

F-to-enter values are included. The process stops when no variables can be deleted or added.

The choice of the *minimum F*-to-enter and the *maximum F*-to-remove affects both the nature of the selection process and the number of variables selected. For instance, if the maximum F-to-remove is much smaller than the minimum F-to-enter, then the process is essentially forward selection. In any case the minimum F-to-enter must be larger than the maximum F-to-remove; otherwise, the same variable will be entered and removed continuously. In many situations it is useful for the investigator to examine the full sequence until all variables are entered. This step can be accomplished by setting the minimum F-to-enter equal to a small value, such as 0.1 (or a corresponding P value of 0.99). In that case the maximum F-to-remove must then be smaller than 0.1. After examining this sequence, the investigator can make a second run, using other F values.

Values that have been found to be useful in practice are a minimum F-to-enter equal to 2.07 and a maximum F-to-remove of 1.07. With these recommended values, for example, the results for the chemical companies data are identical to those of the forward selection method presented in Table 8.3 up to step 4. At step 4 the variables in the equation and their computed F-to-remove are as shown in Table 8.5. Also shown in the table are the variables not in the equation and their computed F-to-enter. Since all the computed F-to-remove values are larger than the minimum F-to-remove of 1.07, none of the variables are removed. Also, since both of the computed F-to-enter values are smaller than the minimum F-to-enter of 2.07, no new variables are entered. The process terminates with four variables in the equation.

There are many situations in which the investigator may wish to include certain variables in the equation. For instance, the theory underlying the subject of the study may dictate that a certain variable or variables be used. The investigator may also wish to include certain variables in order for the results to be comparable with previously published studies. Similarly, the investigator may wish to consider two variables when both provide nearly the same computed F-to-enter. If the variable with the slightly lower F-to-enter is the preferred variable from other viewpoints, the investigator may force it to be included. This preference may arise from cost or simplicity considerations. Most stepwise computer programs offer the user the option of *forcing variables* at the beginning of or later in the stepwise sequence, as desired.

We emphasize that none of the procedures described here are guaranteed or even expected to produce the best possible regression equation. This

TABLE 8.5. Step 4 of the Stepwise Procedure for Chemical Companies Data Example

Variables in Equation	Computed F-to-remove	Variables Not in Equation	Computed F-to-enter
D/E	3.03		
PAYOUTR1	13.03		
NPM1	9.37		
SALESGR5	3.33		
		EPS5	0.37
		ROR5	0.03

comment applies particularly to the intermediate steps where the number of variables selected is more than one but less than $P - 1$. In the majority of applications of the stepwise method, investigators attempt to select two to five variables from the available set of independent variables. Programs that examine the best possible subsets of a given size are also available. These programs are discussed next.

8.7 SUBSET REGRESSION

The availability of the computer makes it possible to compute the multiple R^2 and the regression equation for all possible subsets of variables. For a specified subset size the "best" subset of variables is the one that has the largest multiple R^2. The investigator can thus compare the best subsets of different sizes and choose the preferred subset. The preference is based on a compromise between the value of the multiple R^2 and the subset size. In Section 8.4 three criteria were discussed for making this comparison, namely, adjusted $\overline{R}^2$, C_p, and U_p.

For a small number of independent variables, say three to five, the investigator may indeed be well advised to obtain all possible subsets. This technique allows detailed examination of all the regression models so that an appropriate one can be selected. Programs are available for computing the necessary statistics for all possible subsets (such as RSQUARE in SAS).

If the number of independent variables is larger than five, the number of possible subsets becomes too large for practical purposes. Furnival and Wilson (1974) have devised a technique for finding the subset of a given size (number of independent variables used) having a maximum R^2 without examining all possible subsets. Furthermore, this technique will identify the nearly best subsets of a given size. This technique has been implemented in BMDP9R. The program will further select the subsets that either (1) maximize R^2, (2) maximize adjusted $\overline{R}^2$, or (3) minimize C_p for *different* subset sizes. For a given size these three criteria are equivalent. But if size is not specified, they may select subsets of different sizes.

Table 8.6 includes part of the output produced by running BMDP9R for the chemical companies data. As before, the best single variable is D/E, with a multiple R^2 of 0.224. Included in Table 8.6 are the next two best candidates if only one variable is to be used. Note that if either NPM1 or PAYOUTR1 is chosen instead of D/E, then the value of R^2 drops by about one-half. The relatively low value of R^2 for any of these variables reinforces our understanding that any one independent variable alone does not do well in "explaining" the variation in the dependent variable P/E.

The best combination of two variables is NPM1 and PAYOUTR1, with a multiple R^2 of 0.404. This combination does not include D/E, as would be the case in stepwise regression (Table 8.3). In stepwise regression the two variables selected were D/E and PAYOUTR1 with a multiple R^2 of 0.228. So in this case stepwise regression does not come close to selecting the best combination of two variables. Here stepwise regression resulted in the second-best choice. Also, the third-best combination of two variables is essentially as good as the second-best. The best combination of two variables, NPM1 and PAYOUTR1, as chosen by BMDP9R, is interesting in light of our earlier interpretation of the variables in Section 8.3. Variable NPM1 measures the efficiency of the operation in converting sales to earnings, while PAYOUTR1 measures the intention to plow earnings back into the company or distribute them to stockholders. These are quite different aspects of *current* company behavior. In contrast, the debt-to-equity ratio (D/E) may be, in large part, a *historical* carry-over from past operations or a reflection of management style.

For the best combination of three variables the value of the multiple R^2 is 0.477, only slightly better than the stepwise choice. If D/E is a lot simpler to obtain than SALESGR5, the stepwise selection might be preferred since the loss in the multiple R^2 is negligible. Here the investigator, when given the option of different subsets, might prefer the first (NPM1, PAYOUTR1, SALESGR5) on theoretical grounds, since it is the only option that explicitly includes a measure of growth (SALESGR5). (You should also examine the four-variable combinations in light of the above discussion.)

TABLE 8.6. Best Three Subsets for One, Two, Three, or Four Variables for Chemical Companies Data

Number of Variables	Names of Variables	R^2	Adjusted $\overline{R}^2$	C_p	$U_p \times 1000$
1	D/E*	0.224	0.197	13.0	1.03
1	NPM1	0.123	0.092	18.1	
1	PAYOUTR1	0.106	0.074	18.9	
2	NPM1, PAYOUTR1	0.404	0.360	6.0	0.85
2	D/E, PAYOUTR1*	0.288	0.235	11.8	
2	D/E, NPM1	0.287	0.234	11.8	
3	NPM1, PAYOUTR1, SALESGR5	0.477	0.417	4.3	0.80
3	D/E, PAYOUTR1, NPM1*	0.472	0.410	4.6	
3	NPM1, PAYOUTR1, ROR5	0.428	0.362	6.8	
4	D/E, NPM1, PAYOUTR1, SALESGR5	0.534	0.459	3.4	0.78
4	NPM1, PAYOUTR1, SALESGR5, EPS5*	0.494	0.413	5.4	
4	NPM1, PAYOUTR1, SALESGR5, ROR5	0.484	0.402	5.9	
Best 5	D/E, NPM1, PAYOUTR1, SALESGR5, EPS5*	0.541	0.445	5.1	0.83
All 6	All variables	0.542	0.423	7.0	0.90

*Combinations selected by the stepwise procedure.

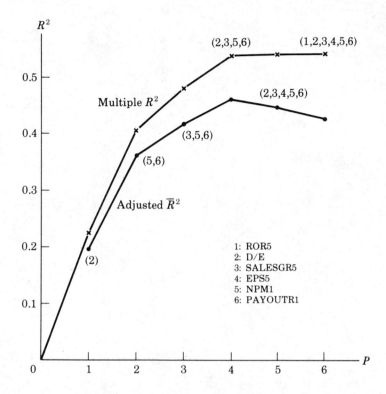

FIGURE 8.1. Multiple R^2 and Adjusted $\bar{R}^2$ Versus P for Best Subset with P Variables for Chemical Companies Data

Summarizing the results in the form of Table 8.6 is advisable. Then graphing the best combination of one, two, three, four, five, or six variables helps the investigator decide how many variables to use. For example, Figure 8.1 shows a plot of the multiple R^2 and the adjusted $\bar{R}^2$ versus the number of variables included in the best combination for the chemical companies data. Note that R^2 is a nondecreasing function. However, it levels off after four variables. The adjusted $\bar{R}^2$ reaches its maximum with four variables (D/E, NPM1, PAYOUTR1, and SALESGR5) and decrease with five and six variables.

Figure 8.2 shows C_p versus P (the number of variables) for the best combinations for the chemical companies data. The same combination of four variables selected by the $\bar{R}^2$ criterion minimizes C_p. A similar graph in Figure 8.3 shows that U_p is also minimized by the same choice of four variables.

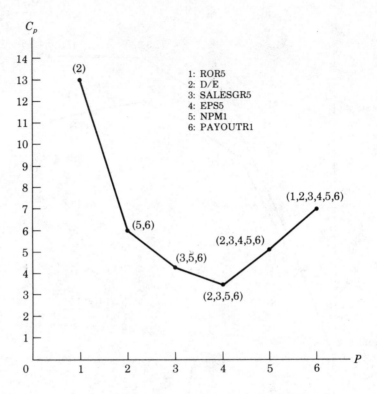

FIGURE 8.2. C_p Versus P for Best Subset with P Variables for Chemical Companies Data

In this particular example all three criteria agree. However, in other situations the criteria may select different numbers of variables. In such cases the investigator's judgment must be used in making the final choice. Even in this case an investigator may prefer to use only three variables, such as NPM1, PAYOUTR1, and SALESGR5. The value of having these tables and figures is that they inform the investigator of how much is being given up, as estimated by the sample on hand.

You should be aware that variable selection procedures are highly dependent on the particular sample being used. For example, data from two different years could give very different results. Also, any tests of hypotheses should be viewed with extreme caution since the significance levels are not valid when variable selection is performed.

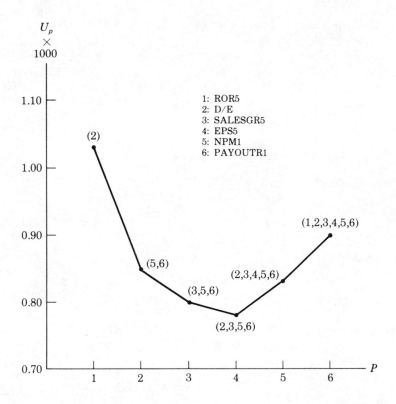

FIGURE 8.3. U_p Versus P for Best Subset with P Variables for Chemical Companies Data

8.8 DISCUSSION OF COMPUTER PROGRAMS

The techniques described in this chapter require extensive computation. It is difficult to imagine situations where variable selection using these techniques is done without the aid of a computer. In Table 8.7 a summary of options suitable for variable selection is presented for the three major packages. Other multiple linear regression options were discussed in Section 7.10 and in Table 7.7.

In Section 8.6 a description of the stepwise regression procedures was presented with results from the analysis of the chemical companies data.

TABLE 8.7. Summary of Computer Output from BMDP, SAS, and SPSS–X for Stepwise or Variable Selection

Output	BMDP	SAS	SPSS–X
Computed criteria			
Adjusted $\overline{R}^2$	2R, 9R	REG	REGRESSION
C_p	9R	RSQUARE and STEPWISE	
U_p			
Variable selection criteria			
Maximize R^2	2R, 9R	STEPWISE	REGRESSION
Maximize adjusted $\overline{R}^2$	9R		
Minimize C_p	9R		
Forcing	2R	STEPWISE	REGRESSION
F-to-enter	2R	STEPWISE	REGRESSION
F-to-remove	2R	STEPWISE	REGRESSION
Swapping variables	2R	STEPWISE	
Selection method			
Forward	2R	STEPWISE	REGRESSION
Backward	2R	STEPWISE	REGRESSION
Stepwise	2R	STEPWISE	REGRESSION
Maximum and minimum R^2		STEPWISE	
Best subset	9R		
All possible subsets	9R	RSQUARE	
Stagewise		SYSREG	
No selection	1R	REG	REGRESSION
Sets of variables entered at each step	2R		

Note: See Table 7.7 for standard regression output.

These results were obtained from running the stepwise regression program BMDP2R. Some further details of specification of the computer run and explanation of some of the output are presented here to help you master the mechanics of the process.

This program example was run on a desktop microcomputer that hosts the 1982 mainframe version of BMDP, with all input at a CRT display terminal. The program statements for the BMDP2R example are as follows:

```
/Problem    title is 'chemical data regression'.
/input      file is chem80.
            format is free.
            cases=30.
            variables=8.
```

```
/variable     names are symb,'p/e',
              ror5,'d/e',salessr5,eps5,npm1,payoutr1.
              label=1.
              use=2 to 8.
/regress      dependent is 'p/e'.
              independent are
              ror5,'d/e',salessr5,eps5,npm1,payoutr1.
              enter=2.1.
              remove=1.0.
/print        data.
              fratio.
/plot         var=ror5,'d/e',salessr5,eps5,npm1,payoutr1.
              resid.
/end
```

The specification statements are shown here in lowercase, allowable with this version of BMDP, but are otherwise identical with the control language used in the mainframe version and published in the manuals. The data file is created before the run at the desktop computer terminal, using a full screen editor. The data are entered in free format with the values for each case on one line, separated by blanks. The data file is named as "chem80" when the screen editor is invoked by the desktop computer's (UNIX) operating system, and the named file is retrieved by the BMDP program from disk storage as a consequence of the input paragraph statement "file is chem80." The data option is included in the print paragraph in order to examine the original input data for entry errors.

The stock symbol (a symbol used in the stock market) for each of the companies (cases) is entered as the first variable—as a label for that case— and the "label=1" sentence is included to tell the computer to read an alphanumeric label of up to four characters for that variable. The variable names 'p/e' and 'd/e' are enclosed in apostrophes, or single quotation marks, as required by BMDP for names including characters other than letters or numbers. The same technique would be necessary for names beginning with a number digit. Note that BMDP also restricts names to a maximum of eight characters in length. (It should be obvious from the examples we include in these chapters that each of the major computer program packages has its particular rules of syntax, formats, conventions, and restrictions for problem specification.)

The manuals published for the packages provide extensive guides to organization and form of the statements. After working with a package, you will develop some semblance of personal style that helps avoid omissions or errors. For example, we find that in writing the input statements in BMDP, we make fewer mistakes by proceeding from the general to the specific. Thus we begin with "Where are the data found"? which corresponds to "file is chem80." Next, "How are the data organized?' corresponds to "format is free." Then "How many cases?" becomes, cases=30," and finally, "How many variables per case?" becomes" variables=8."

The "enter=2.1" and "remove=1.0" statements in the regress paragraph specify an F-to-enter of 2.1 and an F-to-remove of 1.0. Note that the F-to-remove is smaller than the F-to-enter and that they are set at the levels suggested in Section 8.6.

The "fratio" option is included in the print paragraph to obtain a summary table of the computed F-to-enter and F-to-remove values from the analysis. The results of the summary table are given in Table 8.8. The variables ROR5 and EPS5 were not entered in the multiple regression because their F-to-enter values were 0.03 and 0.37, respectively, well below the 2.10 level we specified.

The results of Table 8.8 show that no variables were removed once they were entered, so the results are the same as those we would have obtained with forward selection.

The increase in the squared multiple correlation (RSQ) did not steadily decrease in successive steps. In step 0 the mean of Y is entered, and the printed mean square in the analysis of variance table is simply the variance of the dependent variable Y. When the first independent variable chosen by the program (D/E) was entered in step 1, RSQ was computed as 0.2242. Since it was the first variable entered, the increase in RSQ was also 0.2242. At step 2, RSQ went to 0.2879, for an increase of 0.0637. At step 3, however, RSQ jumped to 0.4717, for an increase of 0.1838. This pattern is not uncommon. The combination of D/E, PAYOUTR1, and NPM1 were much better predictors than just D/E and PAYOUTR1. Note that in the *best-subsets regression* given earlier in Table 8.6, NPM1 and PAYOUTR1 were the best choice when the number of independent variables was two.

Another interesting result is that the order of entry does not correspond to the size of the standardized regression coefficients obtained when the four variables have been entered. The largest standard regression coefficient is 0.623 for PAYOUTR1, followed by 0.492 for NPM1, then 0.280 for SALESGR5, and finally 0.256 for D/E. The order of magnitude of the standardized regression coefficients does, however, relate directly to the size

TABLE 8.8. Summary Table for Stepwise Regression of Chemical Companies Data

	Variable		Multiple		Increase	F-to-
Step	Entered	Removed	R	RSQ	in RSQ	Enter
1	D/E	—	0.4735	0.2242	0.2242	8.0924
2	PAYOUTR1	—	0.5366	0.2879	0.0637	2.4161
3	NPM1	—	0.6868	0.4717	0.1838	9.0463
4	SALESGR5	—	0.7306	0.5338	0.0620	3.3267

of the computed F-to-remove, with larger F-to-remove values corresponding to larger coefficients.

The various plots can be examined for outliers and for nonnormality. In this example no striking outliers were found. The case with the largest positive residual was for the company Air Products and Chemicals.

The REGRESSION program in SPSS–X presents the residuals in a form that makes it very easy to identify cases that are possible outliers. A vertical plot is given with the listing of the values of the cases on the same line. The REGRESSION program combines stepwise features with some of the newer features of residual analysis.

8.9 DISCUSSION AND EXTENSIONS

The process leading to a final regression equation involves several steps and tends to be iterative. The investigator should begin with data screening, eliminating blunders and possibly some outliers and deleting variables with an excessive proportion of missing values. Initial regression equations may be forced to include certain variables as dictated by the underlying situation. Other variables are then added on the basis of selection procedures outlined in this chapter. The choice of the procedure depends on which computers and programs are accessible to the investigator.

The results of this phase of the computation will include various alternative equations. For example, if the forward selection procedure is used, a regression equation at each step is printed. Thus in the chemical companies data the

TABLE 8.9. Regression Coefficients for Chemical Companies Data: First Four Steps

Step	Intercept	D/E	PAYOUTR1	NPM1	SALESGR5
		Regression Coefficients			
1	11.8	−5.86			
2	9.9	−5.35	4.38		
3	5.1	−3.46	8.54	0.36	
4	1.3	−3.17	10.67	0.35	0.19

equations resulting from the first four steps are as given in Table 8.9. Note that the coefficients for given variables could vary appreciably from one step to another. In general, coefficients for variables that are independent of the remaining predictor variables tend to have stable coefficients. The effect of a variable whose coefficient is highly dependent on the other variables in the equation is more difficult to interpret than the effect of a variable with stable coefficients.

A particularly annoying situation occurs when the sign of a coefficient changes from one step to the next. A method for avoiding this difficulty is to impose restrictions on the coefficients (see Chapter 10).

For some of the promising equations in the sequence given in Table 8.9, another program could be run to obtain extensive residual analysis and to determine influential cases (Cook's distance). Examination of this output may suggest further data screening. The whole process could then be repeated.

As discussed in Chapters 6 and 7, some regression programs allow the user to weight each case differently. This option is available in some variable selection programs as well. Deleting an observation is equivalent to giving it a weight of zero. This device can be used to adjust the influence of certain observations without necessarily totally deleting them, namely by giving them small weight relative to the rest of the observations. Such adjustments are cumbersome and require thorough knowledge of the underlying situation. Furthermore, for a large data set the influence of a single observation is less than it is for a small data set, other things being equal.

An alternative approach to variable selection is the so-called *stagewise regression procedure*. In this method the first stage (step) is the same as in forward stepwise regression; i.e., the ordinary regression equation of Y on the most highly correlated independent X variable is computed. Residuals from this equation are also computed. In the second step the X variable that is most highly correlated with these residuals is selected. Here the residuals are considered the "new" Y variables. The regression coefficient of these residuals on the X variable selected in step 2 is computed. The constant and the regression coefficient from the first step are combined with the regression coefficient from the second step to produce the equation with two X variables. This equation is the two-stage regression equation. The process can be continued to any number of desired stages.

The resulting *stagewise regression equation* does not fit the data as well as the least squares equation using the same variables (unless the X variables are independent in the sample). However, stagewise regression has some desirable advantages in certain applications. In particular, econometric models often use stagewise regression to adjust the data source factor, such as a trend or seasonality. Another feature is that the coefficients of the variables already entered are preserved from one stage to the next. For example, for the chemical companies data the coefficient for D/E would be preserved over the successive stages of a stagewise regression. This result can be contrasted with the changing coefficient in the stepwise process summarized in Table 8.9. In behavioral science applications the investigator can determine whether the addition of a variable reflecting a score or an attitude scale improves prediction of an outcome over what results in using only demographic variables.

SUMMARY

Variable selection techniques are helpful in situations in which many independent variables exist and no standard model is available. The methods we described in this chapter constitute one part of the process of making the final selection of one or more equations.

Although these methods are useful in exploratory situations, they should not be used as a substitute for modeling based on the underlying scientific problem. In particular, the variables selected by the methods described here depend on the particular sample on hand, especially when the sample size is small. Also, significance levels are not valid in variable selection situations.

These techniques can, however, be an important component in the modeling process. They frequently suggest new variables to be considered. They also help cut down the number of variables so that the problem becomes less cumbersome both conceptually and computationally. For these reasons variable selection procedures are among the most popular of the packaged options. The ideas underlying variable selection have also been incorporated into other multivariate programs (see, e.g., Chapters 11 and 12).

In the intermediate stages of analysis the investigator can consider several equations as suggested by one or more of the variable selection procedures. The equations selected should be considered tentative until the same variables are chosen repeatedly. Concensus will also occur when the same variables are selected by different subsamples within the same data set or by different investigators using separate data sets.

BIBLIOGRAPHY

*Beale, E. M. L.; Kendall, M. G.; and Mann, D. W. 1967. The discarding of variables in multivariate analysis. *Biometrika* 54 : 357–366.

Bendel, R. B., and Afifi, A. A. 1977. Comparison of stopping rules in forward stepwise regression. *Journal of the American Statistical Association* 72 : 46–53.

Brigham, E. F. 1978. *Fundamentals of financial management*. Hinsdale, Ill.: Dryden Press.

*Efroymson, M. A. 1960. Multiple regression analysis. In *Mathematical methods for digital computers*, ed. Ralston and Wilf. New York: Wiley.

Forsythe, A. B.; Engleman, L.; Jennrich, R.; and May, P. R. A. 1973. Stopping rules for variable selection in multiple regression. *Journal of the American Statistical Association* 68 : 75–77.

*Furnival, G. M., and Wilson, R. W. 1974. Regressions by leaps and bounds. *Technometrics* 16 : 499–512.

Gorman, J. W., and Toman, R. J. 1966. Selection of variables for fitting equations to data. *Technometrics* 8 : 27–51.

Hocking, R. R. 1972. Criteria for selection of a subset regression: Which one should be used. *Technometrics* 14 : 967–970.

———. 1976. The analysis and selection of variables in linear regression. *Biometrics* 32 : 1–50.

*Kendall, M. G., and Stuart, A. 1967. *The advanced theory of statistics*. New York: Hafner.

Mallows, C. L. 1973. Some comments on Cp. *Technometrics* 15 : 661–676.

Mantel, N. 1970. Why stepdown procedures in variable selection. *Technometrics* 12 : 621–626.

Pope, P. T., and Webster, J. T. 1972. The use of an F-statistic in stepwise regression procedures. *Technometrics* 14 : 327–340.

Pyne, D. A., and Lawing, W. D. 1974. *A note on the use of the Cp statistic and its relation to stepwise variable selection procedures*. Technical Report no. 210, Johns Hopkins University.

Theil, H. 1971. *Principles of econometrics*. New York: Wiley.

Wilkinson, L., and Dallal, G. E. 1981. Tests of significance in forward selection regression with an F-to-enter stopping rule. *Technometrics* 23 : 377–380.

PROBLEMS

8.1 Use the depression data set given in Table 3.3. Using CESD as the dependent variable and age, income, and level of education as the independent variables, run a forward stepwise regression program to determine which of the independent variables predict level of depression for women.

8.2 Repeat Problem 8.1, using either the RSQUARE or BMDP9R program, and compare the results.

8.3 *Forbes* gives, each year, the same variables listed in Table 8.1 for the chemical industry. The changes in lines of business and company mergers resulted in a somewhat different list of chemical companies in 1982. We have selected a subset of 13 companies that are listed in both years and whose main product is chemicals. The accompanying table includes data for both years. Do a forward stepwise regression analysis, using P/E as the dependent variable and ROR5, D/E, SALESGR5, EPS5, NPM1, and PAYOUTR1 as independent variables, on both years' data and compare the results. Note that this period showed little growth for this subset of companies, and the variable(s) entered should be evaluated with that idea in mind.

8.4 For adult males it has been demonstrated that age and height are useful in predicting FEV1. Using the data given in Appendix B, determine whether the regression plane can be improved by also including weight.

8.5 Using the data given in Table 8.1, repeat the analyses described in this chapter with $\sqrt{\text{P/E}}$ as the dependent variable instead of P/E. Do the results change much? Does it make sense to use the square root transformation?

8.6 Use the data you generated for Problem 7.7, where X1, X2,..., X9 are the independent variables and Y is the dependent variable. Use the generalized linear hypothesis test to test the hypothesis that $\beta_4 = \beta_5 = \cdots = \beta_9 = 0$. Comment in the light of what you know about the population parameters.

TABLE. Chemical Companies Financial Performance as of 1980 and 1982 (Subset of 13 of the Chemical Companies)

Company	Symbol	P/E	ROR5	D/E	SALESGR5	EPS5	NPM1	PAYOUTR1
1980								
Diamond Shamrock	dia	9	13.0	0.7	20.2	15.5	7.2	0.43
Dow Chemical	dow	8	13.0	0.7	17.2	12.7	7.3	0.38
Stauffer Chemical	stf	8	13.0	0.4	14.5	15.1	7.9	0.41
E. I. du Pont	dd	9	12.2	0.2	12.9	11.1	5.4	0.57
Union Carbide	uk	5	10.0	0.4	13.6	8.0	6.7	0.32
Pennwalt	psm	6	9.8	0.5	12.1	14.5	3.8	0.51
W. R. Grace	gra	10	9.9	0.5	10.2	7.0	4.8	0.38
Hercules	hpc	9	10.3	0.3	11.4	8.7	4.5	0.48
Monsanto	mtc	11	9.5	0.4	13.5	5.9	3.5	0.57
American Cyanamid	acy	9	9.9	0.4	12.1	4.2	4.6	0.49
Celanese	cz	7	7.9	0.4	10.8	16.0	3.4	0.49
Allied Chemical	acd	7	7.3	0.6	15.4	4.9	5.1	0.27
Rohm & Haas	roh	7	7.8	0.4	11.0	3.0	5.6	0.32
1982								
Diamond Shamrock	dia	8	9.8	0.5	19.7	-1.4	5.0	0.68
Dow Chemical	dow	13	10.3	0.7	14.9	3.5	3.6	0.88
Stauffer Chemical	stf	9	11.5	0.3	10.6	7.6	8.7	0.46
E. I. du Pont	dd	9	12.7	0.6	19.7	11.7	3.1	0.65
Union Carbide	uk	9	9.2	0.3	10.4	4.1	4.6	0.56
Pennwalt	psm	10	8.3	0.4	8.5	3.7	3.1	0.74
W. R. Grace	gra	6	11.4	0.6	10.4	9.0	5.5	0.39
Hercules	hpc	14	10.0	0.4	10.3	9.0	3.2	0.71
Monsanto	mtc	10	8.2	0.3	10.8	-0.2	5.6	0.45
American Cyanamid	acy	12	9.5	0.3	11.3	3.5	4.0	0.60
Celanese	cz	15	8.3	0.6	10.3	8.9	1.7	1.23
Allied Corp*	ald	6	8.2	0.3	17.1	4.8	4.4	0.36
Rohm & Haas	roh	12	10.2	0.3	10.6	16.3	4.3	0.45

Data abstracted from *Forbes*, 127, no. 1 (January 5, 1981) and *Forbes*, 131, no. 1 (January 3, 1983).
* Name and symbol changed.

8.7   For the data from Problem 7.7, perform a variable selection analysis, using the methods described in this chapter. Comment on the results in view of the population parameters.

8.8   In Problem 7.7 the population multiple $\mathcal{R}^2$ of Y on X4, X5,..., X9 is zero. However, from the sample alone we don't know this result. Perform a variable selection analysis on X4 to X9, using your sample, and comment on the results.

9 | NONLINEAR REGRESSION ANALYSIS

9.1 WHAT WILL YOU LEARN FROM THIS CHAPTER?

From this chapter you will learn:

- ▶ When nonlinear regression is used (9.2, 9.3).
- ▶ How to interpret the parameters in the regression equation (9.2).
- ▶ How to obtain the appropriate estimated regression equation (9.4).
- ▶ How to perform related tests of hypotheses (9.4).
- ▶ How to obtain confidence intervals (9.4).
- ▶ How to choose the appropriate computational program (9.5).

Note that the material in this chapter is slightly more mathematical than the material in previous chapters. Readers not interested in this subject may skip over the chapter without hindering their understanding of subsequent chapters.

9.2 WHEN IS NONLINEAR REGRESSION USED?

As discussed in Chapters 6 and 7, a regression equation is called linear if it is linear in the parameters α, β_1, β_2, etc. For example, the regression function

TABLE 9.1. Linear and Intrinsically Linear Regression Equations

Original Equation	Transformation	Resulting Equation
Linear in parameters		
$Y = \alpha + \beta\left(\dfrac{1}{X}\right)$	$X^* = \dfrac{1}{X}$	$Y = \alpha + \beta X^*$
$Y = \alpha + \beta \sin X$	$X^* = \sin X$	$Y = \alpha + \beta X^*$
Intrinsically linear		
$Y = \alpha e^{\beta X}$	$Y^* = \log Y,\ \alpha^* = \log \alpha$	$Y^* = \alpha^* + \beta X$
$Y = \alpha \beta^X$	$Y^* = \log Y,\ \alpha^* = \log \alpha,\ \beta^* = \log \beta$	$Y^* = \alpha^* + \beta^* X$

$$\alpha + \beta_1 X + \beta_2 X^2$$

though not linear in X, is linear in α, β_1, and β_2. Therefore it is a linear regression function. Other examples of equations that are not linear in X but are linear in the parameters are given in the first two rows of Table 9.1.

Other regression functions though not linear in their parameters are intrinsically linear; i.e., with appropriate transformations they can be made linear. Examples of intrinsically linear equations are given in the last two rows of Table 9.1. (The * notation in Table 9.1 signifies a transformed variable or parameter.) Ordinary linear regression can then be performed on the transformed equation. An estimate of the original equation can be obtained by performing the reserve transformation. Other transformations discussed in Chapter 6, Section 6.11, may also be used to linearize regression equations.

Some nonlinear equations cannot be transformed to produce linear functions of the parameters. For example, Figures 9.1 and 9.2 illustrate functions that cannot be so transformed and are, therefore, called *nonlinear regression functions*.

In Figure 9.1 a *logistic function* is graphed. The generalized logistic curve is used frequently as a model of growth in biological and other applications. It is useful when the quantity being measured has a natural limit that it eventually reaches. The variable X often represents time or space.

Another useful nonlinear function is the *exponential decay curve* illustrated in Figure 9.2. It represents situations such as radioactive decay, where the amount of radioactivity being emitted is proportional to the amount of radioactive material remaining. Such functions are used frequently in biological, engineering, and business applications.

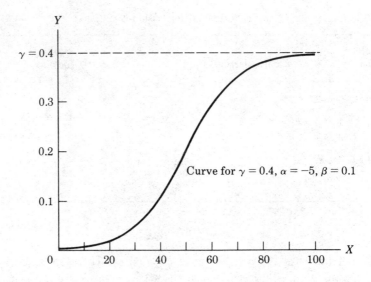

FIGURE 9.1.   Generalized Logistic Growth Curve, where $Y = \gamma/(1 + e^{-(\alpha + \beta X)})$

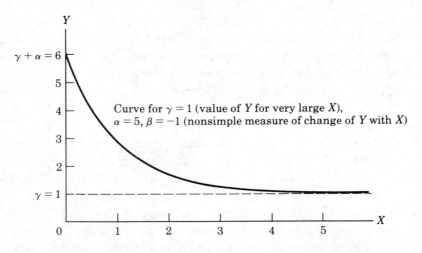

FIGURE 9.2.   Generalized Exponential Decay Curve, where $Y = \gamma + \alpha e^{\beta X}$

In general, whenever the parameters appear in a nonlinear function such that no transformation produces a linear equation, the resulting regression equation is called nonlinear. In addition to exponential functions and logistic functions, trigonometric functions such as sine and cosine often appear in nonlinear regression equations.

9.3 EXAMPLE OF A NONLINEAR REGRESSION EQUATION

In a cooperative study (Hollinger et al. 1982) the issue of prevention of non-A, non-B hepatitis caused by blood transfusions was examined. Samples of donated blood were analyzed and various constituents were measured. A good indicator of whether the donated blood is associated with hepatitis infection in the recipient is the level of alanine aminotransferase (ALT) in the donated blood, so the researchers measured levels of ALT in recipients. The data for this example are called the hepatitis data set. (For the sake of simplicity we use only data on recipients of one unit of blood.) Table 9.2 shows, for various intervals of ALT level, the number of recipients whose donor blood was in that interval and the number of cases of hepatitis.

The proportion of cases of hepatitis as a function of ALT level is graphed in Figure 9.3. For example, the first point in Figure 9.3 shows that at an ALT level of 4.5 (the midpoint of the first interval) the proportion of cases is 0.038. The points obviously do not fit on a straight line, so a function is needed that is a reasonable model and will fit the plotted points.

One possible model is the logistic growth curve shown in Figure 9.1. The general equation for a logistic growth curve is

$$Y = \frac{\gamma}{1 + e^{-(\alpha + \beta X)}}$$

where γ is the value of Y eventually achieved for very large values of X,

$$\frac{\gamma}{1 + e^{-\alpha}} = \text{value of } Y \text{ when } X = 0$$

and β is a measure of the rate of change of Y with X.

The logistic growth curve has been used to fit growth over time for human, animal, and economic data. It has been shown (Goldstein 1979) that if a

TABLE 9.2. Number of Cases of Hepatitis Among Recipients of a Single Unit of Blood, Depending on ALT Level

Donor Range of ALT Levels	Midpoints of ALT Levels	Number of Recipients	Number of Cases with Hepatitis	Proportion of Cases with Hepatitis
≤9	4.5	106	4	0.038
10–19	14.5	95	5	0.053
20–29	24.5	39	3	0.077
30–39	34.5	20	2	0.100
40–49	44.5	6	1	0.167
50–59	54.5	4	1	0.250
60–149	105.0	4	2	0.500

logistic regression function is used, the investigator is assuming that the percentage rate of change of Y decreases proportionately to Y. Percentage rate of change is defined by the *relative* rate of growth (the growth velocity divided by the *size*, or Y). If Y is small relative to γ, the investigator assumes that the relative growth is large, although the *actual* growth rate is small. When $Y = \gamma/2$, the actual growth rate is at a maximum, and as Y approaches γ, the growth rate tends to zero. When Y nears its final magnitude—i.e., γ—both the relative and actual growth rates slow down. If the investigator is willing to assume that the data follow this model, then the logistic function is a reasonable one to fit to those data.

In Figure 9.1, $\gamma = 0.4$; so the maximum growth rate occurs at $X = 50$ and $Y = 0.2$. Note also that the actual growth rate is small for Y close to zero and Y close to 0.4, but the relative growth rate is large for small values of Y.

The logistic growth curve shown in Figure 9.1 is a good model to represent the points of Figure 9.3. Thus if Y represents the proportion of recipients contracting non-A, non-B hepatitis and X is the donor ALT level, then the nonlinear regression equation takes the form

$$Y = \frac{\gamma}{1 + e^{-(\alpha+\beta X)}} + \text{error}$$

where γ, α, and β are unknown parameters. In the next section we discuss methods for estimating the parameters of this and other nonlinear regression equations, as well as some tests of hypotheses regarding them.

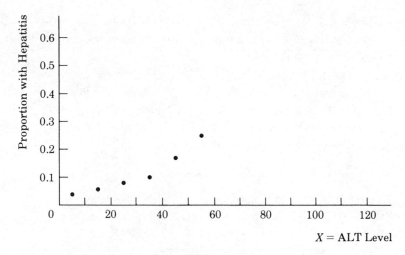

FIGURE 9.3. Proportion of Cases with Hepatitis Versus mid-ALT level

9.4 DESCRIPTION OF METHODS

Any regression model can be written in the form

$$Y = F(X_1, \ldots, X_P; \beta_1, \beta_2, \ldots, \beta_M) + e$$

where e is the error term. For example, the model used to represent the hepatitis data had one X variable (X_1) and three parameters $(\beta_1 = \alpha, \beta_2 = \beta, \text{ and } \beta_3 = \gamma)$. When the function F is linear in the parameters, estimates for the parameters are obtained easily by the least squares method, as was discussed in Chapters 6 and 7. When F is a nonlinear function of the parameters, the estimates can no longer be obtained so easily. Complicated iterative methods are necessary. Usually, because of their complexity, they are performed on a computer. In this section we discuss these methods and explain standard computer output.

Iterative Least Squares Procedure

Various iterative least squares estimation procedures have been developed (Hartley 1961; Sprent 1969; Daniel and Wood 1971; Bard 1974; Gallant

1975; Ralston and Jennrich 1978; Afifi and Azen 1979; Draper and Smith 1981). To show how these computational algorithms work, we must introduce a brief theoretical discussion.

Suppose that a sample of size N is available, where each observation consists of a set of values for Y, $X_1, \ldots, X_P$. Suppose also that $b_1, b_2, \ldots, b_m$ are suggested guesses or estimates of $\beta_1, \beta_2, \ldots, \beta_M$, respectively. For each observation we can compute an estimated value of Y, called $\hat{Y}$, based on these estimates; i.e., we can calculate

$$\hat{Y} = F(X_1, X_2, \ldots, X_P; b_1, b_2 \ldots, b_M)$$

The b_i estimates are considered "good" if the estimated $\hat{Y}$ values are "close" to the actual Y values. In the method of least squares, "close" is measured by the residual sum of squares (RSS), defined as

$$\text{RSS} = \Sigma \, (Y - \hat{Y})^2$$

Each computational algorithm tries to find those b_i estimates that produce the smallest possible value of RSS. They differ in how they go about this minimization process (see above references). Typically, an algorithm requires specifying an initial estimate of each parameter. The algorithm then selects a new set of estimates of the parameters and compares the RSS for the new set with that of the previous set. This process is called an *iteration*. If the new set is an "improvement," i.e., if the new RSS is smaller than the previous RSS, then the algorithm uses it to find a third set of estimates. This iterative process continues until no appreciable "improvement" is obtained, i.e., until the decrease in the RSS is very small; how small is usually built into the computer program itself (see the program manuals).

Initial Estimates

The initial estimates may be obtained graphically, from previous experience with the problem, or simply from intelligent guesses. You should be aware, however, that the choice of the initial estimates can be critical for the estimation process. If a good set of initial estimates is not readily available, you may wish to try several sets to determine whether the program gives the same final estimates.

For the hepatitis example γ represents the value of Y for a very large ALT level (see Figure 9.1). Therefore, from Table 9.2 a good initial estimate of γ is

0.50, because that is the proportion at the largest level of ALT. Next, we estimate from Table 9.2 that $\gamma \simeq 0.03$ when $X = 0$. From Figure 9.3 it is seen that the value of Y when $X = 0$ is

$$Y = \frac{\gamma}{1 + e^{-\alpha}}$$

Substituting 0.5 for γ and 0.03 for Y, we obtain

$$0.03 = \frac{0.5}{1 + e^{-\alpha}}$$

or

$$e^{-\alpha} = \frac{0.5}{0.03} - 1 = 15.667$$

Taking the natural logarithm of each side and solving the equation in α, we see that -2.75 can be used as an initial estimate of α.

We now have initial estimates for two of the three parameters, i.e.,

$$Y \simeq \frac{0.5}{1 + e^{-(-2.75 + \beta X)}}$$

We also know from Table 9.2 that a value of X of 14.5 results in a Y of 0.053. Substituting these values in the above equation, we get

$$0.053 = \frac{0.5}{1 + e^{-(-2.75 + 14.5\beta)}}$$

Solving for β, we obtain an initial value of 0.043.

These initial estimates were used in the BMDP3R nonlinear regression program and produced the final estimated equation as

$$Y = \frac{0.541}{1 + e^{-(-3.17 + 0.054X)}}$$

Thus the initial estimates were rather good guesses since they were fairly close to the final estimates.

The final estimated equation is graphed in Figure 9.4. Note that for low levels of ALT very few recipients of a single unit of blood evidenced hepatitis. In the middle range of levels of ALT the curve is close to a straight-line increase. But the curve flattens out for large values of ALT, with the maximum proportion expected to be equal to 0.541.

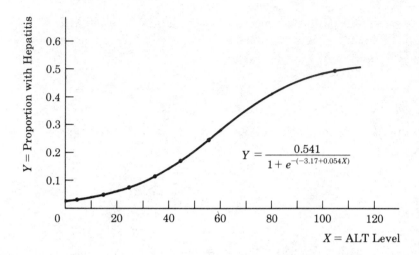

FIGURE 9.4. Logistic Growth Model Fitted to Hepatitis Data and Actual Proportion of Cases

Output from Packaged Programs

Most packaged programs also print an estimate of the variance of the error term. This estimate is called the *residual mean square* or the *error mean square*. The square root of this quantity, often called the *standard error of the estimate*, is a measure of the spread of the Y values around the regression equation. Approximate standard errors of the estimated parameters are also often provided in the output. These values can be used to perform approximate tests or to construct approximate confidence intervals for the parameters. Such procedures use the normal approximation.

For example, suppose we wish to test that $\beta = 0.05$ for the hepatitis data. The program printout shows that the standard error of $\hat{\beta}$ is 0.005. The test statistic is

$$Z = \frac{\text{estimated } \hat{\beta} - \beta}{\text{SE}(\hat{\beta})} = \frac{-0.054 - (0.05)}{0.005} = \frac{-0.004}{0.005} = -0.8$$

Comparing this value with the tabled value in Appendix Table A.1, we compute the appropriate two-sided P value as 0.42 and thus do not reject the null hypothesis that $\beta = 0.05$.

Some programs (P3R or PAR in BMDP) also print the predicted value of Y and its standard error for each observation. These quantities can be used to compute confidence intervals around the regression curve, using the normal approximation. For example, for the hepatitis data at $X = 14.5$ the predicted $\hat{Y} = 0.0455$ with a standard error of 0.0058. Thus an approximate 95% confidence interval for the true mean proportion of recipients contracting non-A, non-B hepatitis when the donor blood ALT level is 14.5 can be computed as

$$0.0455 \pm (1.96)(0.0058)$$

or
$$0.0341 < \text{mean } Y < 0.0569$$

The programs BMDP3R and BMDPAR print the residuals, which also can be obtained from the SAS NLIN procedure by using the print procedure. The residuals can be used to examine how well the equation fits over the range of the observed data and to check for outliers, as was discussed in Chapter 6.

Comparison of Models

In many situations more than one model can present a reasonable approximation to the data. A useful statistic can be computed to compare the relative fit of the various models. This statistic is the simple correlation between the observed Y values and the predicted $\hat{Y}$ values. In the hepatitis example the correlation between Y and $\hat{Y}$ was computed to be 0.97, indicating an excellent fit. The residual sums of squares can also be used to compare different models if none of them involve a transformation of the Y values. The correlation coefficient between Y and $\hat{Y}$ can be useful whether or not such a transformation is made.

9.5 DISCUSSION OF COMPUTER PROGRAMS

As we have seen, there are numerous computer programs that the investigator can use for multiple linear regression. But for nonlinear regression the choice is rather limited. In fact, SPSS-X does not have one, SAS has one, and BMDP has two. The SAS NLIN procedure can be run with the user supplying formulas for certain derivatives, or without them. All the programs require the user to supply the regression model formula and initial values of the

TABLE 9.3. Summary of Computer Output from SAS and BMDP for Nonlinear Regression

Options	SAS	BMDP
Built-in functions		P3R
User-supplied function	NLIN	P3R or PAR
Optional initial estimates equal zero		P3R
User-supplied initial estimates	NLIN	P3R or PAR
No derivatives needed	NLIN	PAR
Derivatives supplied for built-in functions		P3R
User-supplied derivatives	NLIN	P3R* or PAR
Weighted regression	NLIN	P3R or PAR
Constraints on parameters	NLIN	P3R or PAR
Maximum number of iterations	NLIN	P3R or PAR
Specify criterion for convergence	NLIN	P3R or PAR
Estimates of parameters	NLIN	P3R or PAR
Standard error of parameters	NLIN	P3R or PAR
Predicted dependent variable ($\hat{Y}$)	NLIN	P3R or PAR
Standard error of predicted $\hat{Y}$		P3R or PAR

Note: SPSS–X does not have a nonlinear regression program.
*For non-built-in functions (see text discussion).

estimates of the parameters. In NLIN a grid of initial values can be specified. In the remainder of this section we discuss in detail the use of the three programs mentioned in Table 9.3.

Using BMDP3R

The BMDP3R program offers two options for specifying the regression function. One option is for the user to supply formulas for the regression model and its derivatives. The other option is to select one of the built-in functions, in which case the program supplies the derivatives. The built-in functions are rather flexible and can accommodate a variety of other functions with the proper selection of the parameters. For example, the first built-in function has the form

$$f_1 = P_1 e^{(P_2 X)} + P_3$$

The second function is

$$f_2 = \frac{1}{f_1}$$

or
$$f_2 = \frac{1}{P_1 e^{(P_2 X)} + P_3}$$

This function is algebraically equivalent to the model used to fit the hepatitis data, i.e.,

$$Y = \frac{\gamma}{1 + e^{-(\alpha + \beta X)}}$$

In the hepatitis data example we used the built-in function f_2, and the equation fitted by the computer was

$$f_2 = \frac{1}{44.1 e^{-0.054 X} + 1.847}$$

Dividing by 1.847 in both the numerator and denominator yields

$$f_2 = \frac{0.541}{1 + 23.87 e^{-0.054 X}}$$

Since 23.87 equals $e^{\ln(23.87)}$, and since $\ln(23.87) = 3.17$, we can rewrite the equation as

$$f_2 = \frac{0.541}{1 + e^{3.17 - 0.54 X}}$$

or
$$Y = \frac{0.541}{1 + e^{-(-3.17 + 0.054 X)}}$$

Therefore the estimated parameters are as follows:

$$\text{estimate of } \gamma = 0.541$$

$$\text{estimate of } \alpha = -3.17$$
$$\text{estimate of } \beta = 0.054$$

as was given earlier.

If none of the built-in functions can be adapted to the user's desired regression function, BMDPAR or NLIN is an alternative choice because these procedures do not require specification of derivatives. They will, however, on the average require more iterations than BMDP3R. Table 9.3 summarizes this discussion and lists other options.

The instructions required for running BMDP3R with the hepatitis data are very few in number. Following the problem statement, the usual data definition information must be supplied. Next, the information needed to perform the regression analysis must be given. The specifications are reproduced here:

```
/PROBLEM      TITLE IS 'LOGISTIC REGRESSION'.
/INPUT        FORMAT IS FREE.
              CASES = 7.
              VARIABLES ARE 2.
/VARIABLES    NAMES ARE ALT, PROPHEP.
/REGRESSION   DEPENDENT IS PROPHEP.
              INDEPENDENT IS ALT.
              NUMBER IS 2.
              PARAMETERS ARE 3.
/PARAMETERS   INITIAL ARE 31.3,-.043,2.
/END
```

The data on the ALT levels and the proportion of cases with hepatitis are entered with a blank between them and with one case per line after the END paragraph; the data are submitted along with the program so that it is not necessary to say where the data are. The format sentence states that FORMAT IS FREE. This format is a sensible choice of data entry for a small amount of data, such as exists in this data set. The first variable is ALT level midpoints, as given in Table 9.2. The second variable is the proportion of cases with hepatitis, which is called PROPHEP.

In the REGRESSION paragraph the proportion of hepatitis cases is defined to be the dependent variable, while ALT level is the independent variable. The NUMBER IS 2 sentence signifies that we wish to use the second function from the list of built-in functions. The first function is

$$f_1 = P_1 e^{(P_2 X)} + P_3 e^{(P_4 X)} + \cdots$$

and the second function is

$$f_2 = \frac{1}{f_1}$$

The statement PARAMETERS ARE 3 makes the first function become

$$f_1 = P_1 e^{(P_2 X)} + P_3 e^{(0X)}$$

or

$$f_1 = P_1 e^{(P_2 X)} + P_3$$

since $e^{0X} = e^0 = 1$.

Because we want the reciprocal of f_1, the second function is used. The initial values are estimated as

$$\gamma_0 = 0.5$$

$$\alpha_0 = 0.043$$

$$\beta_0 = -2.75$$

so that
$$Y = \frac{0.5}{1 + e^{-(-2.75 + 0.043X)}}$$

This equation now needs to be transformed to the form

$$Y = \frac{1}{P_1 e^{P_2 X} + P_3}$$

First, we write

$$Y = \frac{0.5}{1 + e^{2.75} e^{-0.043X}}$$

Multiplying both the numerator and denominator by 2 results in

$$Y = \frac{1}{2 + 31.3 e^{-0.043X}}$$

since $e^{2.75} = 15.64$. Thus the initial values used in the program are $P_1 = 31.3$, $P_2 = -0.043$, and $P_3 = 2$.

A brief discussion of selected output follows. Procedure P3R prints the mean, standard deviation, and minimum and maximum values of each variable. It also gives the values of the estimated parameters and the residual sums of squares at successive iterations. In the hepatitis example the program rather quickly iterated to a solution. It also prints whether or not the estimates "overshoot" and actually result in a larger residual sum of squares. When this result occurs, the program repeatedly halves the distance it uses to adjust the estimates of the parameters until no increase is obtained or until it has halved the distance five times. In this example halving did not happen until the last step. The results of the iteration are given in Table 9.4, with the data on the estimates rounded to four places.

From Table 9.4 it can be seen that the program did not have appreciably

TABLE 9.4. Successive Iterations for Solution of the Hepatitis Nonlinear Regression

Iteration Number	Incremental Halvings	Residual Sums of Squares	P_1	P_2	P_3
0	0	0.821880-02	31.3000	−0.0430	2.0000
1	0	0.707412-03	40.8199	−0.0538	1.8778
2	0	0.529736-03	43.5352	−0.0539	1.8432
3	0	0.529192-03	44.0437	−0.0541	1.8463
4	0	0.529187-03	44.0980	−0.0541	1.8465
5	0	0.529186-03	44.1036	−0.0541	1.8465
6	0	0.529186-03	44.1042	−0.0541	1.8465
7	0	0.529186-03	44.1042	−0.0541	1.8465
8	0	0.529186-03	44.1043	−0.0541	1.8465
9	5	0.529186-03	44.1043	−0.0541	1.8465

different results after the fifth iteration. The program terminates when the relative change in the sums of squares during five successive iterations is less than 0.0001. (The value 0.529186-03 is read by shifting the decimal point three places to the left, as 0.000529186. This notation is called scientific notation, and it is used to display the significant digits of a value without using a large number of zeros. The representation 0.529186-03 is shorthand for 0.529186 times 10^{-3}.)

The program also presents the residual mean square, which is equal to the residual sum of squares divided by the degrees of freedom (sample size minus the number of fitted parameters). The latter is equal to 4 in this example.

In the printout a separate table is given of the estimated parameters, their asymptotic standard deviations, and a correlation matrix. If the asymptotic standard deviation of a parameter estimate is zero, the manual states that this result indicates a problem with the estimate. It may be an indication that two parameter estimates are highly correlated.

Finally, predicted values of the dependent variable, residuals from these predicted values, and observed values of the dependent variable are printed for all cases. Cook's distance is also printed for each case (see Section 6.8). In the hepatitis example Cook's distance for the last point (ALT=105, PRO-PHEP=0.500) is very large (603.1), indicating that that point has an overwhelming effect on the location of the regression curve. The residuals are all less than 0.02, indicating a very good fit, but the high dependency on one point signifies that we should have confidence in that point before we can trust the estimated logistic equation. Other output can be obtained by using the options given in the manual.

Using BMDPAR

If the investigator wishes to avoid obtaining derivatives or using the built-in functions of P3R, either SAS NLIN or BMDP PAR programs can be used. The output from PAR is similar to the output of P3R, so only the instructions for submission of the program that differ from P3R are given here. The INPUT and VARIABLE paragraph for the hepatitis example are precisely the same.

```
/REGRESSION      DEPENDENT IS PROPHEP.
                 PARAMETERS ARE 3.
/PARAMETER       INITIAL ARE 31.3,-.043,2.
/FUN             F=1/(P1*EXP(P2*ALT)+P3).
/END
```

The independent and number sentence are not given as they are in P3R since the model of the regression equation and the independent variable are given in the FUN paragraph. These statements are like FORTRAN statements, and numerous examples are given in the manual.

The results are obtained by using a numerical method developed by Ralston and Jennrich (1978). A comparison of the results of this method and more conventional methods is given in their article. For the hepatitis example the numerical estimates of the parameters are almost the same as those obtained earlier, but it took 13 iterations instead of the 9 used by P3R. We believe that a user not well versed in mathematics is less apt to make errors in using this program or the similar option in SAS NLIN than in using the derivative options of P3R or SAS NLIN.

Using SAS NLIN

A description of the hepatitis problem run by using SAS NLIN with card input is given next. A data set HEP is set up with variables X and Y entered successively with blanks between them (XblankYblankXblankYblank...). The two @@ (pointer control symbols) indicate this mode of reading the data. The semicolon (;) is used at the start of a new line to indicate the end of the data. The input and data instruction statements are as follows:

```
TITLE           HEPATITIS;
DATA            HEP;
                INPUT X Y @@;
                CARDS;
4.5  .038 14.5  .053 24.5  .077 34.5  .100 44.5  .167
54.5  .250 105.0  .500
;
PROC            NLIN DATA=HEP;
                PARMS A=31.3,B=-.043,C=2;
                MODEL Y=1/(C+A*EXP(B*X));
```

The NLIN procedure requires starting values, as given in the PARMS statement, and a MODEL statement using the same symbols as those in the INPUT statement and the same parameters used in the PARMS statement. The nonlinear equation is given in the MODEL statement, using any valid SAS expression that will yield a numeric result. Note that the program automatically uses the derivative-free method of Ralston and Jennrich (1978) when formulas for the derivatives are not supplied in the statements.

The output from NLIN includes the values of the three parameters at each iteration (14 iterations with the hepatitis data and set of instructions) and the residual sums of squares. Procedure NLIN estimated A as 44.10411, B as -0.05413, and C as 1.84651, results very similar to those of P3R.

The output also includes an analysis of variance table similar to that discussed in Chapters 6 and 7. The estimates of the parameters, their asymptotic standard error, and asymptotic 95% confidence limits are listed as well. In this example the confidence limits are

$$18.924 < A < 69.284$$

$$-0.068 < B < -0.040$$

$$1.647 < C < 2.047$$

The asymptotic standard errors computed by NLIN are very close to the asymptotic standard deviations listed by P3R.

One convenient feature of the NLIN procedure is the ability to define a whole grid of possible initial values of the parameters. For example, if an investigator has two parameters to estimate, A and B, a rectangular grid of possible starting values can be created by using statements such as

```
PARMS    A=0 to 4 by 2;
         B=1 to 3 by .5;
```

which would result in the program trying all possible combinations in the following array:

			B		
A	1	1.5	2	2.5	3
0	0,1	0,1.5	0,2	0,2.5	0,3
2	2,1	2,1.5	2,2	2,2.5	2,3
4	4,1	4,1.5	4,2	4,2.5	4,3

In some equations it is critical to start with initial values that are not too far from the final solution. Often investigators will try numerous starting values to ensure that they have estimated the best-fitting parameters to their model.

Numerous other options are included in the NLIN procedure, and the user is advised to read the manual carefully before using this procedure in order to select the most appropriate options.

SUMMARY

This chapter dealt with regression equations that cannot be transformed so that they are linear in the parameters. Programs for performing nonlinear regression were discussed, and a detailed example was worked out. Because of the complexity of nonlinear regression methods, theory has not been developed to the same extent as it has for linear models. Also, little is known about the robustness of the estimates. Partly for these reasons most investigators will try to first employ linear models for regression analysis. If a good fit is not obtained, they resort to nonlinear methods.

BIBLIOGRAPHY

Afifi, A. A., and Azen, S. P. 1979. *Statistical analysis: A computer oriented approach*. 2nd ed. New York: Academic Press.

Bard, Y. 1974. *Nonlinear parameter estimation*. New York: Academic Press.

Daniel, C., and Wood, F.S. 1980. *Fitting equations to data*. New York: Wiley.

Draper, N. R., and Smith, H. 1981. *Applied regression analysis*. 2nd ed. New York: Wiley.

*Gallant, A. R. 1975. Nonlinear regression. *American Statistician* 29:73–81.

Goldstein, H. 1979. *The design and analysis of longitudinal studies*. New York: Academic Press.

*Hartley, A. O. 1961. Modified Gauss-Newton method for fitting non-linear regression functions. *Technometrics* 3:269–280.

Hollinger, F. B.; Mosley, J. W.; Szmuness, W.; Aach, R. D.; Melnick, J. L.; Afifi, A.; Stevens, C. E.; and Kahn, R. A. 1982. Non-A, non-B, hepatitis following blood transfusions: Risk factors associated with donor characteristics. In *Viral Hepatitis: 1981 International Symposium*, ed. W. Szmuness, H. J. Alter, and J. E. Maynard. Philadelphia: Franklin Institute Press.

*Kowalik, J., and Osborne, M. R. 1968. *Methods for unconstrained optimization problems*. New York: Elsevier.

*Marquardt, D. W. 1963. An algorithm for the least-squares estimation of non-linear parameters. *Journal of the Society of Industrial Applied Mathematics* 11:431–441.

Ralston, M. L., and Jennrich, R. I. 1978. Dud, a derivative-free algorithm for nonlinear least squares. *Technometrics* 20:7–13.

Sprent, P. 1969. *Models in regression and related topics*. London: Methuen.

PROBLEMS

9.1 Rerun the hepatitis data set example, using starting values of 40, −0.05, and 1.9, respectively, to see whether use of these values shortens the number of iterations needed compared with the number given in Section 9.5.

9.2 Fit the nonlinear equation $Y = B(X - A)^2$ to the accompanying data. Test the null hypothesis that $B = 2$. For starting values, try $A = 6$ and $B = 1.5$.

X	Y
10	50.5
11	72.1
12	99.5
14	163.0
16	243.4
16	242.9
17	289.2
18	336.5
20	449.3

9.3 Fit an exponential decay function, $Y = \gamma + \alpha e^{\beta X}$, to the accompanying data. Plot the points on graph paper. By analogy to Figure 9.2, determine starting values for α, β, and γ, and fit the regression equation.

X	Y
0.5	4.59
1.1	3.71
1.8	2.76
2.0	2.53
3.0	2.35
4.0	2.25
5.0	2.08

9.4 In the accompanying table the effects of spraying a new insecticide on cucumber leaves are given. One hundred diabrotica insects were placed on leaves of a cucumber plant in each of six trials. They were sprayed with different concentrations. The number of insects that died within one hour were counted. Fit a logistic regression equation, using the proportion killed as the dependent variable and the concentration as the independent variable. Fit the equation again with the logarithm of concentration as the independent variable, and decide which equation yields the better fit.

Concentration (mg/L)	Number of Deaths	Number of Diabrotica
2	5	100
3	11	100
4	18	100
5	46	100
8	86	100
10	90	100

9.5   From the data in Appendix B we wish to relate Y = height to X = age for boys. Several models are possible:

▶ Linear: $Y = \alpha + \beta X + e$.
▶ Quadratic: $Y = \alpha + \beta_1 X + \beta_2 X^2 + e$.
▶ Cubic: $Y = \alpha + \beta_1 X + \beta_2 X^2 + \beta_3 X^3 + e$.
▶ Logistic: $Y = \gamma/(1 + e^{-(\alpha+\beta X)}) + e$ for $\gamma > 0$.
▶ Exponential: $Y = \alpha + \beta e^{\gamma X} + e$ for $\beta, \gamma < 0$.

Using the data on the first child (boys only), fit a regression equation to each of these five models. (Note that the first three are linear and the last two are nonlinear.) From each equation, predict the height at age = 0 (at birth), 5, 15, 25, and 50 years. (Note that the ages in the sample are restricted to 7–17 years.) Also, compute 95% confidence intervals for the mean height at 10 years from each equation. Comment on the results.

9.6   Continuation of Problem 9.5: Using graphical and numerical methods, compare the five equations and decide which one(s) you prefer.

9.7   Repeat the analysis described in Problems 9.5 and 9.6 for girls. Compare the results for the two sexes.

9.8   Using the data from Appendix B again, now we wish to relate Y = weight to X_1 = height and X_2 = age. Two possible models are

$$Y = \alpha + \beta_1 X_1 + \beta_2 X_2 + e$$

and

$$Y = \alpha + \beta_1 X_1 + \frac{\beta_2}{1 + e^{-(\beta_3+\beta_4 X_2)}} + e$$

The first equation can also be modified by adding quadratic and/or cubic terms for age. For the first child in the data set, fit these and possibly other models for boys and girls separately. Decide on a preferred model and compare the results.

10 RECENT ADVANCES IN REGRESSION ANALYSIS

10.1 WHAT WILL YOU LEARN FROM THIS CHAPTER?

This chapter presents brief descriptions of several topics in regression and specialized uses of regression. From this chapter you will learn:

- How to handle missing values in regression analysis (10.2).
- How to use dummy variables in order to employ nominal data as independent variables in regression analysis (10.3).
- How to interpret the results when dummy variables are used (10.3).
- How to obtain regression equations when constraints are placed on the parameter estimates (10.4).
- How to obtain segmented regression equations (10.4).
- How to determine when to use ridge regression (10.5).
- How to perform ridge regression by using standard computer packages (10.5).
- How to decide whether or not to perform path analysis (10.6).

10.2 MISSING VALUES

Missing observations can occur in various ways. An observation can be missing "at random"; for example, it could simply be lost. In other situations the observation is missing because it falls into a certain range. For example, in archeology fossils of long bones tend to break, so observations of long-bone fossils tend to be missing. In human populations responses to interview data could be missing because of the nature of the possible response or because of a poorly constructed questionnaire.

The effect of nonrandom missing values is difficult to gauge. Data can be missing for many different reasons, and the analysis should take these reasons into account. At the least the investigator should compare the characteristics of the responders and nonresponders to see whether they are similar. If they are, the simplest option is to use only the complete responses. If the nonresponders are different from the responders, the analysis based on the complete data can generalize only to a population represented by the responders. Generalization to the total population must take the characteristics of nonresponders into account.

The subject of randomly missing observations in regression analysis has received wide attention from statisticians. Afifi and Elashoff (1966) present a review of the literature up to 1966, and Hartley and Hocking (1971) present a simple taxonomy of incomplete data problems (see also Little, 1982). When some observations are missing in a random fashion, the simplest option is again to use only the cases with complete data. Since this procedure may result in a reduced sample size, various methods have been developed to fill in the missing data and then perform the regression analysis on the completed data set.

One method for filling in data, called *mean substitution*, replaces the missing value for a given variable by the mean of the observed values of that variable. A more complicated method involves first computing a *regression equation* of a given variable on one or more of the remaining variables and then using that equation to "predict" the missing values. For example, suppose that for a sample of schoolchildren some heights are missing. The investigator could first derive the regressions of height on age for boys and girls and then use these regressions to predict the missing heights. If the children have a wide range of ages, this method presents an improvement over mean substitution (for simple linear regression, see Afifi and Elashoff

$1969a$ and $1969b$). A third method, which applies regression with normally distributed X variables, was studied by Orchard and Woodbury (1972) and further developed by Beale and Little (1975). This method is based on the theoretical principle of *maximum likelihood* used in statistical estimation.

All three methods are available in the BMDPAM program, in addition to estimation based on complete data only. Output from this program can then be used for a regression analysis or other multivariate analyses. We believe that the most understandable analysis is the one based on complete data only. To obtain a complete data set that is as large as possible, though, the investigator may have to delete some variables from the analysis.

If the investigator wishes to make some use of the incomplete cases, mean substitution can be useful when the intercorrelations among the variables are small. This method, however, may result in biases of unknown magnitude in the resulting estimates. The regression method of filling in the data is preferable when higher correlations exist. Some biases are incurred here as well. The maximum likelihood method is known to have theoretically appealing features if the data are, indeed, normally distributed. Little is known about the properties of this method of dealing with missing values when the underlying distribution is not normal.

In the SPSS–X REGRESSION program the investigator has the option of using only the cases with no missing values on all variables, mean substitution, or two other options. One option is to include cases with missing values, with the missing value code considered a valid observation. If the data are continuous, missing values are often coded as extreme values, such as 99 or blanks. If so, then this option would not appear to be very useful. If the investigator is using dummy variables (as discussed in the next section), then considering the missing values as a separate response to a question along with other responses can be an interesting option. Sometimes, in survey data, the fact that a respondent did not answer a particular question can be a useful piece of information to analyze.

The SPSS–X program also offers the option of computing each correlation, using cases with complete data on each pair of variables being correlated irrespective of whether or not the cases have missing data on other variables. This and other methods are also available in BMDP8D for computing correlation and covariances. These methods of using available pairs may result in computational problems (see the program manuals) and should be used with caution. Their use is only recommended when the number of completed cases is small and the investigator is trying to use as much of the data as possible.

A practical piece of advice is to use the various options available and then compare the results. If a consensus is obtained, then further credence is given to the results.

10.3 DUMMY VARIABLES

Often X variables desired for inclusion in a regression model are not continuous. Such variables could be either nominal or ordinal. Ordinal measurements represent variables with an underlying scale. For example, the severity of a burn could be classified as mild, moderate, or severe. These burns are commonly called first-, second-, and third-degree burns, respectively. The X variable representing these categories may be coded 1, 2, or 3. The data from ordinal variables could be analyzed by using the numerical coding for them and treated as though they were continuous (or interval data). This method takes advantage of the underlying order of the data. On the other hand, this analysis assumes equal intervals between successive values of the variable. Thus in the burn example, we would assume that the difference between a first- and a second-degree burn is equivalent to the difference between a second- and third-degree burn. As was discussed in Chapter 5, an appropriate choice of scale may be helpful in assigning numerical values to the outcome that would reflect the varying lengths of intervals between successive responses.

In contrast, the investigator may ignore the underlying order for ordinal variables and treat them as though they were nominal variables. In this section we discuss methods for using one or more nominal X variables in regression analysis along with the role of interactions.

A Single Binary Variable

We will begin with a simple example to illustrate the technique. Suppose that the dependent variable Y is yearly income in dollars and the independent variable X is the sex of the respondent (male or female). To represent sex, we create a *dummy* (or *indicator*) *variable* $D = 0$ if the respondent is male and $D = 1$ if the respondent is female. (All the major computer packages offer the option of recoding the data in this form.) The sample regression equation can then be written as $Y = A + BD$. The value of Y is

$$Y = A \qquad \text{if} \qquad D = 0$$

and
$$Y = A + B \quad \text{if} \quad D = 1$$

Since our best estimate of Y for a given group is that group's mean, A is estimated as the average income for males ($D = 0$) and $A + B$ is the average income for females ($D = 1$). The regression coefficient B is therefore $B = \overline{Y}_{\text{females}} - \overline{Y}_{\text{males}}$. In effect, males are considered the *reference group*; and females' income is measured by how much it differs from males' income.

A Second Method of Coding Dummy Variables

An alternative way of coding the dummy variables is

$$D^* = -1 \quad \text{for males}$$

and
$$D^* = +1 \quad \text{for females}$$

In this case the regression equation would have the form

$$Y = A^* + B^*D^*$$

The average income for males is now

$$A^* - B^* \quad (\text{when } D^* = -1)$$

and for females it is

$$A^* + B^* \quad (\text{when } D^* = +1)$$

Thus
$$A^* = \tfrac{1}{2}(\overline{Y}_{\text{males}} + \overline{Y}_{\text{females}})$$

and
$$B^* = \overline{Y}_{\text{females}} - \tfrac{1}{2}(\overline{Y}_{\text{males}} + \overline{Y}_{\text{females}})$$

or
$$B^* = \tfrac{1}{2}(\overline{Y}_{\text{females}} - \overline{Y}_{\text{males}})$$

In this case neither males nor females are designated as the reference group.

A Nominal Variable with Several Categories

For another example we consider an X variable with $k > 2$ categories. Suppose income is now to be related to the religion of the respondent. Religion is classified as Catholic (C), Protestant (P), Jewish (J), and other

(O). The religion variable can be represented by the dummy variables that follow:

Religion	D_1	D_2	D_3
C	1	0	0
P	0	1	0
J	0	0	1
O	0	0	0

Note that to represent the four categories, we need only three dummy variables. In general, to represent k categories, we need $k - 1$ dummy variables. Here the three variables represent C, P, and J, respectively. For example, $D_1 = 1$ if Catholic and zero otherwise; $D_2 = 1$ if Protestant and zero otherwise; and $D_3 = 1$ if Jewish and zero otherwise. The "other" group has a value of zero on each of the three dummy variables.

The estimated regression equation will have the form

$$Y = A + B_1 D_1 + B_2 D_2 + B_3 D_3$$

The average incomes for the four groups are

$$\overline{Y}_C = A + B_1$$

$$\overline{Y}_P = A + B_2$$

$$\overline{Y}_J = A + B_3$$

$$\overline{Y}_O = A$$

Therefore,

$$B_1 = \overline{Y}_C - A = \overline{Y}_C - \overline{Y}_O$$

$$B_2 = \overline{Y}_P - A = \overline{Y}_P - \overline{Y}_O$$

$$B_3 = \overline{Y}_J - A = \overline{Y}_J - \overline{Y}_O$$

Thus the group "other" is taken as the reference group to which all the others are compared. Although the analysis is independent of which group is chosen as the reference group, the investigator should select the group that makes the interpretation of the B's most meaningful. The mean of the reference group is the constant A in the equation. An example of program specification for dummy variables is given in Problem 10.2.

If none of the groups represent a natural choice of a reference group, an alternative choice of assigning values to the dummy variables is as follows:

Religion	D_1^*	D_2^*	D_3^*
C	1	0	0
P	0	1	0
J	0	0	1
O	-1	-1	-1

As before, Catholic has a value of 1 on D_1^*, Protestant a value of 1 on D_2^*, and Jewish a value of 1 on D_3^*. However, "other" has a value of -1 on each of the three dummy variables. Note that zero is also a possible value of these dummy variables.

The estimated regression equation is now

$$Y = A^* + B_1^* D_1^* + B_2^* D_2^* + B_3^* D_3^*$$

The group means are

$$\overline{Y}_C = A^* + B_1^*$$

$$\overline{Y}_P = A^* + B_2^*$$

$$\overline{Y}_J = A^* + B_3^*$$

$$\overline{Y}_O = A^* - B_1^* - B_2^* - B_3^*$$

In this case the constant

$$A^* = \tfrac{1}{4}(\overline{Y}_C + \overline{Y}_P + \overline{Y}_J + \overline{Y}_O)$$

the unweighted average of the four group means. Thus

$$B_1^* = \overline{Y}_C - \tfrac{1}{4}(\overline{Y}_C + \overline{Y}_P + \overline{Y}_J + \overline{Y}_O)$$

$$B_2^* = \overline{Y}_P - \tfrac{1}{4}(\overline{Y}_C + \overline{Y}_P + \overline{Y}_J + \overline{Y}_O)$$

and $$B_3^* = \overline{Y}_J - \tfrac{1}{4}(\overline{Y}_C + \overline{Y}_P + \overline{Y}_J + \overline{Y}_O)$$

Note that with this choice of dummy variables there is no reference group. Also, there is no slope coefficient corresponding to the "other" group. As before, the choice of the group to receive -1 values is arbitrary. That is, any of the four religious groups could have been selected for this purpose.

One Nominal and One Interval Variable

Another example, which includes one dummy variable and one continuous variable, is the following: Suppose an investigator wants to relate vital capacity (Y) to height (X) for men and women (D) in a restricted age group. One model for the population regression equation is

$$Y = \alpha + \beta X + \delta D + e$$

where $D = 0$ for females and $D = 1$ for males. This equation is a multiple regression equation with $X = X_1$ and $D = X_2$. This equation breaks down to an equation for females,

$$Y = \alpha + \beta X + e$$

and one for males,

$$Y = \alpha + \beta X + \delta + e$$

or
$$Y = (\alpha + \delta) + \beta X + e$$

Figure 10.1a illustrates both equations.

Note that this model forces the equation for males and females to have the same slope β. The only difference between the two equations is in the intercept: α for females and $(\alpha + \delta)$ for males.

Interaction

A model that does not force the lines to be parallel is

$$Y = \alpha + \beta X + \delta D + \gamma(XD) + e$$

This equation is a multiple regression equation with $X = X_1$, $D = X_2$, and $XD = X_3$. The variable X_3 can be generated on the computer as the product of the other two variables. With this model the equation for females $(D = 0)$ is

$$Y = \alpha + \beta X + e$$

and for males it is

$$Y = \alpha + \beta X + \delta + \gamma X + e$$

or
$$Y = (\alpha + \delta) + (\beta + \gamma)X + e$$

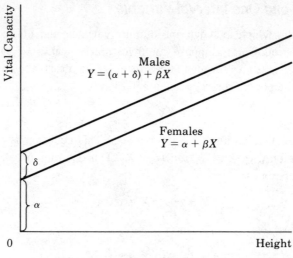

a. Lines are Parallel (No Interaction)

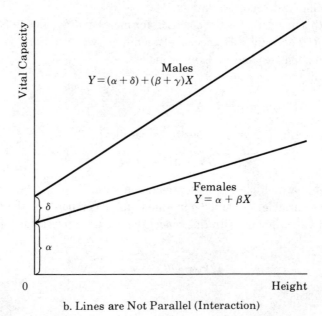

b. Lines are Not Parallel (Interaction)

FIGURE 10.1.   Vital Capacity Versus Height for Males and Females

The equations for this model are illustrated in Figure 10.1b for the case where γ is a positive quantity.

Thus with this model we allow both the intercept and slope to be different for males and females. The quantity δ is the difference between the two intercepts, and γ is the difference between the two slopes. Note that some investigators would call the product XD the *interaction* between height and sex. With this model we can test the hypothesis of no interaction, i.e., $\gamma = 0$, and thus decide whether parallel lines are appropriate. This test can be done by using the methods discussed in Chapter 7, since this is an ordinary multiple regression model.

Extensions

We can use these ideas and apply them to situations where there may be several X variables and several D (or dummy) variables. For example, we can estimate vital capacity by using age and height for males and females and for smokers and nonsmokers. The selection of the model—i.e., whether or not to include interaction—and the interpretation of the resulting equations must be done with caution. The previous methodology for variable selection applies here as well, with appropriate attention given to interpretation. For example, if a nominal variable such as religion is split into three new dummy variables, with "other" used originally as a reference group, a stepwise program may enter only one of these three dummy variables. Suppose the variable D_1 is entered, where $D_1 = 1$ signifies Catholic and $D_1 = 0$ signifies Protestant, Jewish, or other. In this case the reference group becomes Protestant, Jewish, or other. Care must be taken in the interpretation of the results to report the proper reference group. Sometimes, investigators will force in the three dummy variables D_1, D_2, D_3 if any one of them is entered in order to keep the reference group as it was originally chosen. This feature can be accomplished easily by using the subsets-of-variables option in BMDP2R.

The investigator is advised to write out the separate equation for each subgroup, as was shown in the previous examples. This technique will help clarify the interpretation of the regression coefficients and the implied reference group (if any). Also, it may be advisable to select a meaningful reference group prior to selecting the model equations rather than rely on the computer program to do it for you.

Another alternative to using dummy variables is to find separate regression equations for each level of the nominal or ordinal variables. For example, in the prediction of FEV1 from age and height, a better prediction can be achieved if an equation is found for males and a separate one for females.

Females are not simply "short" men. If the sample size is adequate, it is a good procedure to check whether it is necessary to include an interaction term. If the slopes "seem" equal, no interaction term is necessary; otherwise, such terms should be included. Formal tests exist for this purpose, but they are beyond the scope of this book (see Graybill 1976).

10.4 CONSTRAINTS ON PARAMETERS

As mentioned in Chapter 9, some standard nonlinear regression programs offer the user the option of restricting the range of possible values of the parameter estimates. In addition, some programs (e.g., BMDP3R, BMDPAR, and SAS REG) offer the option of imposing *linear constraints* on the parameters. These constraints take the form

$$C_1\beta_1 + C_2\beta_2 + \cdots + C_P\beta_P = C$$

where $\beta_1, \beta_2, \ldots, \beta_P$ are the parameters in the regression equation and $C_1, C_2, \ldots, C_P$ and C are constants supplied by the user. The program finds estimates of the parameters restricted to satisfy this constraint as well as any other supplied.

Although some of these programs are intended for nonlinear regression, they also provide a convenient method of performing a *linear* regression with constraints on the parameters. For example, suppose that the coefficient of the first variable in the regression equation was demonstrated from previous research to have a specified value, such as $B_1 = 2.0$. Then the constraint would simply be

$$C_1 = 1, \qquad C_2 = \ldots = C_P = 0$$

and $$C = 2.0 \qquad \text{or} \qquad 1\beta_1 = 2.0$$

Another example of an inequality constraint is the situation when coefficients are required to be nonnegative. For example, if $\beta_2 \geq 0$, this constraint can also be supplied to the program.

The use of linear constraints offers a simple solution to the problem known as *spline regression* or *segmented-curve regression*. For instance, in economic applications we may want to relate the consumption function Y to the level of aggregate disposable income X. A possible nonlinear relationship is a linear function up to some level X_0, i.e., for $X \leq X_0$, and another linear function for $X > X_0$. As illustrated in Figure 10.2, the equation for $X \leq X_0$ is

$$Y = \alpha_1 + \beta_1 X + e$$

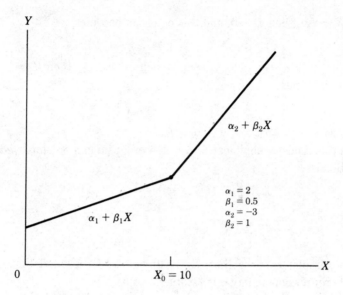

FIGURE 10.2. Segmented Regression Curve with $X = X_0$ as the Change Point

and for $X > X_0$ it is

$$Y = \alpha_2 + \beta_2 X + e$$

The two curves must meet at $X = X_0$. This condition produces the linear constraint

$$\alpha_1 + \beta_1 X_0 = \alpha_2 + \beta_2 X_0$$

Equivalently, this constraint can be written as

$$\alpha_1 + X_0 \beta_1 - \alpha_2 - X_0 \beta_2 = 0$$

To formulate this problem as a multiple linear regression equation, we first define a dummy variable D such that

$$D = 0 \quad \text{if} \quad X \leq X_0$$

$$D = 1 \quad \text{if} \quad X > X_0$$

The segmented regression equation can be combined into one multiple linear equation as

$$Y = \gamma_1 + \gamma_2 D + \gamma_3 X + \gamma_4 DX + e$$

When $X \leq X_0$, then $D = 0$, and this equation becomes

$$Y = \gamma_1 + \gamma_3 X + e$$

Therefore $\gamma_1 = \gamma_3 = \alpha_1$ and $\gamma_3 = \beta_1$. When $X > X_0$, then $D = 1$, and the equation becomes

$$Y = (\gamma_1 + \gamma_2) + (\gamma_3 + \gamma_4) X + e$$

Therefore $\gamma_1 + \gamma_2 = \alpha_2$ and $\gamma_3 + \gamma_4 = \beta_2$.

With this model a nonlinear regression program can be employed with the restriction

$$\gamma_1 + X_0\gamma_3 - (\gamma_1 + \gamma_2) - X_0(\gamma_3 + \gamma_4) = 0$$

or

$$-\gamma_2 - X_0\gamma_4 = 0$$

or

$$\gamma_2 + X_0\gamma_4 = 0$$

For example, suppose that X_0 is known to be 10 and that the fitted multiple regression equation is estimated to be

$$Y = 2 - 5D + 0.5X + 0.5DX + e$$

Then we estimate $\alpha_1 = \gamma_1$ as 2, $\beta_1 = \gamma_3$ as 0.5, $\alpha_2 = \gamma_1 + \gamma_2$ as $2 - 5 = -3$, and $\beta_2 = \gamma_3 + \gamma_4$ as $0.5 + 0.5 = 1$. These results are pictured in Figure 10.2. Further examples can be found in Draper and Smith (1981).

Where there are more than two segments with known values of X at which the segments intersect, a similar procedure using dummy variables could be employed. When these points of intersection are unknown, more complicated estimation procedures are necessary. Quandt (1972) presents a method for estimating two regression equations when an unknown number of the points belong to the first and second equation. The method involves numerical maximization techniques and is beyond the scope of this book.

Another kind of restriction occurs when the value of the dependent variable is constrained to be above (or below) a certain limit. For example, it may be physically impossible for Y to be negative. Tobin (1958) derived a procedure for obtaining a multiple linear regression equation satisfying such a constraint.

10.5 RIDGE REGRESSION

In Section 7.9 we discussed the problem of multicollinearity, i.e., when the independent variables are highly intercorrelated. In that section it was mentioned that the concept of tolerance offers the basis for a method of removing redundant variables. Other methods for excluding variables were described in Chapter 8 (variable selection).

If a successful variable selection process is used, the problem of multicolinearity is often implicitly overcome. However, there are situations where the investigator wishes to use several independent variables that are highly intercorrelated. For example, such will be the case when the independent variables are the prices of related commodities. Another situation where multicollinearity may occur is when the investigator measures several highly related physiological variables on animals. In those situations the estimated regression coefficients can be unstable in the sense that the standard errors of the standardized regression coefficients are large. One effect of this instability is that a minor change in the data set may result in a large change in the estimates. In this section we present a recently developed solution to this problem and show how it can be implemented on the computer.

Theoretical Background

One solution to the problem of multicollinearity is the so-called *ridge regression procedure* (Marquardt and Snee 1975; Hoerl and Kennard 1970; Gunst and Mason 1980). In effect, ridge regression artificially reduces the correlations among the independent variables in order to obtain more stable estimates of the regression coefficients. Note that although such estimates are biased, they may, in fact, produce a smaller mean square error for the estimates.

To explain the concept of ridge regression, we must follow through a theoretical presentation. Readers not interested in this theory may skip to the discussion of how to perform the analysis, using ordinary least squares regression programs.

We will restrict our presentation to the standardized form of the observations, i.e., where the mean of each variable is subtracted from the observation and the difference is divided by the standard deviation of the variable. The resulting least squares regression equation will automatically have a zero intercept and standardized regression coefficients. These coefficients are functions of only the correlation matrix among the X variables, namely,

$$
\begin{bmatrix}
1 & r_{12} & r_{13} & \cdots & r_{1P} \\
r_{12} & 1 & r_{23} & \cdots & \\
r_{13} & r_{23} & 1 & \cdots & \vdots \\
\vdots & \vdots & \vdots & & \vdots \\
r_{1P} & r_{2P} & r_{3P} & \cdots & 1
\end{bmatrix}
$$

The instability of the least squares estimates stems from the fact that some of the independent variables can be predicted accurately by the other independent variables. (For those familiar with matrix algebra, note that this feature results in the correlation matrix having a nearly zero determinant.) The ridge regression method artificially inflates the diagonal elements of the correlation matrix by adding a positive amount k to each of them. The correlation matrix is modified to

$$
\begin{bmatrix}
(1+k) & r_{12} & r_{13} & \cdots & r_{1P} \\
r_{12} & (1+k) & r_{23} & \cdots & \\
r_{13} & r_{23} & (1+k) & \cdots & \vdots \\
\vdots & \vdots & \vdots & & \\
r_{1P} & r_{2P} & r_{3P} & \cdots & (1+k)
\end{bmatrix}
$$

The value of k can be any positive number and is usually determined empirically, as will be described later.

Example of Ridge Regression

For $P = 2$—i.e., with two independent variables X_1 and X_2—the ordinary least squares *standardized* coefficients are computed as

$$
b_1 = \frac{r_{1Y} - r_{12}r_{2Y}}{1 - r_{12}^2}
$$

and

$$
b_2 = \frac{r_{2Y} - r_{12}r_{1Y}}{1 - r_{12}^2}
$$

The ridge estimators turn out to be

$$
b_1^* = \frac{r_{1Y} - [r_{12}/(1+k)]r_{2Y}}{1 - [r_{12}/(1+k)]^2} \left(\frac{1}{1+k} \right)
$$

and

$$
b_2^* = \frac{r_{2Y} - [r_{12}/(1+k)]r_{1Y}}{1 - [r_{12}/(1+k)]^2} \left(\frac{1}{1+k} \right)
$$

Note that the main difference between the ridge and least squares coefficients is that r_{12} is replaced by $r_{12}/(1+k)$, thus artificially reducing the correlation between X_1 and X_2.

For a numerical example, suppose that $r_{12} = 0.9$, $r_{1Y} = 0.3$, and $r_{2Y} = 0.5$. Then the standardized least squares estimates are

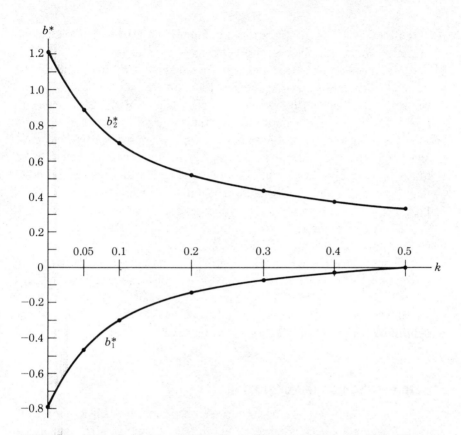

FIGURE 10.3. Ridge Regression Coefficients for Various Values of k (Ridge Trace)

$$b_1 = \frac{0.3 - (0.9)(0.5)}{1 - (0.9)^2} = -0.79$$

and
$$b_2 = \frac{0.5 - (0.9)(0.3)}{1 - (0.9)^2} = 1.21$$

For a value of $k = 0.4$ the ridge estimates are

$$b_1^* = \frac{0.3 - [0.9/(1 + 0.4)](0.5)}{1 - [0.9/(1 + 0.4)]^2} \left(\frac{1}{1 + 0.4}\right) = -0.026$$

and
$$b_2^* = \frac{0.5 - [0.9/(1 + 0.4)](0.3)}{1 - [0.9/(1 + 0.4)]^2} \left(\frac{1}{1 + .4}\right) = 0.374$$

Note that both coefficients are reduced in magnitude from the original slope coefficients. We computed several other ridge estimates for $k = 0.05$ to 0.5. In practice, these estimates are plotted as a function of k, and the resulting graph, as shown in Figure 10.3, is called the *ridge trace*.

The ridge trace should be supplemented by a plot of the residual mean square of the regression equation. From the ridge trace together with the residual mean square graph, it becomes apparent when the coefficients begin to stabilize. That is, an increasing value of k produces a small effect on the coefficients. The value of k at the point of stabilization (say $k*$) is selected to compute the final ridge coefficients. The final ridge estimates are thus a compromise between bias and inflated coefficients.

For the above example, as seen from Figure 10.3, the value of $k* = 0.2$ seems to represent that compromise. Values of k greater than 0.2 do not produce appreciable changes in the coefficients, although this decision is a subjective one on our part. Unfortunately, no objective criterion has been developed to determine $k*$ from the *sample*. Proponents of ridge regression agree that the ridge estimates will give better predictions in the *population*, although they do not fit the *sample* as well as the least squares estimates.

Implementation on the Computer

In practice, ordinary least squares regression programs such as those discussed in Chapter 7 can be used to obtain ridge estimates. As usual, we denote the dependent variable by Y and the P independent variables by X_1, $X_2, \ldots, X_P$. First, the Y and X variables are standardized by creating new variables in which we subtract the mean and divide the difference by the standard deviation for each variable. Then, P additional dummy observations are added to the data set. In these dummy observations the value of Y is always set equal to zero.

For a specified value of k the values of the X variables are defined as follows:

1. Compute $\sqrt{k(N-1)}$. Suppose $N = 20$ and $k = 0.2$; then $\sqrt{.2(19)} = 1.95$.

2. The *first* dummy observation has $X_1 = \sqrt{k(N-1)}$, $X_2 = 0, \ldots, X_P = 0$, and $Y = 0$.

3. The *second* dummy observation has $X_1 = 0$,
$X_2 = \sqrt{k(N-1)}$, $X_3 = 0, \ldots, X_P = 0$, and $Y = 0$,
etc.

4. The Pth dummy observation has $X_1 = 0$, $X_2 = 0$,
$X_3 = 0, \ldots, X_{P-1} = 0$, $X_P = \sqrt{k(N-1)}$, and $Y = 0$.

For a numerical example, suppose $P = 2$, $N = 20$, and $k = 0.2$. The two additional dummy observations are as follows:

X_1	X_2	Y
1.95	0	0
0	1.95	0

With the N regular observations and these P dummy observations, we use an ordinary least squares regression program in which the intercept is *forced* to be *zero*. The regression coefficients obtained from the output are automatically the ridge coefficients for the specified k. The resulting residual mean square should also be noted. Repeating this process for various values of k will provide the information necessary to plot the ridge trace of the coefficients, similar to Figure 10.3, and the residual mean square. The usual range of interest for k is between zero and one. Most computer packages allow the user to perform the above procedure for all desired values of k in one run. An example of the control statements for such a run using BMDP2R is given in Problem 10.4.

10.6 PATH ANALYSIS

Recall that regression and correlation analyses have two major purposes: prediction and description. For predictive purposes the techniques outlined in Chapters 6 through 9 cover the most commonly used methods in current statistical practice. For descriptive purposes many applied statisticians prefer geometric rather than algebraic presentations. That is, they prefer pictures to equations. A technique that accomplishes this graphic explanation of the interrelationships among several variables is called *path analysis*. A brief introduction to the subject with references for more detailed treatment follows.

To perform a path analysis, the investigator begins with an assumed model, indicating the direction and sources of causality among the variables. For example, suppose that an investigator wishes to examine the relationships among height (X_1), daily caloric intake (X_2), and weight (Y) for adult males. First, the investigator decides which variables could have a causal effect on other variables. Such a decision is based on previous experience and knowledge of the situation. In this example it is intuitively obvious that height and caloric intake both affect weight. This relationship can be displayed graphically as follows:

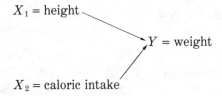

$$X_1 = \text{height}$$
$$Y = \text{weight}$$
$$X_2 = \text{caloric intake}$$

The arrows indicate that X_1 and X_2 affect Y and not the other way around. This graph also indicates that X_1 and X_2 are independent since they are not connected. However, it is known that height and caloric intake are interrelated, and we therefore must add to the graph a double-arrowed curve indicating this relationship:

$$X_1$$
$$Y$$
$$X_2$$

This curve linking X_1 and X_2 does not imply causal direction; it simply shows that X_1 and X_2 are not statistically independent.

To make the above diagram into a path diagram, we place, on the arrows and the curve, numbers that measure the strength of the relationship. The curve linking X_1 and X_2 is labeled with the correlation coefficient between X_1 and X_2. The labels on the arrows (or *paths*) from X_1 and X_2 to Y are the standardized regression coefficients in the multiple regression equation of Y on X_1 and X_2. Finally, to complete this diagram, we include a variable U in the graph to symbolize other variables affecting weights that have not been included. The label accompanying the arrow from U to Y is the square root of 1 minus the square of the multiple correlation between Y and (X_1, X_2). The final diagram, with hypothetical numerical values, looks like this:

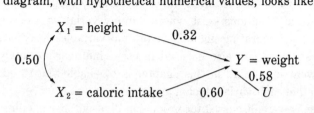

In this diagram the following hypothetical numerical values were used:

0.50 = correlation coefficient between X_1 and X_2

0.32 = standardized regression coefficient of X_1 in the multiple regression of Y on X_1 and X_2

0.60 = standardized regression coefficient of X_2 in the multiple regression of Y on X_1 and X_2

$0.58 = \sqrt{1 - R}$, where R is the multiple correlation coefficient of Y on X_1 and X_2

This path diagram indicates that caloric intake is the major contributor to weight and that height and caloric intake are intercorrelated. Variables other than height and caloric intake also affect weight. The effect of these variables not included in the analysis is measured by the quantity 0.58. This quantity, $\sqrt{1 - R^2}$, was given in Chapter 7 as the proportion of the standard deviation of Y *not* explained by the X variables in the regression equation.

A comprehensive treatment of path analysis written at a similar level to the level of this book can be found in Li (1975).

SUMMARY

In this chapter we reviewed methods for handling some special problems in regression analysis and introduced some recent advances in the subject. We reviewed the literature on missing values and recommended that the largest possible complete sample be selected from the data set and then used in the analysis. In certain special circumstances it may be useful to replace missing values with estimates from the sample, and methods for this technique were referenced.

We gave a detailed discussion of the use of dummy variables in regression analysis. Several situations exist where dummy variables are very useful. These include incorporating nominal or ordinal variables in equations. The ideas explained here can also be used in other multivariate analyses. One reason we went into detail in this chapter is to enable you to adapt these methods to other multivariate techniques.

Two methods were discussed for producing "biased" parameter estimates, i.e., estimates that do not average out to the true parameter value. The first method is placing constraints on the parameters. This method is useful when you wish to restrict the estimates to a certain range or when natural constraints on some function of the parameters must be satisfied. The second biased regression technique is ridge regression. This method is used only when you must employ variables that are known to be highly intercorrelated. You should also consider some of the methods we discussed in Chapter 7, such as removing one of the intercorrelated variables. In addition, the tolerance option available in many programs will often prevent an intercorrelated variable from entering the equation if a variable selection method is used.

BIBLIOGRAPHY

Missing Data in Regression Analysis

Afifi, A. A., and Elashoff, R. M. 1966. Missing observations in multivariate statistics. I: Review of the literature. *Journal of the American Statistical Association* 61:595–604.

———. 1969a. Missing observations in multivariate statistics. III: Large sample analysis of simple linear regression. *Journal of the American Statistical Association* 64:337–358.

———. 1969b. Missing observations in multivariate statistics. IV: A note on simple linear regression. *Journal of the American Statistical Association* 64:359–365.

*Beale, E. M. L., and Little, R. J. A. 1975. Missing values in multivariate analysis. *Journal of the Royal Statistical Society* 37:129–145.

Hartley, H. O., and Hocking, R. R. 1971. The analysis of incomplete data. *Biometrics* 27:783–824.

Little, R. J. A. 1982. Models for nonresponse in sample surveys. *Journal of the American Statistical Association* 77:237–250.

*Orchard, T., and Woodbury, M. A. 1972. A missing information principle: Theory and application. *Proceedings of the 6th Berkeley Symposium on Mathematical Statistical Problems* 1:697–715.

Path Analysis

Duncan, O. D. 1966. Path analysis in sociological examples. *American Journal of Sociology* 72:1–16.

Goldsmith, J. R., and Berglund, K. 1974. Epidemiological approach to multiple factor interactions in pulmonary disease: The potential usefulness of path analysis. *Annals of the New York Academy of Science* 221:361–375.

Li, C. C. 1975. *Path analysis—A primer*. Pacific Grove, Calif.: Boxwood Press.

*Wright, S. 1934. The method of path coefficients. *Annals of Mathematical Statistics* 5:161–215.

Ridge Regression

Fennessey, J., and D'Amico, R. 1980. Collinearity, ridge regression, and investigator judgement. *Social Methods and Research* 8:309–340.

Gunst, R. F., and Mason, R. L. 1980. *Regression analysis and its application*. New York: Dekker.

Hoerl, A. E., and Kennard, R. W. 1970. Ridge regression: Application to non-orthogonal problems. *Technometrics* 12:55–82.

Marquardt, D. W., and Snee, R. D. 1975. Ridge regression in practice. *American Statistician* 29:3–20.

Segmented (Spline) Regression

Quandt, R. E. 1958. The estimation of the parameters of a linear regression system obeying two separate regimes. *Journal of the American Statistical Association* 53:873–880.

*———. 1960. Tests of the hypothesis that a linear regression system obeys two separate regimes. *Journal of the American Statistical Association* 55:324-331.

*———. 1972. New approaches to estimating switching regressions. *Journal of the American Statistical Society* 67:306–310.

Other Regression

Box, G. E. P. 1966. Use and abuse of regression. *Technometrics* 8:625–629.

*Draper, N.R., and Smith, H. 1981. *Applied regression analysis*. 2nd ed. New York: Wiley.

*Graybill, F. A. 1976. *Theory and application of the linear model*. N. Scituate, Mass.: Duxbury Press.

*Tobin, J. 1958. Estimation of relationships for limited dependent variables. *Econometrica* 26:24–36.

PROBLEMS

10.1   In the depression data set described in Chapter 3, data on educational level, age, sex, and income are presented for a sample of adults from Los Angeles County. Fit a regression plane with income as the dependent variable and the other variables as independent variables. Use a dummy variable for the variable "sex" that was originally coded 1,2 by stating sex=sex − 1. Which sex is the reference group?

10.2   Repeat Problem 10.1, but now use a dummy variable for education. Divide the education level into three categories: did not complete high school, completed at least high school, and completed at least a bachelor's degree. An example of a program specification using BMDP2R follows:

```
/Problem title is 'dummy edu'.
/input   format is free.
         variables are 37.
         cases are 294.
         file is depress.
/variable names are id,sex,age,marital,educ,employ,
         income,relig,c1,c2,c3,c4,c5,c6,c7,c8,c9,c10,
         c11,c12,c13,c14,c15,c16,c17,c18,c19,c20,
         cesd,cases,drink,health,regdoc,treat,
         beddays,acuteill,chronill,d1,d2.
         add=2.
         use=sex,age,educ,income,d1,d2.
/transformation sex=sex-1.
         if (educ eq 1 or educ eq 2) then (d1=1. d2=0.).
         if (educ eq 3 or educ eq 4) then (d1=0. d2=1.).
         if (educ gt 4) then (d1=0. d2=0.).
/regress        depend=income.indep=age,sex,d1,d2.
         enter=.001.remove=.000.
/end
```

Compare the interpretation you would make of the effects of education on income in this problem and in Problem 10.1.

10.3   In the depression data set, determine whether religion has an effect on income when used as an independent variable along with age, sex, and educational level.

10.4   Draw a ridge trace for the accompanying data.

Case	Variable			
	X1	X2	X3	Y
1	0.46	0.96	6.42	3.46
2	0.06	0.53	5.53	2.25
3	1.49	1.87	8.37	5.69
4	1.02	0.27	5.37	2.36
5	1.39	0.04	5.44	2.65
6	0.91	0.37	6.28	3.31
7	1.18	0.70	6.88	3.89
8	1.00	0.43	6.43	3.27
Mean	0.939	0.646	6.340	3.360
Standard deviation	0.475	0.566	0.988	1.100

The following control statements can be used for BMDP2R. Very low F-to-enter and -remove values have been chosen so that all variables will surely enter. The sentence "type = 0" is included to obtain the zero intercept. To run multiple problems with different values of k, include a "for" statement with the desired values of k, and provide a % sign ahead of the input paragraph to indicate that everything beyond is repeated.

```
/Problem        title = 'ridge',
for k = 0,0.1,0.3,0.5,1.0,2.0,3.0,
%/input         file = 'ridge data',
                format = free,
                cases = 11,
                var = 4,
/var            names = x1,x2,x3,y,
/transf         if (kase lt 9) then (x1 = (x1 - .939)/.475),
                    x2 = (x2 - .646)/0.566,
                    x3 = (x3 - 6.340)/ 0.988,
                y = (y - 3.360)/ 1.100,
                if (kase eq 9) then (x1 = (7 * k) ** 0.5),
                if (kase eq 10) then (x2 = (7 * k)** 0.5),
                if (kase eq 11) then (x3 = (7 * k)** 0.5),
/regress        dependent = y,
                enter = 0.001,remove = 0.0,
                type = 0,
/end
```

The details of the transform paragraph are given in the program manual.

10.5 Use the data in Appendix B. For the parents we wish to relate Y = weight to X = height for both men and women in a single equation. Using dummy variables, write an equation for this purpose, including an interaction term. Interpret the parameters. Run a regression analysis, and test whether the rate of change of weight versus height is the same for men and women. Interpret the results with the aid of appropriate graphs.

10.6 Continuation of Problem 10.5: Do a similar analysis for the first boy and girl. Include age and age squared in the regression equation. Compare the results with those obtained in Problem 9.8.

10.7 Unlike the real data used in Problem 10.5, the accompanying data are "ideal" weights published by the Metropolitan Life Insurance Company for American men and women. Compute Y = midpoint of weight range for medium-framed men and women for the various heights shown in the table. Pretending that the results represent a real sample, repeat the analysis requested in Problem 10.5, and compare the results of the two analyses.

	Men		
Height	**Small Frame**	**Medium Frame**	**Large Frame**
5 ft 2 in.	128–134	131–141	138–150
5 ft 3 in.	130–136	133–143	140–153
5 ft 4 in.	132–138	135–145	142–156
5 ft 5 in.	134–140	137–148	144–160
5 ft 6 in.	136–142	139–151	146–164
5 ft 7 in.	138–145	142–154	149–168
5 ft 8 in.	140–148	145–157	152–172
5 ft 9 in.	142–151	148–160	155–176
5 ft 10 in.	144–154	151–163	158–180
5 ft 11 in.	146–157	154–166	161–184
6 ft 0 in.	149–160	157–170	164–188
6 ft 1 in.	152–164	160–174	168–192
6 ft 2 in.	155–168	164–178	172–197
6 ft 3 in.	158–172	167–182	176–202
6 ft 4 in.	162–176	171–187	181–207

Height	Women		
	Small Frame	Medium Frame	Large Frame
4 ft 10 in.	102–111	109–121	118–131
4 ft 11 in.	103–113	111–123	120–134
5 ft 0 in.	104–115	113–126	122–137
5 ft 1 in.	106–118	115–129	125–140
5 ft 2 in.	108–121	118–132	128–143
5 ft 3 in.	111–124	121–135	131–147
5 ft 4 in.	114–127	124–138	134–151
5 ft 5 in.	117–130	127–141	137–155
5 ft 6 in.	120–133	130–144	140–159
5 ft 7 in.	123–136	133–147	143–163
5 ft 8 in.	126–139	136–150	146–167
5 ft 9 in.	129–142	139–153	149–170
5 ft 10 in.	132–145	142–156	152–173
5 ft 11 in.	135–148	145–159	155–176
6 ft 0 in.	138–151	148–162	158–179

Note: Figures include 5 lb of clothing for men, 3 lb for women, and shoes with 1-in heels for both.

10.8 Use the data described in Problem 7.7. Since some of the X variables are intercorrelated, it may be useful to do a ridge regression analysis of Y on X1 to X9. Perform such an analysis, and compare the results to those of Problems 7.10 and 8.7.

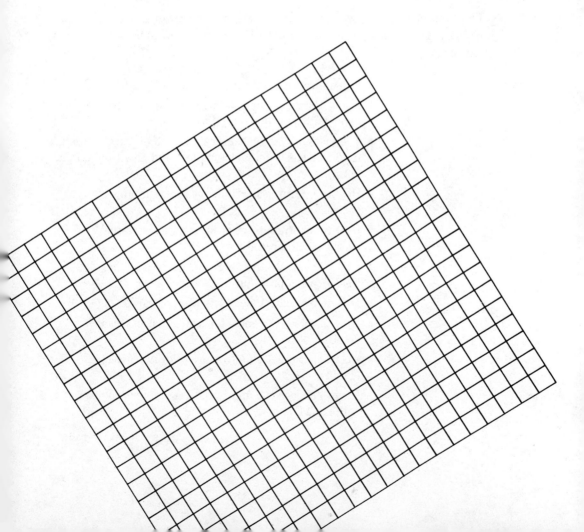

Three

MULTIVARIATE ANALYSIS

11 | DISCRIMINANT ANALYSIS

11.1 WHAT WILL YOU LEARN FROM THIS CHAPTER?

From this chapter you will learn how to classify an individual into one of two or more populations on the basis of the values of one or more variables. In particular, you will learn:

▶ When discriminant analysis is used (11.2, 11.3).

▶ About the basic concepts underlying classification of individuals (11.4).

▶ The meaning of the classical method of classification, the Fisher discriminant function, and how to obtain it (11.5).

▶ How to interpret discriminant function programs (11.6).

▶ How to incorporate prior information into the classification procedure (11.7).

▶ How to evaluate the degree of success of the classification procedure (11.8).

▶ How to evaluate the contribution of variables (11.9).

▶ How to select variables for use in classification (11.10).

▶ About classification into more than two groups (11.11).

▶ How to choose the appropriate computer program and options (11.12).

11.2 WHEN IS DISCRIMINANT ANALYSIS USED?

Discriminant analysis techniques are used to classify individuals into one of two or more alternative groups (or populations) on the basis of a set of measurements. The populations are known to be distinct, and each individual belongs to one of them. These techniques can also be used to identify which variables contribute to making the classification. Thus as in regression analysis, we have two uses, prediction and description.

As an example, consider an archeologist who wishes to determine which of two possible tribes created a particular statue found in a dig. The archeologist takes measurements for several characteristics of the statue and must decide whether these measurements are more likely to have come from the distribution characterizing the statues of one tribe or from the other tribe's distribution. These distributions are based on data from statues known to have been created by members of one tribe or the other. The problem of classification is therefore to guess who made the newly found statue on the basis of measurements obtained from statues whose identities are certain.

The measurements on the new statue may consist of a single observation, such as its height. However, we would then expect a low degree of accuracy in classifying the new statue since there may be quite a bit of overlap in the distribution of heights of statues from the two tribes. If, on the other hand, the classification is based on several characteristics, we would have more confidence in the prediction. The discriminant analysis methods described in this chapter are multivariate techniques in the sense that they employ several measurements.

As another example, consider a loan officer at a bank who wishes to decide whether to approve an applicant's automobile loan. This decision is made by determining whether the applicant's characteristics are more similar to those of persons who in the past repaid loans successfully or to those of persons who defaulted. Information on these two groups, available from past records, would include factors such as age, income, marital status, outstanding debt, and homeownership.

A third example, which is described in detail in the next section, comes from the depression data set (Chapters 1 and 3). We wish to predict whether an individual living in the community is more or less likely to be depressed on the basis of readily available information on the individual.

11.3 DATA EXAMPLE

As described in Chapter 1, the depression data set was collected for individuals residing in Los Angeles County. To illustrate the ideas described in this chapter, we will develop a method for estimating whether an individual is likely to be depressed. For the purposes of this example "depression" is defined by a score of 16 or greater on the CESD scale (see the code book given in Table 3.2). This information is given in the variable called "cases." We will base the estimation on demographic and other characteristics of the individual. The variables used are education and income. We may also wish to determine whether we can improve our prediction by including information on illness, sex, or age. Additional variables are an overall health rating, number of bed days in the past two months (0 if less than eight days, 1 if eight or more), acute illness (1 if yes in the past two months, 0 if no), and chronic illness (0 if none, 1 if one or more).

TABLE 11.1. Means and Standard Deviations for Nondepressed and Depressed Adults in Los Angeles County

Variable	Group I, Nondepressed (N = 244)		Group II, Depressed (N = 50)	
	Mean	Standard Deviation	Mean	Standard Deviation
Sex (male = 1, female = 2)	1.59*	0.49	1.80*	0.40
Age (in years)	45.2	18.1	40.4	17.4
Education (1 to 7, 7 high)	3.55	1.33	3.16	1.17
Income (thousands of dollars per year)	21.68*	15.98	15.20*	9.84
Health index (1 to 4, 1 = excellent)	1.71*	0.80	2.06*	0.98
Bed days (0 : less than 8 days per year; 1 : 8 or more days)	0.17*	0.39	0.42*	0.50
Acute conditions (0 = no, 1 = yes)	0.28	0.45	0.38	0.49
Chronic conditions (0 = none, 1 = one or more)	0.48	0.50	0.62	0.49

* For a test of equal means, P less than 0.01, assuming a normal distribution.

The first step in examining the data is to obtain descriptive measures of each of the groups. Table 11.1 lists the means and standard deviations for each variable in both groups. Note that in the depressed group, group II, we have a higher percentage of females, a lower average age, a lower educational level, and lower incomes. The standard deviations in the two groups are similar except for income, where they are slightly different. Note also that the health characteristics of the depressed group are generally worse than those of the nondepressed, even though the members of the depressed group tend to be younger on the average. Because sex is coded males = 1 and females = 2, the average sex of 1.80 indicates that 80% of the depressed group are females. Similarly, 59% of the nondepressed individuals are female.

Suppose that we wish to predict whether or not individuals are depressed, on the basis of their incomes. Examination of Table 11.1 shows that the mean value for depressed individuals is significantly lower than that for the nondepressed. Thus, intuitively, we would classify those with lower incomes as depressed and those with higher incomes as nondepressed. Similarly, we may classify the individuals on the basis of age alone, or sex alone, etc. However, as in the case of regression analysis, the use of several variables simultaneously can be superior to the use of any one variable. The methodology for achieving this result will be explained in the next sections.

11.4 BASIC CONCEPTS OF CLASSIFICATION

In this section we present the underlying concepts of classification and give an example illustrating their use. We also briefly discuss interpretation of the coefficients.

Principal Ideas

Suppose that an individual may belong to one of two populations. We begin by considering how an individual can be classified into one of these populations on the basis of a measurement of one characteristic, say X. Suppose that we have a representative sample from each population, enabling us to estimate the distributions of X and their means. Typically, these distributions can be represented as in Figure 11.1.

FIGURE 11.1. Hypothetical Frequency Distributions of Two Populations Showing Percentage of Cases Incorrectly Classified

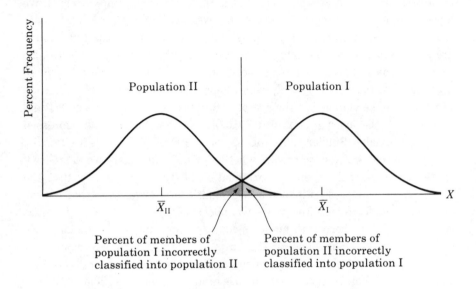

From the figure it is intuitively obvious that a low value of X would lead us to classify an individual into population II and a high value would lead us to classify an individual into population I. To define what is meant by *low* or *high*, we must select a dividing point. If we denote this dividing point by C, then we would classify an individual into population I if $X \geq C$. For any given value of C we would be incurring a certain percentage of error. If the individual came from population I but the measured X were less than C, we would incorrectly classify the individual into population II, and vice versa. These two types of errors are illustrated in Figure 11.1. If we can assume that the two populations have the same variance, then the usual value of C is

$$C = \frac{\overline{X}_\mathrm{I} + \overline{X}_\mathrm{II}}{2}$$

This value ensures that the two probabilities of error are equal.

The idealized situation illustrated in Figure 11.1 is rarely found in practice. In real life situations the degree of overlap of the two distributions is frequently large, and the variances are rarely precisely equal. For example, in the depression data the income distributions for the depressed and non-

FIGURE 11.2. Distribution of Income for Depressed and Nondepressed Individuals Showing Effects of a Dividing Point at an Income of $18,440

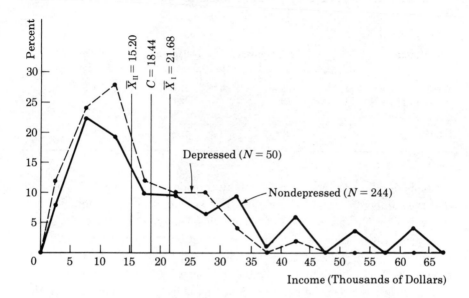

depressed individuals do overlap to a large degree, as illustrated in Figure 11.2. The usual dividing point is

$$C = \frac{15.20 + 21.68}{2} = 18.44$$

As can be seen from Figure 11.2, the percentage errors are rather large. The exact data on the errors are shown in Table 11.2. These numbers were obtained by first checking whether each individual's income was greater than or equal to 18.44×10^3 and then determining whether the individual was correctly classified. For example, of the 244 nondepressed individuals, 123 had income greater than or equal to 18.44×10^3 and were therefore correctly classified as not depressed (see Table 11.2). Similarly, of the 50 depressed individuals, 31 were correctly classified. The total number of correctly classified individuals is 123 + 31 = 154, amounting to 52.4% of the total sample of 294 individuals, as shown in Table 11.2. Thus although the mean incomes were significantly different from each other $(P < 0.01)$, income alone is not very successful in identifying whether an individual is depressed.

TABLE 11.2. Classification of Individuals as Depressed or Not Depressed on the Basis of Income Alone

	Classified as		
Actual Status	**Not Depressed**	**Depressed**	**% Correct**
Not depressed (N = 244)	123	121	50.4
Depressed (N = 50)	19	31	62.0
Total N = 294	142	152	52.4

Combining two or more variables may provide better classification. Note that the number of variables used must be less than N_I plus N_{II} minus 1. For two variables X_1 and X_2 concentration ellipses may be illustrated as shown in Figure 11.3 (see Section 7.5 for an explanation of concentration ellipses). Figure 11.3 also illustrates the univariate distributions of X_1 and X_2 separately. The univariate distribution of X_1 is what is obtained if the values of X_2 are ignored. On the basis of X_1 alone, and its corresponding dividing point C_1, a relatively large amount of error of misclassification would be encountered. Similar results occur for X_2 and its corresponding dividing point C_2. To use both variables simultaneously, we need to divide the plane of X_1 and X_2 into two regions, each corresponding to one population, and classify the individuals accordingly. A simple way of defining the two regions is to draw a straight line through the points of intersection of the two concentration ellipses, as shown in Figure 11.3.

The percentage of individuals from population II incorrectly classified is shown in the crosshatched areas. The shaded areas show the percentage of individuals from population I who are misclassified. The errors incurred by using two variables are often much smaller than those incurred by using either variable alone. In the illustration in Figure 11.3 this result is, in fact, the case.

The dividing line was represented by R. A. Fisher (1936) as an equation $Z = C$, where Z is a linear combination of X_1 and X_2 and C is a constant defined as follows:

$$C = \frac{\overline{Z}_I + \overline{Z}_{II}}{2}$$

FIGURE 11.3. Classification into Two Groups on the Basis of Two Variables

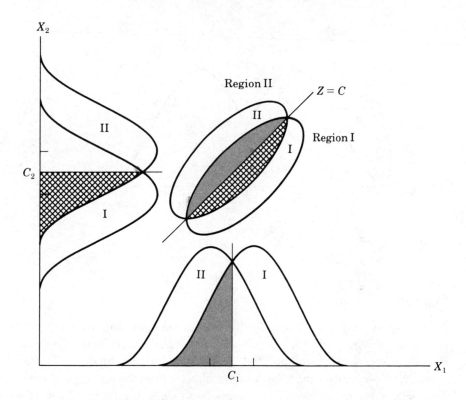

where $\overline{Z}_I$ is the average value of Z in population I and $\overline{Z}_{II}$ is the average value of Z for population II.

In this book we will call Z the *Fisher discriminant function*, written as

$$Z = a_1 X_1 + a_2 X_2$$

for the two-variable case. The formulas for computing the coefficients a_1 and a_2 can be found in Fisher (1936), Lachenbruch (1975), or Afifi and Azen (1979).

For each individual from each population, the value of Z is calculated. When the frequency distributions of Z are graphed separately for each population, the result is as illustrated in Figure 11.4. In this case, the bivariate classification problem with X_1 and X_2 is reduced to a univariate situation using the single variable Z.

FIGURE 11.4. Frequency Distributions of Z for Populations I and 11

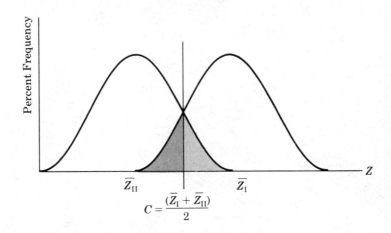

$$C = \frac{(\bar{Z}_I + \bar{Z}_{II})}{2}$$

Example

As an example of this technique for the depression data, it may be better to use both income and age to classify depressed individuals. Program BMDP7M was used to obtain the Fisher discriminant function. Unfortunately, this equation cannot be obtained directly from the output, and some intermediate computations must be made, as explained in Section 11.6. The result is

$$Z = 0.0209(\text{age}) + 0.0336(\text{income})$$

The mean Z value for each group can be obtained as follows, using the means from Table 11.1:

$$\text{mean } Z = 0.0209(\text{mean age}) + 0.0336(\text{mean income})$$

Thus $\quad \bar{Z}_{\text{not depressed}} = 0.0209(45.2) + 0.0336(21.68) = 1.67$

and $\quad \bar{Z}_{\text{depressed}} = 0.0209(40.4) + 0.0336(15.20) = 1.36$

The dividing point is therefore

$$C = \frac{1.67 + 1.36}{2} = 1.515$$

FIGURE 11.5. Classification of Individuals as Depressed or Not Depressed, on the Basis of Income and Age

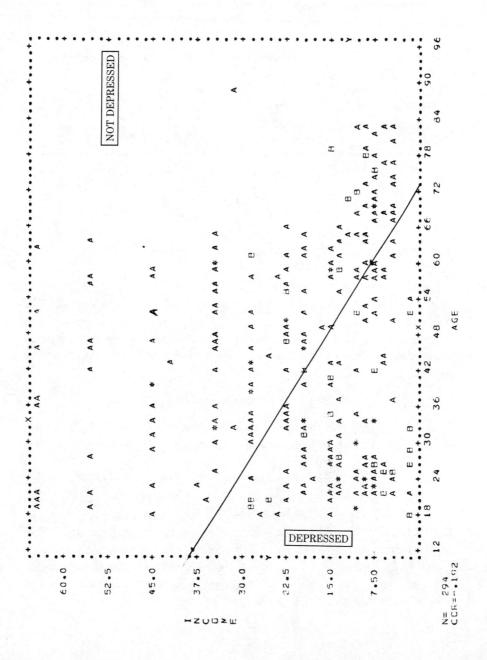

TABLE 11.3. Classification of Individuals as Depressed or Not Depressed on the Basis of Income and Age

	Classified As		
Actual Status	Not Depressed	Depressed	% Correct
Not depressed (N = 244)	154	90	63.1
Depressed (N = 50)	20	30	60.0
Total N = 294	174	120	62.6

An individual is then classified as depressed if his or her Z value is less than 1.52.

For two variables it is possible to illustrate the classification procedure as shown in Figure 11.5. This figure was obtained from the output of a BMDP6D program (see Chapter 6). In Figure 11.5 each A denotes the income and age of a nondepressed person, and each B denotes the same information for a depressed person. The dividing line is a graph of the equation $Z = C$, i.e.,

$$0.0209(\text{age}) + 0.0336(\text{income}) = 1.515$$

An individual falling in the region above the dividing line is classified as not depressed. Note that, indeed, very few depressed persons fall far above the dividing line.

To measure the degree of success of the classification procedure for this sample, we must count how many of each group are correctly classified. The computer program produces these counts automatically; they are shown in Table 11.3. Note that 63.1% of the nondepressed are correctly classified. This value is compared with 50.4%, which results when income alone is used (Table 11.2). The percentage of depressed correctly classified is comparable in both tables. Combining age with income improved the overall percentage of correct classification from 52.4% to 62.6%.

Interpretation of Coefficients

In addition to its use for classification, the Fisher discriminant function is helpful in indicating the direction and degree to which each variable contributes to the classification. The first thing to examine is the sign of each coefficient: If it is positive, the individuals with larger values of the corresponding variable tend to belong to population I, and vice versa. In the depression data example, both coefficients are positive, indicating that large values of both variables are associated with a lack of depression. In more complex examples, comparisons of those variables having positive coefficients with those having negative coefficients can be revealing. To quantify the magnitude of the distribution, the investigator may find standardized coefficients helpful, as explained in Section 11.6.

The concept of discriminant functions applies as well to situations where there are more than two variables, say $X_1, X_2, \ldots, X_P$. As in multiple linear regression, it is often sufficient to select a small number of variables. Variable selection will be discussed in Section 11.10. In the next section we present some necessary theoretical background for discriminant function analysis.

11.5 THEORETICAL BACKGROUND

In deriving his linear discriminant function, R. A. Fisher (1936) did not have to make any distributional assumptions for the variables used in classification. Fisher denoted the discriminant function by

$$Z = a_1 X_1 + a_2 X_2 + \cdots + a_P X_P$$

As in the previous section, we denote the two-mean values of Z by $\overline{Z}_{\mathrm{I}}$ and $\overline{Z}_{\mathrm{II}}$. We also denote the pooled sample variance of Z by S_Z^2 (this statistic is similar to the pooled variance used in the standard two-sample t test; e.g., see Dixon and Massey 1969). To measure how "far apart" the two groups are in terms of values of Z, we compute

$$D^2 = \frac{(\overline{Z}_{\mathrm{I}} - \overline{Z}_{\mathrm{II}})^2}{S_Z^2}$$

Fisher selected the coefficients $a_1, a_2, \ldots, a_P$ so that D^2 has the maximum possible value.

The term D^2 can be interpreted as the squared distance between the means of the standardized value of Z. A larger value of D^2 indicates that it is easier to discriminate between the two groups. The quantity D^2 is called the *Mahalanobis distance*. Both a_i and D^2 are functions of the group means and the pooled variances and covariances of the variables. The manuals for the statistical package programs make frequent use of these formulas, and you will find a readable presentation of them in Lachenbruch (1975) or Klecka (1980).

Some distributional assumptions make it possible to develop further statistical procedures relating to the problem of classification. These procedures include tests of hypotheses for the usefulness of some or all of the variables and methods for estimating errors of classification.

The variables used for classification are denoted by $X_1, X_2, \ldots, X_P$. The standard model makes the assumption that for each of the two populations these variables have a multivariate normal distribution. It further assumes that the covariance matrix is the same in both populations. However, the mean values for a given variable may be different in the two populations. We further assume that we have a random sample from each of the populations. The sample sizes are denoted by N_I and N_{II}.

Alternatively, we may think of the two populations as subpopulations of a single population. For example, in the depression data the original population consists of all adults over 18 years old in Los Angeles County. Its two subpopulations are the depressed and nondepressed. A single sample was collected and later diagnosed to form two subsamples.

11.6 INTERPRETATION

In this section we present various methods for interpreting discriminant functions. Specifically, we discuss the regression analogy, computations of the coefficients, standardized coefficients, and posterior probabilities.

Regression Analogy

A useful connection exists between regression and discriminant analyses. For the regression interpretation we think of the classification variables

$X_1, X_2, \ldots, X_P$ as the independent variables. The dependent variable is a dummy variable indicating the population from which each observation comes. Specifically,

$$Y = \frac{N_{II}}{N_I + N_{II}}$$

if the observation comes from population I, and

$$Y = -\frac{N_I}{N_I + N_{II}}$$

if the observation comes from population II. For instance, for the depression data $Y = 50/(244 + 50)$ if the individual is not depressed and $Y = -244/(244 + 50)$ if the individual is depressed.

When the usual multiple regression analysis is performed, the resulting regression coefficients are proportional to the discriminant function coefficients $a_1, a_2, \ldots, a_P$ (see Lachenbruch 1975). The value of the resulting multiple correlation coefficient R is related to the Mahalanobis D^2 by the following formula:

$$D^2 = \frac{R^2}{1 - R^2} \frac{[N_I + N_{II}][N_I + N_{II} - 2]}{N_I \times N_{II}}$$

Hence from a multiple regression program it is possible to obtain the coefficients of the discriminant function and the value of D^2. The $\overline{Z}$'s for each group can be obtained by multiplying each coefficient by the corresponding variable's sample mean. The dividing point C can then be computed as

$$C = \frac{\overline{Z}_I + \overline{Z}_{II}}{2}$$

As in regression analysis, some of the independent variables (or classification variables) may be dummy variables (see Section 10.3). In the depression example we may, for instance, use sex as one of the classification variables by treating it as a dummy variable. Research has shown that even though such variables do not follow a normal distribution, their use in linear discriminant analysis can still help improve the classification.

TABLE 11.4. Classification Function and Discriminant Coefficients for Age and Income from BMDP7M

| Variables | Classification Function | | Discriminant Function |
	Group I, Not Depressed	Group II, Depressed	
Age	0.1634	0.1425	$0.0209 = a_1$
Income	0.1360	0.1024	$0.0336 = a_2$
Constant	−5.8641	−4.3483	$1.5158 = C$

Computing the Fisher Discriminant Function

In the discriminant analysis programs that will be discussed in Section 11.12, some computations must be performed to obtain the values of the discriminant coefficients. Some programs (such as BMDP7M and the SPSS–X DISCRIMINANT procedure) print what is called a "classification function" for each group. Other programs (such as the SAS DISCRIM procedure) call these functions the "linearized discriminant functions." For each population the coefficients are printed for each variable. The discriminant function coefficients $a_1, a_2, \ldots, a_P$ are then obtained by subtraction.

As an example, we again consider the depression data using age and income. The classification functions are shown in Table 11.4.

The coefficient a_1 for age is $0.1634 - 0.1425 = 0.0209$. For income, $a_2 = 0.1360 - 0.1024 = 0.0336$. The dividing point C is also obtained by subtraction, but in *reverse order*. Thus $C = -4.3483 - (-5.8641) = 1.5158$. (This agrees closely with the previously computed value $C = 1.515$, which we use throughout this chapter.) For more than two variables the same procedure is used to obtain $a_1, a_2, \ldots, a_P$, and C.

Renaming the Groups

If we wish to put the depressed persons into group I and the nondepressed into group II, we can do so by reversing the zero and one values for the "cases" variable. In the data used for our example "cases" equals 1 if a person is depressed and zero if not depressed. However, we can make cases equal zero if a person is depressed and 1 if a person is not depressed. The BMDP TRANSFORM paragraph for this conversion is as follows:

```
/TRANSFORM
X = CASES.
IF (X EQ 0) THEN CASES = 1.
IF (X EQ 1) THEN CASES = 0.
```

Note that this reversal does not change the classification functions but simply changes their order so that *all* the signs in the linear discriminant function are changed. The new constant and discrimination function are

$$-1.515 \quad \text{and} \quad -0.0209(\text{age}) - 0.0336(\text{income}), \quad \text{respectively.}$$

The ability to discriminate is exactly the same, and the number of individuals correctly classified is the same.

Standardized Coefficients

As in the case of regression analysis, the values of $a_1, a_2, \ldots, a_P$ are not directly comparable. However, an impression of the relative effect of each variable on the discriminant function can be obtained from the *standardized discriminant coefficients*. This technique involves the use of the pooled (or within-group) covariance matrix from the computer output. In the original example this covariance matrix is as follows:

	Age	Income
Age	324.8	−57.7
Income	−57.7	228.6

Thus the pooled standard deviations are $\sqrt{324.8} = 18.02$ for age and $\sqrt{228.6} = 15.10$ for income. The standardized coefficients are obtained by multiplying the a_i's by the corresponding pooled standard deviations. Hence the standardized discriminant coefficients are

$$(0.0209)(18.02) = 0.377 \quad \text{for age}$$

and

$$(0.0336)(15.10) = 0.505 \quad \text{for income}$$

It is therefore seen that income has a slightly larger effect on the discriminant function than age. (Note that these are not the same as the standardized canonical discriminant functions discussed in Section 11.11.)

Posterior Probabilities

Thus far the classification procedure assigned an individual to either group I or group II. Since there is always a possibility of making the wrong classification, we may wish to compute the probability that the individual has come from one group or the other. We can compute such a probability under the multivariate normal model discussed in Section 11.5. The formula is

$$\text{probability of belonging to population I} = \frac{1}{1 + \exp(-Z + C)}$$

where $\exp(-Z + C)$ indicates e raised to the power $(-Z + C)$, as discussed in Truett, Cornfield, and Kannell (1967). The probability of belonging to population II is 1 minus the probability of belonging to population I.

For example, suppose that an individual from the depression study is 42 years old and earns $\$24 \times 10^3$ income per year. For that individual the discriminant function is

$$Z = 0.0209(42) + 0.0336(24) = 1.718$$

Since $C = 1.515$—and therefore Z is greater than C—we classify the individual as not depressed (in population I). To determine how likely this person is to be not depressed, we compute the probability

$$\frac{1}{1 + \exp(-1.718 + 1.515)} = 0.55$$

The probability of being depressed is $1 - 0.55 = 0.45$. Thus this individual is only slightly more likely to be not depressed than to be depressed.

Several packaged programs compute the probabilities of belonging to both groups for each individual in the sample. In some programs these probabilities are called the *posterior probabilities* since they express the probability of belonging to a particular population posterior to (i.e., after) performing the analysis.

Posterior probabilities offer a valuable method of interpreting classification results. The investigator may wish to classify only those individuals whose probabilities clearly favor one group over the other. Judgment could be withheld for individuals whose posterior probabilities are close to 0.5. In the next section another type of probability, called prior probability, will be defined and used to modify the dividing point.

Finally, we note that the discriminant function presented here is a sample estimate of the population discriminant function. We would compute the latter if we had the actual values of the population parameters. If the populations were both multivariate normal with equal covariance matrices, then the population discriminant classification procedure would be optimal; i.e., no other classification procedure would produce a smaller total classification error (see Anderson 1958).

11.7 ADJUSTING THE VALUE OF THE DIVIDING POINT

In this section we indicate how prior probabilities and costs of misclassification can be incorporated into the choice of the dividing point C.

Incorporating Prior Probabilities into the Choice of C

Thus far, the dividing point C was used as the point producing an equal percentage of errors of both types, i.e., the probability of misclassifying an individual from population I into population II, or vice versa. This use can be seen in Figure 11.4. But the choice of the value of C can be made to produce any desired ratio of these probabilities of errors. To explain how this choice is made, we must introduce the concept of *prior probability*. Since the two populations constitute an overall population, it is of interest to examine their relative size. The prior probability of population I is the probability that an individual selected at random actually comes from population I. In other words, it is the proportion of individuals in the overall population who fall in population I. This proportion is denoted by q_I.

In the depression data the definition of a depressed person was originally designed so that 20% of the population would be designated as depressed and 80% nondepressed. Therefore the prior probability of not being depressed (population I) is $q_I = 0.8$. Likewise, $q_{II} = 1 - q_I = 0.2$. Without knowing any of the characteristics of a given individual, we would thus be inclined to classify him or her as nondepressed, since 80% were in that group. In this case we would be correct 80% of the time. This example offers an intuitive interpretation of prior probabilities. Note, however, that we would be always wrong in identifying depressed individuals.

The theoretical choice of the dividing point C is made so that the total probability of misclassification is minimized. This total probability is defined as $q_I \cdot$ (probability of misclassifying an individual from population I into population II) plus $q_{II} \cdot$ (probability of misclassifying an individual from population II into population I), or

$$q_I \cdot \text{Prob(II given I)} + q_{II} \cdot \text{Prob(I given II)}$$

Under the multivariate normal model mentioned in Section 11.5 the choice of the dividing point C is

$$C = \frac{\overline{Z}_I + \overline{Z}_{II}}{2} + \ln \frac{q_{II}}{q_I}$$

where ln stands for the natural logarithm. Note that if $q_I = q_{II} = \frac{1}{2}$, then $q_{II}/q_I = 1$ and $\ln q_{II}/q_I = 0$. In this case C is

$$C = \frac{\overline{Z}_I + \overline{Z}_{II}}{2} \qquad \text{if} \qquad q_I = q_{II}$$

Thus in the previous sections we have been implicitly assuming that $q_I = q_{II} = \frac{1}{2}$.

For the depression data we have seen that $q_I = 0.8$, and therefore the theoretical dividing point should be

$$C = 1.515 + \ln(0.25) = 1.515 - 1.386 = 0.129$$

In examining the data, we see that using this dividing point classifies all of the nondepressed individuals correctly and all of the depressed individuals incorrectly. Therefore the probability of classifying a nondepressed individual (population I) as depressed (population II) is zero. On the other hand, the probability of classifying a depressed individual (population II) as nondepressed (population I) is 1. Therefore the total probability of misclassification is $(0.8)(0) + (0.2)(1) = 0.2$. When $C = 1.515$ was used, the two probabilities of misclassification were 0.369 and 0.400, respectively (see Table 11.3). In that case the total probability of misclassification is $(0.8)(0.379) + (0.2)(0.400) = 0.383$. This result verifies that the theoretical dividing point did produce a small value of this total probability of misclassification.

In practice, however, it is not appealing to identify none of the depressed individuals. If the purpose of classification were preliminary screening, we would be willing to incorrectly label some individuals as depressed in order to avoid missing too many of those who are truly depressed. In practice, we would choose various values of C and for each value determine the two probabilities of misclassification. The desired choice of C would be made when some balance of these two is achieved.

Incorporating Costs into the Choice of C

One method of weighting the errors is to determine the relative costs of the two types of misclassification. For example, suppose that it is four times as serious to falsely label a depressed individual as nondepressed as it is to label a nondepressed individual as depressed. These costs can be denoted as

$$\text{cost(II given I)} = 1$$

and

$$\text{cost(I given II)} = 4$$

The dividing point C can then be chosen to minimize the total cost of misclassification, namely

$$q_{\text{I}} \cdot \text{Prob(II given I)} \cdot \text{cost(II given I)} + q_{\text{II}} \cdot \text{Prob(I given II)} \cdot \text{cost(I given II)}$$

The choice of C that achieves this minimization is

$$C = \frac{\overline{Z}_{\text{I}} + \overline{Z}_{\text{II}}}{2} + K$$

where

$$K = \ln \frac{q_{\text{II}} \cdot \text{cost(I given II)}}{q_{\text{I}} \cdot \text{cost(II given I)}}$$

In the depression example the value of K is

$$K = \ln \frac{0.2(4)}{0.8(1)} = \ln 1 = 0$$

In other words, this numerical choice of cost of misclassification and the use of prior probabilities counteract each other so that $C = 1.515$, the same value obtained without incorporating costs and prior probabilities.

Finally, it is important to note that incorporating the prior probabilities and costs of misclassification alters only the choice of the dividing point C. It does not affect the computation of the coefficients $a_1, a_2, \ldots, a_P$ in the discriminant function. If the computer program does not allow the option of incorporating those quantities, you can easily modify the dividing point as was done in the above example. In Section 11.12, we show how costs of misclassification can be incorporated into a program that allows only prior probabilities to be specified.

11.8 HOW GOOD IS THE DISCRIMINANT FUNCTION?

A *measure of goodness* for the classification procedure consists of the two probabilities of misclassification, probability(II given I) and probability(I given II). Various methods exist for estimating these probabilities. One method, called the *empirical method*, was used in the previous examples. That is, we applied the discriminant function to the same samples used for deriving it and computed the proportion incorrectly classified from each group (see Tables 11.2 and 11.3). This process is a form of validation of the discriminant function. Although this method is intuitively appealing, it does produce biased estimates. In fact, the resulting proportions underestimate the true probabilities of misclassification, because the same sample is used for deriving and validating the discriminant function.

Ideally, we would like to derive the function from one sample and apply it to another sample to estimate the proportion misclassified. This procedure is called *cross-validation*, and it produces unbiased estimates. The investigator can achieve cross-validation by randomly splitting the original sample from each group into two subsamples: one for deriving the discriminant function and one for cross-validating it.

The investigator may be hesitant to split the sample if it is small. An alternative method sometimes used in this case, which imitates splitting the samples, is called the *jackknife procedure*. In this method we exclude one observation from the first group and compute the discriminant function on the basis of the remaining observations. We then classify the excluded observation. This procedure is repeated for each observation in the first sample. The proportion of misclassified individuals is the jackknife estimate of

Prob(II given I). A similar procedure is used to estimate Prob(I given II). This method produces nearly unbiased estimators. Some programs offer this option.

If we accept the multivariate normal model, theoretical estimates of the probabilities are also available and require only an estimate of D^2. The formulas are

$$\text{estimated Prob(II given I)} = \text{area to left of } \left(\frac{K - 1/2D^2}{D}\right) \text{ under standard}$$

normal curve

and

$$\text{estimated Prob(I given II)} = \text{area to left of } \left(\frac{-K - 1/2D^2}{D}\right) \text{ under}$$

standard normal curve

where
$$K = \ln \frac{q_{II} \cdot \text{cost(I given II)}}{q_I \cdot \text{cost(II given I)}}$$

If $K = 0$, these two estimates are each equal to the area to the left of $(-D/2)$ under the standard normal curve. For example, in the depression example $D^2 = 0.319$ and $K = 0$. Therefore $D/2 = 0.282$, and the area to the left of -0.282 is 0.389. From this method we estimate both Prob(II given I) and Prob(I given II) as 0.39. This method is particularly useful if the discriminant function is derived from a regression program, since D^2 can be easily computed from R^2 (see Section 11.6).

Unfortunately, this last method also underestimates the true probabilities of misclassification. An *unbiased estimator* of the population Mahalanobis D^2 is

$$\text{unbiased } D^2 = \frac{N_I + N_{II} - P - 3}{N_I + N_{II} - 2} D^2 - P\left(\frac{1}{N_I} + \frac{1}{N_{II}}\right)$$

In the depression example we have

$$\text{unbiased } D^2 = \frac{50 + 244 - 2 - 3}{50 + 244 - 2}(0.319) - 2\left(\frac{1}{50} + \frac{1}{244}\right) = 0.316 - 0.048$$

$$= 0.268$$

The resulting area is computed in a similar fashion to the last method. Since unbiased $D/2 = 0.259$, the resulting area to the left of $-D/2$ is 0.398. In comparing this result with the estimate based on the biased D^2, we note that the difference is small because (1) only two variables are used and (2) the sample sizes are fairly large. On the other hand, if the number of variables P were close to the total sample size $(N_\text{I} + N_\text{II})$, the two estimates could be very different from each other.

Whenever possible, it is recommended that the investigator obtain at least some of the above estimates of errors of misclassification and the corresponding probabilities of correct prediction.

To evaluate how well a particular discriminant function is performing, the investigator may also find it useful to compute the probability of correct prediction based on pure *guessing*. The procedure is as follows: Suppose that the prior probability of belonging to population I is known to be q_I. Then $q_\text{II} = 1 - q_\text{I}$. One way to classify individuals using these probabilities alone is to imagine a coin that comes up heads with probability q_I and tails with probability q_II. Every time an individual is to be classified, the coin is tossed. The individual is classified into population I if the coin comes up heads and into population II if it is tails. Overall, a proportion q_I of all individuals will be classified into population I.

Next, the investigator computes the total probability of correct classification. Recall that the probability that a person comes from population I is q_I, and the probability that any individual is classified into population I is q_I. Therefore the probability that a person comes from population I and is *correctly* classified into population I is q_I^2. Similarly, q_II^2 is the probability that an individual comes from population II and is correctly classified into population II. Thus the total probability of correct classification using only knowledge of the prior probabilities is $q_\text{I}^2 + q_\text{II}^2$. Note that the lowest possible value of this probability occurs when $q_\text{I} = q_\text{II} = 0.5$, i.e., when the individual is equally likely to come from either population. In that case $q_\text{I}^2 + q_\text{II}^2 = 0.5$.

Using this method for the depression example, with $q_\text{I} = 0.8$ and $q_\text{II} = 0.2$, gives us $q_\text{I}^2 + q_\text{II}^2 = 0.68$. Thus we would expect more than two-thirds of the individuals to be correctly classified if we simply flipped a coin that comes up heads 80% of the time. Note, however, that we would be wrong on 80% of the depressed individuals, a situation we may not be willing to tolerate. (In this context you might recall the role of costs of misclassification discussed in Section 11.7.)

11.9 TESTING FOR THE CONTRIBUTIONS OF CLASSIFICATION VARIABLES

Can we classify individuals by using variables available to us better than we can by chance alone? One answer to this question assumes the multivariate normal model presented in Section 11.4. The question can be formulated as an hypothesis-testing problem. The null hypothesis being tested is that none of the variables improve the classification based on chance alone. Equivalent null hypotheses are that the two population means for each variable are identical, or that the population D^2 is zero. The test statistic for the null hypothesis is

$$F = \frac{N_\mathrm{I} + N_\mathrm{II} - P - 1}{P(N_\mathrm{I} + N_\mathrm{II} - 2)} \times \frac{N_\mathrm{I} N_\mathrm{II}}{N_\mathrm{I} + N_\mathrm{II}} \times D^2$$

with degrees of freedom of P and $N_\mathrm{I} + N_\mathrm{II} - P - 1$ (see Rao 1965). The P value is the tail area to the right of the computed test statistic. We point out that $N_\mathrm{I} N_\mathrm{II} D^2 / (N_\mathrm{I} + N_\mathrm{II})$ is known as the two-sample Hotelling T^2, derived originally for testing the equality of two sets of means (see Morrison 1976).

For the depression example using age and income, the computed F value is

$$F = \frac{244 + 50 - 2 - 1}{2(244 + 50 - 2)} \times \frac{244 \times 50}{244 + 50} \times 0.319 = 6.60$$

with 2 and 291 degrees of freedom. The P value for this test is less than 0.005. Thus these two variables together significantly improve the prediction based on chance alone. Equivalently, there is statistical evidence that the population means are not identical in both groups. It should be noted that most existing computer programs do not print the value of D^2. However, from the printed value of the above F statistic we can compute D^2 as follows:

$$D^2 = \frac{P(N_\mathrm{I} + N_\mathrm{II})(N_\mathrm{I} + N_\mathrm{II} - 2)}{(N_\mathrm{I} N_\mathrm{II})(N_\mathrm{I} + N_\mathrm{II} - P - 1)} F$$

Another useful test is whether one additional variable improves the discrimination. Suppose that the population D^2 based on $X_1, X_2, \ldots, X_P$ variables is denoted by pop D_P^2. We wish to test whether an additional variable X_{p+1} will significantly increase the pop D^2; i.e., we test the hypothesis that pop $D_{p+1}^2 =$ pop D_P^2. The test statistic under the multivariate normal model is also an F statistic and is given by

$$F = \frac{(N_I + N_{II} - P - 2)(N_I N_{II})(D_{p+1}^2 - D_P^2)}{(N_I + N_{II})(N_I + N_{II} - 2) + N_I N_{II} D_P^2}$$

with 1 and $(N_I + N_{II} - P - 2)$ degrees of freedom (see Rao 1965).

For example, in the depression data we wish to test the hypothesis that age improves the discrimination function when combined with income. The D_1^2 for income alone is 0.183, and D_2^2 for income and age is 0.319. Thus with $P = 1$

$$F = \frac{(50 + 244 - 1 - 2)(50 \times 244)(0.319 - 0.183)}{(294)(292) + 50 \times 244 \times 0.183} = 5.48$$

with 1 and 291 degrees of freedom. The P value for this test is equal to 0.02. Thus age significantly improves the classification when combined with income.

A generalization of this last test allows for the checking of the contribution of several additional variables simultaneously. Specifically, if we start with $X_1, X_2, \ldots, X_P$ variables, we can test whether $X_{P+1}, \ldots, X_{P+Q}$ variables improve the prediction. We test the hypothesis that pop $D_{P+Q}^2 = $ pop D_P^2. For the multivariate normal model the test statistic is

$$F = \frac{(N_I + N_{II} - P - Q - 1)}{Q} \times \frac{N_I N_{II}(D_{P+Q}^2 - D_P^2)}{(N_I + N_{II})(N_I + N_{II} - 2) + N_I N_{II} D_P^2}$$

with Q and $(N_I + N_{II} - P - Q - 1)$ degrees of freedom.

The last two formulas for F are useful in variable selection, as will be shown in the next section.

11.10 VARIABLE SELECTION

Recall that there is an analogy between regression analysis and discriminant function analysis. Therefore much of the discussion of variable selection given in Chapter 8 applies to selecting variables for classification into two groups. In fact, the computer programs discussed in Chapter 8 may be used here as well. These include, in particular, stepwise regression programs and subset regression programs. In addition, some computer programs are available for performing stepwise discriminant analysis. They employ the same concepts discussed in connection with stepwise regression analysis.

In discriminant function analysis, instead of testing whether the value of multiple R^2 is altered by adding (or deleting) a variable, we test whether the value of pop D^2 is altered by adding or deleting variables. The F statistic given in Section 11.9 is used for this purpose. As before, the user may specify a value for the F-to-enter and F-to-remove values. For F-to-enter Costanza and Afifi (1979) recommend using a value corresponding to a P of 0.15. No recommended value from research can be given for the F-to-remove value, but a reasonable choice may be a P of 0.30.

11.11 CLASSIFICATION INTO MORE THAN TWO GROUPS

A comprehensive discussion of classification into more than two groups is beyond the scope of this book. However, several books do include a detailed discussion of this subject, including Tatsuoka (1971), Lachenbruch (1975), Morrison (1976), and Afifi and Azen (1979). In this section we summarize the classification procedure for a multivariate normal model and discuss an example.

We now assume that an individual is to be classified into one of k populations, for $k \geq 2$, on the basis of the values of P variables $X_1, X_2, \ldots, X_P$. We assume that in each of the k populations the P variables have a multivariate normal distribution, with the same covariance matrix. A typical packaged computer program will compute, from a sample from each population, a classification function. In applications the variable values from a given individual are substituted in each classification function, and the individual is classified into the population corresponding to the highest classification function.

Now we will consider an example. In the depression data set values of CESD can vary from 0 to 60 (see Chapter 3). In the previous runs persons were classified as depressed if their CESD scores were greater than or equal to 16; otherwise, they were classified as not depressed. Another possibility is to divide the individuals into $k = 3$ groups: those who deny any symptoms of depression (CESD score = 0), those who have CESD scores between 1 and 15 inclusive, and those who have scores of 16 or greater. Arbitrarily, we call these three groups lowdep (1), meddep (2), and highdep (3).

The variables considered for entry in a stepwise fashion are sex, age, education, income, health, bed days, and chronic illness. These variables are

TABLE 11.5. Partial Printout from BMDP7M for Classification into More Than Two Groups, Using the Depression Data with $k = 3$ Groups

```
APPROXIMATE F-STATISTIC 4.347 DEGREES OF FREEDOM 12.00  572.00

            F-MATRIX        DEGREES OF FREEDOM = 6   286

            lowdep          medder
medder      2.36
hishder     7.12            5.58

CLASSIFICATION FUNCTIONS

                    GROUP = lowdep       medder            hishder
        VARIABLE
        2 sex          7.07529          7.49617           8.14165
        3 age           .16774           .13935            .11698
        5 educat       2.54993          2.82551           2.68116
        7 income        .10533           .09005            .06537
       32 health       2.13954          2.75024           3.10425
       35 beddays      -.97394          -.80246            .46685

        CONSTANT     -17.62107        -18.54811         -18.81630

        VARIABLE     COEFFICIENTS FOR CANONICAL VARIABLES

        2 sex          -.73103          -.01977
        3 age           .03167           .02531
        5 educat       -.00617          -.65481
        7 income        .02758          -.00067
       32 health       -.57524          -.68822
       35 beddays     -1.13644          1.13117

        CONSTANT        .49631          2.17769
```

also used in Section 11.12, where the control statements for $k = 2$ are given for BMDP7M. The method of entering variables used in this program is similar to that described for forward stepwise regression in Chapter 8.

Partial results for this example are given in Table 11.5. Note that not all the variables entered the discriminant function.

The approximate F statistic given in Table 11.5 tests the null hypothesis that the means of the three groups are equal for all variables simultaneously. From Appendix Table A.4 we observe that the P value is very small, indicating that the null hypothesis should be rejected.

The F matrix part of Table 11.5 gives F statistics for testing the equality of means for each pair of groups. The F value (2.36) for the lowdep and meddep groups is barely significant at the 5% level. The other F statistics indicate a significant difference between the highdep group and each of the other two. This result partially justifies our previous analysis of two groups, where highdep is the depressed group and lowdep plus meddep constitute the nondepressed group.

The classification functions are also shown in Table 11.5. To classify a new individual into one of the three groups, we first evaluate each of the three classification functions, using that individual's scores on all the variables that entered the function. Then the individual is assigned to the group for which the computed classification function is the highest. If we are interested in the groups taken two at a time, the corresponding pair of classification functions could be subtracted from each other to produce a discriminant function, as explained in Section 11.6.

In the example considered here we divided a continuous variable into three subsets to get the three groups. Often when we are working with nominal data, three or more separate groups exist; examples include classification by religion or by race of individuals. Thus the capability that discriminant function analysis has of handling more than two groups is a useful feature in some applications.

Recall that there is an analogy between regression and discriminant analysis for two groups. It is unfortunate that this analogy does *not* extend to the case of k greater than two. There is, however, a correspondence between canonical correlation and classification into several populations.

In Chapter 15 we discuss the subject of canonical correlations and show how this analysis applies to the classification problem. Here we present an interpretation of the so-called *canonical variables* (called *canonical discriminant functions* in SPSS–X). These functions are derived in such a way that they best exhibit the differences among the groups.

The canonical scores for a given individual are calculated in a manner similar to that for the classification functions. The coefficients necessary for making the calculations are shown in the lower part of Table 11.5. Note that there are two canonical variables. In general, the number of canonical variables is equal to either the number of variables or $k - 1$, whichever is smaller.

When there are two or more canonical variables, some packaged programs produce a plot of the values of the first two canonical variables for each individual in the total sample. In this plot each group is identified by a different letter or number. Such a plot is a valuable tool in seeing how separate the groups are since this plot illustrates the maximum possible

FIGURE 11.6. Plot of the Canonical Variables for the Depression Data Set with $k = 3$ Groups

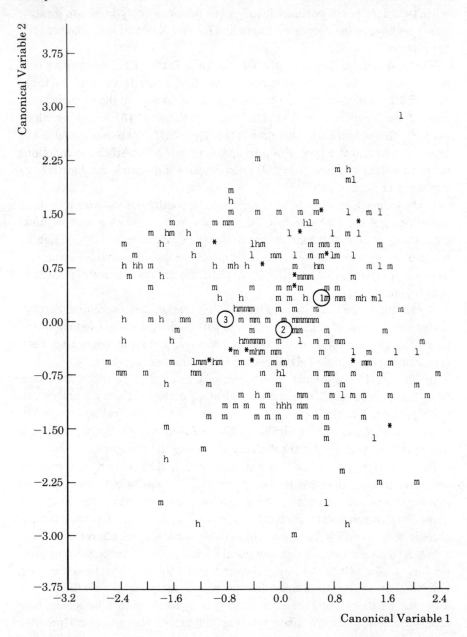

separation among the groups. The same plot can also be used to identify outliers or blunders.

Figure 11.6 shows a plot of the two canonical variables for this example. The symbols l, m, and h indicate the three groups. The figure also indicates the position of the mean values (1, 2, and 3) of the canonical variables for each of the three groups. These means show that the main variation is exhibited in canonical variable 1. The plot also shows a great deal of overlap between the individuals in the three groups. Finally, no extreme outliers are evident, although the case in the upper right-hand corner may merit further examination.

Various options and extensions for classification are offered in standard packaged programs, some of which are presented in the next section. In particular, the stepwise procedure of variable selection is available with most standard programs. However, unless the investigator is familiar with the complexities of a given option, we recommend the use of the default options. For most standard packaged programs the variable to be entered is selected to maximize a test statistic called Wilks' lambda. The one exception to the procedure is that we recommend modifying the values of the F-to-enter and F-to-remove as discussed in the previous section.

When the number of groups k is greater than two, the investigator may wish to examine the discrimination between the groups taken two at a time. This examination will serve to highlight specific differences between any two groups, and the results may be easier to interpret. Another possibility is to contrast each group with the remaining groups taken together.

11.12 DISCUSSION OF COMPUTER PROGRAMS

In this section we present the features of packaged programs for discriminant analysis, and we give an example of a program run for the depression data set.

Features of Packaged Programs

Table 11.6 summarizes items available in the output of standard packaged programs. The BMDP package has one program, P7M; SPSS-X has

TABLE 11.6 . Summary of Computer Output from BMDP, SAS, and SPSS–X for Discriminant Function Analysis

Output	BMDP	SAS	SPSS–X
Means	BMDP7M	DISCRIM and STEPDISC	DISCRIMINANT
Standard deviations	BMDP7M	DISCRIM and STEPDISC	DISCRIMINANT
Pooled covariance	BMDP7M	DISCRIM and STEPDISC	DISCRIMINANT
Pooled correlation	BMDP7M	DISCRIM and STEPDISC	DISCRIMINANT
Classification function	BMDP7M	DISCRIM and STEPDISC	DISCRIMINANT
Standardized classification function			DISCRIMINANT
D^2		Generalized distance (DISCRIM)	
F statistics	BMDP7M	STEPDISC	DISCRIMINANT
Wilks' lambda	BMDP7M	STEPDISC	DISCRIMINANT
Stepwise	BMDP7M	STEPDISC	DISCRIMINANT
Forcing variables	BMDP7M	STEPDISC	DISCRIMINANT
Classification table	BMDP7M	DISCRIM	DISCRIMINANT
Jackknife estimates	BMDP7M		
Cross-validation with subsamples	BMDP7M	*	DISCRIMINANT
Canonical plots	BMDP7M		DISCRIMINANT[+]
Prior probabilities	BMDP7M	DISCRIM	DISCRIMINANT
Costs	‡	‡	‡
Posterior probabilities	BMDP7M	DISCRIM	DISCRIMINANT

[+] Called canonical discriminant score.
‡ See the discussion in this section.
* The same procedure described in the BMDP manual can be used with SAS.

DISCRIMINANT; and SAS has four procedures, CANDISC, DISCRIM, NEIGHBOR, and STEPDISC. Of the SAS procedures only DISCRIM and STEPDISC are included in Table 11.6. The CANDISC procedure performs a canonical discriminant analysis that can be better understood after reading Chapters 13 and 15. The NEIGHBOR procedure classifies observations into groups on the basis of the so-called nearest-neighbor rule or the k-nearest-neighbor rule. This procedure is one of several nonparametric methods. For a discussion of these techniques, see Lachenbruch (1975) or Hand (1981).

An examination of the variable means is useful in obtaining a feel for how the groups differ from each other. When there are only two groups, it is useful to compute the Mahalanobis D^2 for a single variable or a group of variables. Unfortunately, most programs do not print these values of D^2. For a single variable X, say, the value of D^2 is easily computed from the group means $\overline{X}_I$ and $\overline{X}_{II}$ and the pooled variance S^2 as

$$D^2 \text{ (for single variable } X) = \frac{(\overline{X}_I - \overline{X}_{II})^2}{S^2}$$

These grouped means and pooled variances are available from all programs. For a subset of variables each program prints an F statistic to test the equality of the group means. The value of this F can be used to compute D^2, as shown in Section 11.9. Note that BMDP7M prints the Mahalanobis D^2 for each individual. This value is a measure of the distance between the point representing an individual and the point representing the estimated population mean. A small value of D^2 indicates that the individual probably belongs to that population.

In Section 11.7 we discussed how prior probabilities and costs of misclassification can be used to adjust the value of the dividing point C for two groups. If the costs are assumed equal, we may have the program do this adjustment automatically by supplying the prior probabilities as input. If the costs are not equal, we may trick the program into incorporating them into the prior probabilities, as the following example illustrates. Suppose $q_I = 0.4$, $q_{II} = 0.6$, cost(II given I) = 5, and cost(I given II) = 1. Then

$$\text{adjusted } q_I = q_I \cdot \text{cost(II given I)} = (0.4)(5) = 2$$

and $\quad$ $$\text{adjusted } q_{II} = q_{II} \cdot \text{cost(I given II)} = (0.6)(1) = .6$$

Since the prior probabilities must add up to 1, we further adjust q_I and q_{II} such that their sum is 1, i.e.,

$$\text{adjusted } q_I = \frac{2}{2.6} \quad \text{and} \quad \text{adjusted } q_{II} = \frac{0.6}{2.6}$$

The packaged programs BMDP7M, SAS CANDISC, and SPSS–X DISCRIMINANT compute canonical variables. As mentioned in Section 11.11, these variables are linear combinations of the variables chosen to represent the maximum separation possible among the groups. When the number of groups is two, only one canonical variable exists, and its value for a given individual is proportional to the value of the Fisher discriminant function. The program may then produce a histogram of this variable for each of the two groups.

For the two-groups case we remind you of the analogy between regression and discrimination. Thus it is possible to perform discriminant analysis for two groups using any of the regression programs discussed in Part 2 of this book. In particular, an all-subsets regression, such as BMDP9R, can be used to perform variable selection for discriminant function analysis. Note that the discriminant function coefficients resulting from two different programs may be different, but they will be proportional to each other.

Example

An example of the input statements used to run the stepwise discriminant function program BMDP7M is presented here for the depression data set. In this example eight variables are considered as possible variables for the discriminant function: sex, age, education, income, health status, bed days, acute illness, and chronic illness. A description of these variables is given in the code book in Table 3.2. Note that the first four variables are typical demographic data and the last four are measures of health and illness. On the basis of income and age alone, we are able to classify 62.6% correctly (see Table 11.3), and this run will enable us to see whether we can improve on that percentage.

One confusing use of terminology emerges in this run. In BMDP programs the word *case* is used to indicate the number of observations. But note that we have used the same word to signify whether or not a person is depressed, following common medical terminology use of the word *case* as someone who is ill.

The program input statements for BMDP7M are as follows:

```
/problem      title is 'disc cases of depression'.
/input        file is depress.
              format is '(8x,f1.0,f2.0,1x,f1.0,1x,
              f2.0,23x,f1.0,1x,f1.0,2x,f1.0,f1.0,f1.0)'.
              cases are 294.
              variables are 9.
/variable     names are sex,age,educat,income,cases,health,
              beddays,acuteill,chronill.
              group is cases.
/group        code(5) = 0,1.
              names(5) = notdep,depress.
/disc         enter = 1,1.
              remove = 0,0.
/end
```

TABLE 11.7. Means and Standard Deviations for Depression Data

	Not Depressed		Depressed	
Variable	Mean	Standard Deviation	Mean	Standard Deviation
Sex	1.59	0.49	1.80	0.41
Age	45.24	18.15	40.38	17.40
Education	3.55	1.33	3.16	1.17
Income	21.68	15.98	15.20	9.84
Health	1.71	0.80	2.06	0.98
Bed days	0.17	0.38	0.42	0.50
Acute illness	0.29	0.45	0.38	0.49
Chronic illness	0.48	0.50	0.62	0.49

The "group" sentence must be used to signify how the two separate groups are defined. This is the fifth variable and takes on the value 0 if normal and 1 if depressed. The F-to-enter and F-to-remove values have been set very low in order to see what will enter. It is also possible to obtain additional printout by using the "print" paragraph, but here we will just use the default printing option.

The program first prints the means and standard deviations for both groups (50 depressed and 244 not-depressed persons). These results are shown in Table 11.7.

Note from the table that the standard deviations are not grossly different between the two groups but that the data cannot be considered to be multivariate normal. Some of the variables only take on two values, 1 and 2, or 0 and 1. Also, income appears to be skewed, as indicated by the large standard deviation relative to the mean and the lack of negative incomes. Simply looking at the mean values indicates that depressed people are more apt to be female, to be younger, to have less income and education, and to be in poorer health.

At step 0 the largest F-to-enter is for bed days, so it enters at step 1. Income enters next at step 2, sex at step 3, age at step 4, and health at step 5. Then no more variables enter since all of the F-to-enter levels are less than 1. At the last step the F-to-remove levels are 3.91 for sex, 6.02 for age, 6.95 for income, 4.20 for health, and 8.65 for bed days. It is not possible to attach precise P values to these F-to-remove levels because of the lack of

TABLE 11.8. Classification and Discriminant Functions

	Classification Function		
Variables	**Not Depressed**	**Depressed**	**Discriminant Function**
Sex	7.268	7.962	−0.694
Age	0.132	0.108	0.024
Income	0.181	0.151	0.030
Health	1.793	2.240	−0.447
Bed days	−0.140	1.119	−1.259
Constant	−12.932	−13.724	−0.792

normality and the use of the stepwise procedure, but the P values appear to be high enough so that these variables should be left in.

The classification function is given in Table 11.8 from the printout, and the discriminant function is obtained by subtraction (see Section 11.6).

Note that the magnitudes of the coefficients for age and income are quite similar to those in Table 11.4 even with the addition of three other variables.

The value of F at the last step is given as 7.60 with 5 and 288 degrees of freedom. This value tests the hypothesis that the population $D^2 = 0$ and is highly significant ($P < 0.001$). We can compute D^2 (see Section 11.9) for five variables with $N_I = 244$ and $N_{II} = 50$ as

$$D^2 = \frac{(5)(294)(292)}{(50)(244)(294 - 5 - 1)}(7.60) = 0.9285$$

Thus $-D/2$ is -0.4818, which results in an area of about 0.31 to the left of this value in Appendix Table A.1. The unbiased estimate of D^2 is

$$\frac{244 + 50 - 5 - 3}{244 + 50 - 2}(0.9285) - 5\left(\frac{1}{50} + \frac{1}{244}\right) = 0.7889$$

(see Section 11.8). This estimate modifies the above area to 0.33, rather close to 0.31.

The program prints the posterior probability of belonging to the not-depressed and the depressed group for each individual. For the first individual in the sample, who is a 68-year-old woman with a $4000-per-year

TABLE 11.9. Classification of Individuals as Depressed or Not Depressed on the Basis of Sex, Age, Income, Health, and Bed Days

Actual Status	N	Classified As		% Correct
		Not depressed	Depressed	
Not depressed	244	175	69	71.7
Depressed	50	16	34	68.0
Total	294	191	103	71.1

reported income, a health status of 2, and no bed days (see Table 3.3),

$$\text{Prob(not depressed)} = \frac{1}{1 + \exp(-Z + C)} = \frac{1}{1 + \exp(0.530 - 0.792)} = 0.565$$

where $C = -0.792$ and

$$Z = -0.694(2) + 0.024(68) + 0.030(4) - 0.447(2) - 1.259(0) = -0.530$$

The probability of being not depressed is close to one-half, but the program classifies this individual as not depressed. The first individual classified as depressed is the fifth person in Table 3.3, who is a 33-year-old female with an income of $35,000, a health status of 1, and bed days of 1. This woman actually had a CESD score of only 6, so she was not depressed. Overall, the program classified correctly 71.1% of the time, as Table 11.9 indicates. These results appear to be an improvement over the results listed in Table 11.3.

Using the method described in Section 11.9, we can test whether the population D^2 for all five variables is equal to the population D^2 for two variables (age and income). With $P = 2$ and $Q = 3$ we compute, using $D_5^2 = 0.929$ and $D_2^2 = 0.319$,

$$F = \frac{244 + 50 - 2 - 3 - 1}{3} \times \frac{(244)(50)(0.929 - 0.319)}{(244 + 50)(244 + 50 - 2) + (244)(50)(0.319)}$$

$$= 7.96$$

FIGURE 11.7. Histogram of Canonical Variable from BMDP7M

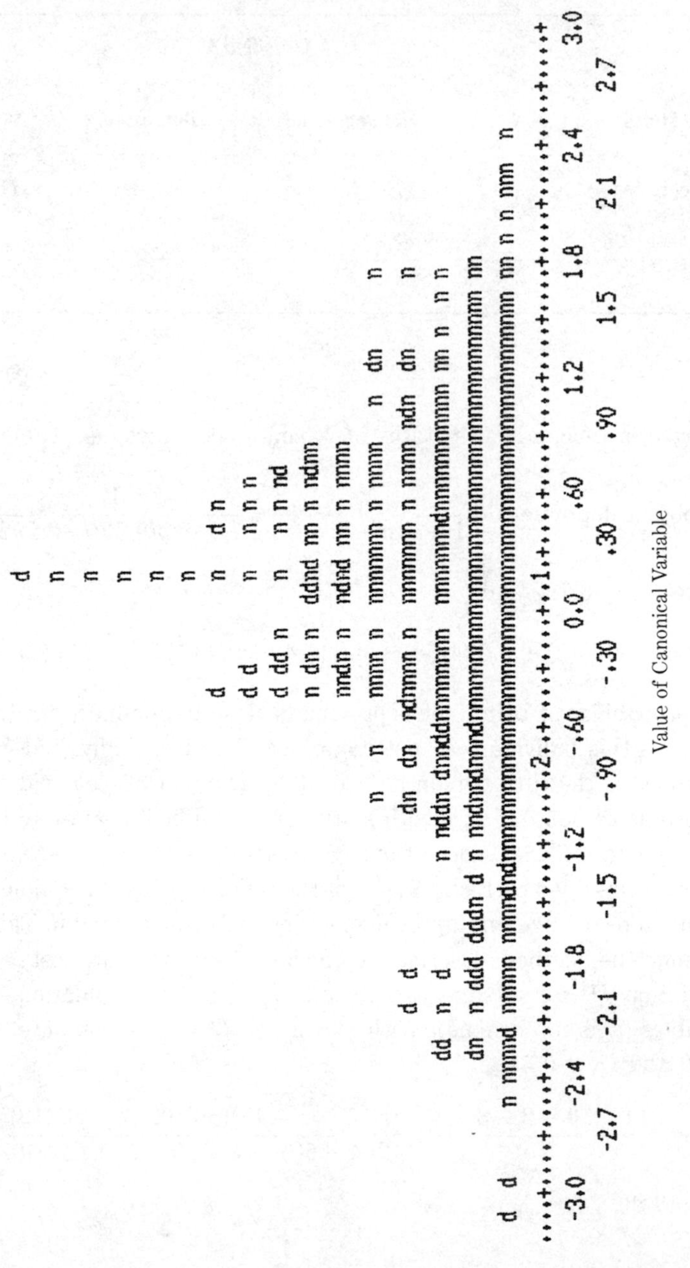

with 3 and 288 degrees of freedom. Since this result corresponds to a P value less than 0.005, we conclude that sex, health, and bed days significantly improve the prediction based on age and income alone.

The program next lists the canonical variable for each individual and makes a histogram of the values, with the not-depressed labeled n and the depressed labeled d, as shown in Figure 11.7. Note that the upper end of the graph is dominated by the letter n, as should be the case. When better discrimination is possible, such a figure would consist of two overlapping but clearly discernable histograms.

The person the program classifies as most depressed according to the value of the canonical variable and the posterior probability is an 18-year-old female with a reported income of $2000 per year, a health status of 3, and one reported bed day. This woman had a CESD score of 39, and so the program correctly classified her as depressed.

Further runs could be made by introducing other variables, considering transformations on some variables, such as the logarithm of income, and setting costs and prior probabilities equal to desired values.

SUMMARY

In this chapter we discussed discriminant function analysis, a technique dating back to at least 1936. However, its popularity began with the introduction of large-scale computers in the 1960s. The method's original concern was to classify an individual into one of several populations. It is also used for explanatory purposes to identify the relative contributions of a single variable or a group of variables to the classification.

In this chapter our main emphasis was on the case of two populations. We gave some theoretical background and presented an example of the use of a packaged program for this situation. We also discussed the case of more than two groups. We return to this subject in Chapter 15.

Interested readers may pursue the subject of classification further by consulting the references given in the Bibliography. In particular, Lachenbruch (1975) presents a comprehensive discussion of the subject.

BIBLIOGRAPHY

Afifi, A. A., and Azen, S. P. 1979. *Statistical analysis: A computer oriented approach*. 2nd ed. New York: Academic Press.

*Anderson, T. W. 1958. *An introduction to multivariate statistical analysis*. New York: Wiley.

Costanza, M. C., and Afifi, A. A. 1979. Comparison of stopping rules for forward stepwise discriminant analysis. *Journal of the American Statistical Association* 74:777–785.

Dixon, W. J., and Massey, F. J. 1983. *Introduction to statistical analysis*. 4th ed. New York: McGraw-Hill.

Fisher, R. A. 1936. The use of multiple measurements in taxonomic problems. *Annals of Eugenics* 7:179–188.

Hand, D. J. 1981. *Discrimination and classification*. New York: Wiley.

Klecka, W. R. 1980. *Discriminant analysis*. Beverly Hills: Sage.

Lachenbruch, P. A. 1975. *Discriminant analysis*. New York: Hafner Press.

*Morrison, D. F. 1976. *Multivariate statistical methods*. New York: McGraw-Hill.

*Rao, C. R. 1965. *Linear inference and its application*. New York: Wiley.

*Tatsuoka, M. M. 1971. *Multivariate analysis: Techniques for educational and psychological research*. New York: Wiley.

Truett, J.; Cornfield, J.; and Kannell, W. 1967. Multivariate analysis of the risk of coronary heart disease in Framingham. *Journal of Chronic Diseases* 20:511–524.

PROBLEMS

11.1 Using the depression data set, perform a stepwise discriminant function analysis with age, sex, log (income), bed days, and health as possible variables. Compare the results with those given in Section 11.12.

11.2 For the data shown in Table 8.1, divide the chemical companies into two groups: group I consists of those companies with a P/E less than 9, and group II consists of those companies with a P/E greater than or equal to 9. Group I should be considered mature or troubled firms, and group II should be considered growth firms. Perform a discriminant function analysis, using ROR5, D/E, SALESGR5, EPS5, NPM1, and PAYOUTR1. Assume equal prior probabilities and costs of misclassification. Test the hypothesis that the population D^2

= 0. Produce a graph of the posterior probability of belonging to group I versus the value of the discriminant function. Use of the sentence "POST." in the PRINT paragraph of BMDP7M will result in both these quantities being printed. Estimate the probabilities of misclassification by several methods.

11.3   Continuation of Problem 11.2: Test whether D/E alone does as good a classification job as all six variables.

11.4   Continuation of Problem 11.2: Choose a different set of prior probabilities and costs of misclassification that seem reasonable to you and repeat the analysis.

11.5   Continuation of Problem 11.2: Perform a variable selection analysis, using stepwise and best-subset programs. Compare the results with those of the variable selection analysis given in Chapter 8.

11.6   Continuation of Problem 11.2: Now divide the companies into three groups: group I consists of those companies with a P/E of 7 or less, group II consists of those companies with a P/E of 8 to 10, and group III consists of those companies with a P/E greater than or equal to 11. Perform a stepwise discriminant function analysis, using these three groups and the same variables as in Problem 11.2. Comment.

11.7   In this problem you will modify the data set created in Problem 7.7 to make it suitable for the theoretical exercises in discriminant analysis. Generate the sample data for X1, X2,..., X9 as in Problem 7.7 (Y is not used here). Then for the first 50 cases, add 6 to X1, add 3 to X2, add 5 to X3, and leave the values for X4 to X9 as they are. For the last 50 cases, leave all the data as they are. Thus the first 50 cases represent a random sample from a multivariate normal population called population I with the following means: 6 for X1, 3 for X2, 5 for X3, and zero for X4 to X9. The last 50 observations represent a random sample from a multivariate normal population (called population II) whose mean is zero for each variable. The population Mahalanobis D^2's for each variable separately are as follows: 1.44 for X1, 1 for X2, 0.5 for X3, and zero for each of X4 to X9. It can be shown that the population D^2 is as follows: 3.44 for X1 to X9; 3.44 for X1, X2, and X3; and zero for X4 to X9. For all nine variables the population discriminant function has the following coefficients; 0.49 for X1, 0.5833 for X2, −0.25 for X3, and zero for each of X4 to X9. The population errors of misclassification are

$$\text{Prob(I given II)} = \text{Prob(II given I)} = 0.177$$

Now perform a discriminant function analysis on the data you constructed, using all nine variables. Compare the results of the sample with what you know about the populations.

11.8   Continuation of Problem 11.7: Perform a similar analysis, using only X1, X2, and X3. Test the hypothesis that these three variables do as well as all nine in classifying the observations. Comment.

11.9   Continuation of Problem 11.7: Do a variable selection analysis for all nine variables. Comment.

11.10   Continuation of Problem 11.7: Do a variable selection analysis, using variables X4 to X9 only. Comment.

12 LOGISTIC REGRESSION

12.1 WHAT WILL YOU LEARN FROM THIS CHAPTER?

In Chapter 11 we presented a method of classifying individuals into one of two possible populations, a method developed originally for continuous variables. From this chapter you will learn another method, which applies to discrete or continuous variables. In particular, you will learn:

- ▶ When logistic regression is used (12.2, 12.3).
- ▶ The meaning of the multiple logistic regression model (12.4).
- ▶ How to choose between logistic regression and discriminant function analysis (12.4).
- ▶ How to interpret the results of a logistic regression analysis (12.5, 12.6).
- ▶ How to evaluate the resulting equation (12.7).
- ▶ How to obtain and adjust risk functions (12.8).
- ▶ About what computer programs to use to obtain various output for the logistic regression model (12.9).

12.2 WHEN IS LOGISTIC REGRESSION USED?

Logistic regression can be used whenever an individual is to be classified into one of *two* populations. Thus it is an alternative to the discriminant analysis presented in Chapter 11. In the past most of the applications of logistic regression were in the medical field. It has been used, for example, to calculate the risk of developing heart disease as a function of certain personal and behavioral characteristics. In principle, limitation to the medical field is unnecessary. For example, the so-called logistic equation has been used to represent the proportion of insects killed with increasing doses of insecticide. Another area where the logistic model is used is transfer of technology. For example, the electronic hand-held calculator represented a competitor to the slide rule for scientific and engineering calculations. Until the advent of electronic hand-held calculators, almost every engineer carried a slide rule. They began to switch technology gradually, particularly as the price of the electronic calculators shifted downward. The proportion of engineers carrying an electronic calculator could be approximated by a logistic curve as a function of time. Now it is the rare engineer who still carries a slide rule.

The linear discriminant function gives rise to the logistic posterior probability when the multivariate normal model is assumed. Logistic regression represents an alternative method of classification when the multivariate normal model is not justified. As we will discuss in this chapter, the logistic regression analysis is applicable for any combination of discrete and continuous variables. (If the multivariate normal model is applicable, the methods discussed in Chapter 11 will result in a better classification procedure and will require less computer time to analyze.)

12.3 DATA EXAMPLE

The same depression data set described in Chapter 11 will be used in this chapter. Recall that the discriminant function based on age and income is

$$Z = 0.0209(\text{age}) + 0.0336(\text{income})$$

with a dividing point $C = 1.515$. Assuming equal prior probabilities, the posterior probability of being not depressed is

$$\text{Prob(not depressed)} = \frac{1}{1 + \exp[1.515 - 0.0209(\text{age}) - 0.0336(\text{income})]}$$

as given in Section 11.6. For a given individual with a discriminant function

FIGURE 12.1. Logistic Function for Depression Data Set

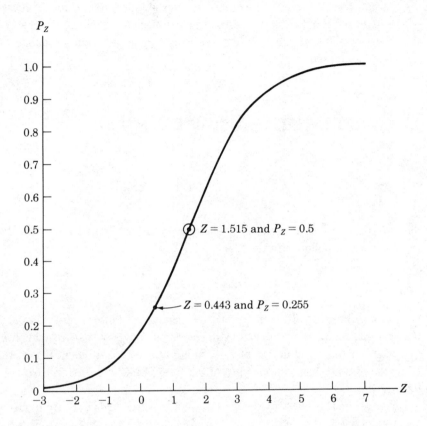

value of Z, we can write this posterior probability as

$$P_Z = \frac{1}{1 + e^{C-Z}}$$

As a function of Z, the probability P_Z has the logistic form shown in Figure 12.1. Note that P_Z is always positive; in fact, it must lie between zero and 1 because it is a probability. The minimum age is 18 years, and the minimum income is $\$2 \times 10^3$. These minimums result in a Z value of 0.443 and a probability $P_Z = 0.255$, as graphed. When $Z = 1.515$, the dividing point C, then $P_Z = 0.5$. Larger values of Z occur when age is older and/or income is higher. It is clear that a value of Z greater than $Z = 1.515$ produces a P_Z greater than 0.5, and hence we classify the individual as not depressed.

The same data and other data from this depression set will be used later in the chapter to illustrate a different method of computing the probability of not being depressed.

12.4 BASIC CONCEPTS OF LOGISTIC REGRESSION

The *logistic function*

$$P_Z = \frac{1}{1 + e^{C-Z}}$$

may be transformed to produce a new interpretation. Specifically, we define the *odds* as the following ratio:

$$\text{odds} = \frac{P_Z}{1 - P_Z}$$

Computing the odds is a commonly used technique of interpreting probabilities (see Fleiss 1981). For example, in sports we may say that the odds are 3 to 1 that one team will defeat another in a game. This statement means that the favored team has a probability of $3/(3 + 1)$ of winning, since $0.75/(1 - 0.75) = 3/1$.

Note that as the value of P_Z varies from 0 to 1, the odds vary from 0 to ∞. When $P_Z = 0.5$, the odds are 1. On the odds scale the values from 0 to 1 correspond to values of P_Z from 0 to 0.5. On the other hand, values of P_Z from 0.5 to 1.0 result in odds of 1 to ∞. Taking the logarithm of the odds will cure this asymmetry. When $P_Z = 0$, ln odds $= -\infty$; when $P_Z = 0.5$, ln odds $= 0.0$; and when $P_Z = 1.0$, ln odds $= +\infty$. The term *logit* is sometimes used instead of ln (odds).

By taking the natural logarithm of the odds and performing some algebraic manipulation, we obtain

$$\ln \left(\frac{P_Z}{1 - P_Z} \right) = -C + Z$$

In words, the logarithm of the odds is linear in the discriminant function Z. Since $Z = a_1 X_1 + a_2 X_2 + \cdots + a_p X_p$ (from Chapter 11), ln(odds) is seen to be linear in the original variables. If we rewrite $- C + Z$ as $a + b_1X_1 + b_2X_2 + \cdots + b_PX_P$, the equation relating ln(odds) to the discriminant function is

$$\ln \text{ (odds)} = a + b_1X_1 + b_2X_2 + \cdots + b_PX_P$$

This equation is in the same form as the multiple linear regression equation (see Chapter 7), where $a = -C$ and $b_i = a_i$ for $i = 1$ to P. For this reason the logistic function has been called the *multiple logistic regression equation*, and the coefficients in the equation can be interpreted as regression coefficients.

The fundamental assumption in logistic regression analysis is that ln(odds) is linearly related to the independent variables. No assumptions are made regarding the distributions of the X variables. In fact, one of the major advantages of this method is that the X variables may be discrete or continuous.

The model assumed is

$$\ln \text{ (odds)} = \alpha + \beta_1X_1 + \beta_2X_2 + \cdots + \beta_PX_P$$

In terms of the probability of belonging to population I, the equation can be written as

$$\frac{\text{probability of belonging}}{\text{to population I}} = \frac{1}{1 + \exp[-(\alpha + \beta_1X_1 + \beta_2X_2 + \cdots + \beta_PX_P)]}$$

This equation is called the *logistic regression equation*.

As mentioned earlier, the technique of linear discriminant analysis can be used to compute estimates of the parameters $\alpha, \beta_1, \beta_2, \cdots, \beta_p$. However, the method of *maximum likelihood* produces estimates that depend only on the logistic model. The maximum likelihood estimates should, therefore, be more robust than the linear discriminant function estimates. However, if the distribution of the X variables is, in fact, multivariate normal, then the discriminant analysis method requires a smaller sample size to achieve the same precision as the maximum likelihood method (see Efron 1975). The maximum likelihood estimates of the probabilities of belonging to one population or the other are preferred to the discriminant function estimates when the X's are nonnormal (see Halperin, Blackwelder, and Verter 1971 or Press and Wilson 1978). The estimates of the coefficients or the probabilities derived from the two methods will rarely be substantially different from each other, whether or not the multivariate normality assumption is satisfied. An exception to this statement is given in O'Hara et al. (1982); they demonstrate that the use of discriminant function analysis may lead to underestimation of the coefficients when all the X's are categorical and the probability of the outcome is small.

Most logistic regression programs use the method of maximum likelihood to compute estimates of the parameters. The procedure is iterative, and a theoretical discussion of it is beyond the scope of this book. When the number of variables is large (say more than nine), the computer time required may be prohibitively long. Some programs, such as BMDPLR, allow the investigator to use an approximation to the maximum likelihood method.

12.5 INTERPRETATION: CONTINUOUS VARIABLES

To show the similarity between discriminant function and logistic regression analysis, we performed a logistic regression on age and income in the depression data set, using the BMDPLR program. The program requires the user to state, for each variable, whether it is categorical or interval. Variables called interval are used as they are supplied to the program. The program, however, assigns dummy values to variables called categorical. For this example both age and income are called interval. The estimates obtained from the program are as follows:

Term	Coefficient	Standard Error
Age	0.020	0.009
Income	0.041	0.014
Constant	−0.028	0.487

So the estimate of α is −0.028, of β_1 is 0.020, and of β_2 is 0.041. The equation for ln(odds), or logit, is estimated by −0.028 + 0.020(age) + 0.041(income). The coefficients 0.020 and 0.041 are interpreted in the same manner they are interpreted in a multiple linear regression equation, where the dependent variable is ln $[P_I/(1 - P_I)]$ and where P_I is the logistic regression equation, estimated as

$$\frac{\text{probability of not}}{\text{being depressed}} = \frac{1}{1 + \exp\{-[-0.028 + 0.020(\text{age}) + 0.041(\text{income})]\}}$$

Recall that the coefficients for age and income in the discriminant function are 0.0209 and 0.0336, respectively. These estimates are within one standard error of 0.020 and 0.041, respectively, the estimates obtained from the BMDPLR program. The constant −0.028 corresponds to a dividing point of + 0.028. This value is different from the dividing point of 1.515 obtained by the discriminant analysis. The explanation for this discrepancy is that the logistic regression program implicitly uses prior probability estimates obtained from the sample. In this example these prior probabilities are 244/294 and 50/294, respectively. When these prior probabilities are used, the discriminant function dividing point is $1.515 + \ln q_{II}/q_{I} = 1.515 - 1.585 = -0.070$. This value is closer to the value 0.028 obtained by the logistic regression program than is 1.515.

Note also that for each coefficient printed by the program, an *asymptotic standard error* is also printed. These errors can be used to test hypotheses and obtain confidence intervals. For example, to test the hypothesis that the coefficient for age is zero, we compute the test statistic

$$Z = \frac{0.020 - 0}{0.009} = 2.22$$

For large samples an approximate P value can be obtained by comparing this Z statistic to percentiles of the standard normal distribution.

For the first individual, whose age is 68 and whose income is $\$4 \times 10^{3}$, for example, we compute

$$\frac{\text{probability of not}}{\text{being depressed}} = \frac{1}{1 + \exp\{-[-0.028 + 0.020(68) + 0.041(4)]\}}$$

$$= \frac{1}{1 + \exp(-1.496)} = 0.817$$

The program proceeds by computing this probability for every individual in the sample.

If the logistic regression equation is used for classification, a *cutoff point* on the probability of not being depressed must be found. This cutoff point is denoted by P_C. So we would classify an individual as not depressed if that individual's probability is greater than or equal to P_C. For a series of cutoff points P_C, the program computes the percentage of individuals correctly classified from each group. These percentages are shown in Figure 12.2. For example, a cutoff point $P_C = 0.7$ results in correct classification of approximately 94% of the nondepressed and 22% of the depressed. A cutoff point of

FIGURE 12.2.   Percentage of Individuals Correctly Classified by Logistic Regression Analysis According to Whether They are Depressed or Not

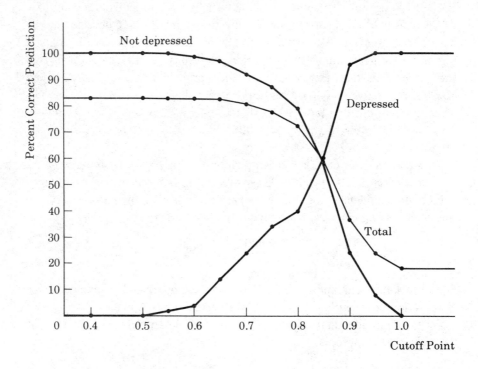

$P_C = 0.83$ results in both percentages of correct classification being approximately 60%. As was discussed in Chapter 11, the choice of the cutoff point depends on the relative costs of misclassification. The BMDPLR program allows the user to supply these costs as input in order to choose the value of P_C that minimizes the total cost.

   As was mentioned in Section 7.8, it is sometimes useful to incorporate *interactions* of two or more variables into the logistic regression model. Interactions are simply represented as the products of variables in the model. For example, if we wish to examine the possibility of including the effect of the age-income interaction, we state that the interaction is part of the model (for BMDPLR). [In other programs the user may have to include the product (age) × (income) as a variable.] The program output is as follows:

Term	Coefficient	Standard Error	Coefficient ÷ Standard Error
Age	0.039	0.015	2.630
Income	0.104	0.043	2.426
Age × income	−0.002	0.001	−1.655
Constant	−0.847	0.711	−1.192

To test whether the interaction should be included, we compare the value −1.655 with percentiles of the normal distribution. The two-sided P value is barely less than 0.10. There is therefore some evidence for including the interaction term. Another method for evaluating the effect of interaction is to examine the correct classification chart, shown in Figure 12.2. In this case there was very little improvement due to including the interaction term.

12.6 INTERPRETATION: CATEGORICAL VARIABLES

One of the important benefits of the logistic model is that it allows the use of categorical X variables. Any number of such variables can be used in the model. The simplest situation is one in which we have a single X variable with two possible values. For example, for the depression data we can attempt to predict who is depressed on the basis of the individual's sex. Table 12.1 shows the individuals classified by depression and sex.

TABLE 12.1. Classification of Individuals by Depression Level and Sex

Sex	Depression		Total
	Yes	No	
Female (1)	40	143	183
Male (0)	10	101	111
Total	50	244	294

If the individual is a female, then the odds of being depressed are 40/143. Similarly, for males the odds of being depressed are 10/101. The *ratio* of these odds is

$$\text{odds ratio} = \frac{40/143}{10/101} = \frac{40 \times 101}{10 \times 143} = 2.825$$

The odds of a female being depressed are 2.825 times that of a male. Note that we could just as well compute the odds ratio of *not* being depressed. In this case we have

$$\text{odds ratio} = \frac{143/40}{101/10} = 0.354$$

The concept of odds ratio is used extensively in biomedical applications (see Fleiss 1981). It is a measure of association of a binary variable (risk factor) with the occurrence of a given event (disease)(see Reynolds 1977 for applications in behavioral science).

To represent a variable such as sex, we customarily use a dummy variable: $X = 0$ if male and $X = 1$ if female. The logistic regression equation can then be written as

$$\text{Prob(not depressed)} = \frac{1}{1 + e^{-\alpha - \beta X}}$$

The sample estimates of the parameters are

$$a = \text{estimate of } \alpha = 2.313$$

$$b = \text{estimate of } \beta = -1.039$$

We note that the estimate of β is the natural logarithm of the odds ratio; or

$$-1.039 = \ln 0.354$$

Equivalently,

$$\text{odds ratio } e^b = e^{-1.039} = 0.354$$

Also, the estimate of α is the natural logarithm of the odds for males ($X = 0$); or

$$2.313 = \ln \frac{101}{10}$$

When there is only a single dichotomous variable, it is not worthwhile to perform a logistic regression analysis. However, in a multivariate logistic equation the value of the coefficient of a dichotomous variable can be related to the odds ratio in a manner similar to that outlined above. For example, for the depression data, if we include age, sex, and income in the same logistic model, the estimated equation is

$$\text{Prob(not depressed)} = \frac{1}{1 + \exp\{-[0.676 + 0.021(\text{age}) + 0.037(\text{income}) - 0.929(\text{sex})]\}}$$

Since sex is a 0, 1 variable, its coefficient can be given an interesting interpretations. The quantity $e^{-0.929} = 0.395$ may be interpreted as the odds ratio of not being depressed if female after adjusting for the linear effects of age and income. It is important to note that such an interpretation is valid only when we do *not* include the interaction of the dichotomous variable with any of the other variables. Further discussion on this subject may be found in Breslow and Day (1980) and Schlesselman (1982). In particular, the case of several dummy variables is discussed in some detail in these books.

If the ones and zeros are reversed for the "cases" variable in the depression data, then the depressed persons will have a value of zero and the nondepressed a value of one. Rerunning the logistics program will then produce a regression equation for the probability of being depressed. This regression equation will be same as it was before, except that it will be multiplied by − 1:

$$\text{Prob(depressed)} = \frac{1}{1 + \exp\{-[-0.676 - 0.021(\text{age}) - 0.037(\text{income}) + 0.929(\text{sex})]\}}$$

Again, the quantity $e^{0.929} = 2.532$ may be interpreted as the odds ratio of being depressed if female after adjusting for the linear effects of age and income.

If a categorical variable takes on more than two values, we may create several dummy variables to represent it, as was discussed in Chapter 10. The BMDPLR program creates such variables automatically. However, the program assigns the values − 1 and + 1 to each dummy variable if the standard default is used and not the values 0 and 1. The user may force the computer to use any two values by reading the dummy variables as "interval" variables after transforming the variable to the desired values. Alternatively, using the sentence

```
DVAR = PART,
```

in the REGRESSION paragraph will result in the usual dummy variables, with the first category the reference group (see the program manual).

12.7 HOW GOOD IS THE EQUATION?

Several approaches have been proposed for testing how well the derived multiple logistic regression equation fits the data. They all rely on the idea of comparing an observed number of individuals with the number expected if the model were valid. These observed (O) and expected (E) numbers are combined to form a χ^2 statistic called the *goodness-of-fit* χ^2. Large values of the test statistic indicate a poor fit of the model. Equivalently, small P values are an indication of poor fit. In this section we present two different approaches to the goodness-of-fit test.

The *classical approach* begins with identifying different combinations of values of the variables used in the equation, called *patterns*. For example, if there are two dichotomous variables, sex and employment status, there are four distinct patterns: male employed, male unemployed, female employed, female unemployed. For each of these combinations we count the observed number of individuals (O) in populations I and II. Similarly, for each of these individuals we compute the probability of being in population I or II on the basis of the logistic equation. The sum of these probabilities for a given pattern is denoted by E. The goodness-of-fit test statistic is computed as

$$\text{goodness-of-fit } \chi^2 = \Sigma 2 \times O \times \ln\left(\frac{O}{E}\right)$$

where the summation is extended over all the distinct patterns. The BMDPLR program computes this statistic and the corresponding P value. The investigator is warned, though that *this statistic may give a misleading impression when the number of distinct patterns is too small or too large*. A large number of distinct patterns occur when continuous variables are used.

Another goodness-of-fit approach is described by Lemeshow and Hosmer (1982). In this approach the probability of belonging to population I is calculated for every individual in the sample, and the resulting numbers are arranged in increasing order. The range of probability values is then divided into ten groups (deciles). For each decile the observed numbers of individuals in population I are computed (O). Also, the expected numbers (E) are calculated by adding the logistic probabilities for all individuals in each decile. The goodness-of-fit statistic is calculated as

$$\text{goodness-of-fit } \chi^2 = \Sigma \, \frac{(O - E)^2}{E}$$

where the summation extends over the two populations and the ten deciles. The BMDPLR program prints this statistic and its P value as well.

Both of these goodness-of-fit tests are approximate and require large sample sizes. A large goodness-of-fit chi-square (or a small P value) indicates that the fit may not be good.

A different sort of statistic is also computed to test whether inclusion of one or more variables *improves* the prediction. The BMDPLR program prints an "improvement chi-square" that tests whether one new variable entered in a stepwise fashion significantly improves prediction. The LOGIST procedure in SAS computes this statistic, called the "*model chi-square*," for the total set of variables in the equation. This statistic may be used to test the hypothesis that the set of variables is useless in classifying individuals. A large value of this statistic (or a small P value) is an indication that the variables are *useful* in classification. This test is analogous to testing the hypothesis that the population Mahalanobis D^2 is zero in discriminant function analysis (see Chapter 11).

The test for whether a *single* variable improves the prediction forms the basis for a stepwise logistic regression procedure. The principle is the same as that used in stepwise linear regression or stepwise discriminant function analysis. The chi-square value or an approximate F value is used for entering or deleting variables. The BMDPLR program prints an improvement chi-square that tests whether a new variable entered in a stepwise fashion significantly improves prediction. A large value of chi-square (or a small P value) indicates that the variable should be added.

12.8 APPLICATIONS OF LOGISTIC REGRESSION

Multiple logistic regression equations are often used to estimate the probability of a certain event occurring to a given individual. Examples of such events are failure to repay a loan, the occurrence of a heart attack, or death from lung cancer. In such applications the period of time in which such an event is to occur must be specified. For example, the event might be a heart attack occurring within ten years from the start of observation.

For estimation of the equation a sample is needed in which each individual has been observed for the specified period of time and values of a set of relevant variables have been obtained at or up to the start of the observation. Such a sample can be selected and used in the following two ways:

1. A sample is selected in a random manner and observed for the specified period of time. This sample is called a *cross-sectional sample*. From this single sample two subsamples result, namely, those who experience the event and those who do not. The methods described in this chapter are then used to obtain the estimated logistic regression equation. This equation can be applied directly to a new member of the population from which the original sample was obtained. This application assumes that the population is in a steady state; i.e., no major changes occur that alter the relationship between the variables and the occurrence of the event. Use of the equation with a different population may require an adjustment, as described in the next paragraph.

2. The second way of obtaining a sample is to select two random samples, one for which the event occurred and one for which the event did not occur. This sample is called a *case-control sample*. Values of the predictive variables must be obtained in a retrospective fashion, i.e., from past records or recollection. The data can be used to estimate the logistic regression equation. This method has the advantage of enabling us to specify the number of individuals with or without the event. In application of the equation the regression coefficients $b_1, b_2, \ldots,$

b_P are valid. However, the constant a must be adjusted to reflect the true population proportion of individuals with the event. To make the adjustment, we need estimated mean values of the variable X_1, $X_2, \ldots, X_P$ and the probability P of the event in the population to which it is to be applied. The adjusted constant is obtained by solving the following equation for $a*$:

$$P = \frac{1}{1 + \exp[-(a* + b_1\overline{X}_1 + b_2\overline{X}_2 + \cdots + b_P\overline{X}_P)]}$$

The solution is

$$a* = \ln\left(\frac{P}{1-P}\right) - (b_1\overline{X}_1 + b_2\overline{X}_2 + \cdots + b_P\overline{X}_P)$$

With this adjustment the probability that the event occurs to a person whose values are average corresponds to the population probability of the event.

For example, in the depression data suppose that we wish to estimate the probability of not being depressed on the basis of sex, age, and income. From the depression data we obtain an equation for the probability of not being depressed as follows:

$$\text{Prob(not depressed)} =$$

$$\frac{1}{1 + \exp\{-[0.676 + 0.021(\text{age}) + 0.037(\text{income}) - 0.929(\text{sex})]\}}$$

This equation is derived from data on an urban population with a nondepression rate of approximately 80%. Since the sample has 244/294, or 83%, nondepressed individuals, we must adjust the constant. Estimates of the variable means in the population are

$$\text{mean age} = 44 \text{ years for adults over 18}$$

$$\text{mean income} = \$20 \times 10^3$$

We therefore obtain the adjusted constant as

$$a* = \ln\left(\frac{0.8}{1-0.8}\right) - (+0.021 \times 44 + 0.037 \times 20 - 0.929 \times 0.5) = 1.3863 -$$

$$1.1995 = 0.187$$

Therefore the equation used for estimating the probability of not being depressed is

$$\text{Prob(not depressed)} = \frac{1}{1 + \exp\{-[0.187 + 0.021(\text{age}) + 0.037(\text{income}) - 0.929(\text{sex})]\}}$$

Frequently, we may wish to include predictive variables measured in different studies. Suppose, for example, that the variables X_1 and X_2 are measured in one study and variables X_3 and X_4 in another. From the first study we obtain an equation such as

$$P(\text{event}) = \frac{1}{1 + \exp[-(a_1 + b_1 X_1 + b_2 X_2)]}$$

Similarly, the equation from the second study may be written as

$$P(\text{event}) = \frac{1}{1 + \exp[-(a_2 + b_3 X_3 + b_4 X_4)]}$$

If the event is defined in the same way and if we can assume that the interactions of X_1 and X_2 with X_3 and X_4 are negligible, we may combine the two equations into one. The information needed is the probability of the event in the population, say P, and estimates $\overline{X}_i$ of the means of the variables. The constant is then estimated by

$$a^* = \ln\left(\frac{P}{1-P}\right) - (b_1\overline{X}_1 + b_2\overline{X}_2 + b_3\overline{X}_3 + b_4\overline{X}_4)$$

The combined equation for predicting the probability of an event is

$$P(\text{event}) = \frac{1}{1 + \exp[-(a^* + b_1 X_1 + b_2 X_2 + b_3 X_3 + b_4 X_4)]}$$

Note that in using this equation, we are making the assumption that the coefficients of X_1 and X_2 are not affected by including X_3 and X_4 in the equation.

Another method of sampling is to select *pairs* of individuals *matched* on certain characteristics. One member of the pair has the event and the other does not. Data from such matched studies may be analyzed by the logistic regression programs described in the next section after making certain adjustments. Holford, White, and Kelsey (1978) describe these adjustments for one-to-one matching; Breslow and Day (1980) give a theoretical discussion of the subject. Woolson and Lachenbruch (1982) describe adjustments that allow a least squares approach to be used for the same estimation.

12.9 DISCUSSION OF COMPUTER PROGRAMS

Of the three major packages, only SAS and BMDP include logistic regression programs as such. The SAS package has two logistic regression programs, but they are both author-supported, not SAS-supported (see Reinhardt 1980). Their features and those of BMDPLR are summarized in Table 12.2.

The PREDICT procedure of SAS first performs a discriminant function analysis. Then it uses the results from that analysis to obtain starting values for the logistic regression analysis.

The LOGIST procedure can be used if the forward stepwise or backward elimination procedure is desired. One unique feature of the LOGIST procedure is the D statistic. If the independent variables follow a multivariate normal distribution and if N is large, then D can be viewed as approximately

TABLE 12.2. Summary of Computer Output from BMDP and SAS for Logistic Regression Analysis

Output and Options	BMDPLR	SAS LOGIST	SAS PREDICT
Means	Yes	Yes	Yes
Assigns categories	− 1, + 1, or 0,1	0,1	
Beta coefficient	Yes	Yes	Yes
Intercept	Yes	Yes	Yes
Improvement chi-square	Yes	Yes	
Model chi-square		Yes	
Goodness-of-fit chi-square	Yes		
Classification table	Yes	Yes	
Probability-of-correct-prediction plot	Yes	Yes	Yes
Cost matrix	Yes		
Stepwise	Yes	Yes	
D statistic		Yes	
Covariance matrix of estimates	Correlation matrix	Yes	Yes
Standard Errors of estimates	Yes	Yes	
Individual probabilities of event	Yes	Yes	

equal to R^2. Thus it provides a method of judging how well the variables predict the outcome. For each variable a D statistic (or partial correlation coefficient squared) between that variable and the outcome variable is also printed. However, the normality assumptions should be kept in mind when viewing these statistics.

It is also possible to perform logistic regression by using the FUNCAT procedure in SAS. For special instructions for use of this program for logistic regression, the user should consult the program manual. We note that the signs for the coefficients will be reversed if this procedure is used. Similarly, the procedure LOGLINEAR in SPSS—X can be used to obtain estimates of the coefficients of the logistic equation.

It is important to note that the computational method for each of these programs is iterative, and hence the program can be expensive if the number of variables is large. It may be advisable for the investigator to perform some initial variable selection, using discriminant analysis programs. In BMDP two computation options are offered. The default option, called the ACE method, can be much less expensive than the MLE option.

To illustrate the use of BMDPLR, we give below the control statements needed to predict nondepressed or depressed individuals, with age, sex and income as independent variables, as discussed in Section 12.6. Note that the first three paragraphs and the "group" paragraph are essentially the same as those in the discriminant function example in Section 11.12. A "transform" paragraph is added to change sex to a 0,1 variable from a 1,2 variable. Since it is called by the same name, no new variable is considered to be added.

```
/problem        title is 'logistic regression depress'.
/input          file is depress.
                format is '(8x,f1,f2,3x,f2,23x,f1)'.
                cases are 294.
                variables are 4.
/variables      names are sex,age,income,cases.
/transform      sex=sex - 1.
/group          codes(4)=0,1.
                names(4)=notdep,depress.
/regress        depend is cases.
                interval is sex,age,income.
                enter=.2,.2.
                remove=1,1.
/end
```

TABLE 12.3. Partial Output of BMDPLR for Steps 2 and 3

Term	Coefficient	Standard Error	Coefficient ÷ Standard Error
Step 2			
Income	0.031	0.014	2.286
Sex	−0.903	0.382	−2.360
Constant	1.660	0.417	3.975
Step 3			
Age	0.021	0.009	2.318
Income	0.037	0.014	2.595
Sex	−0.929	0.386	−2.409
Constant	0.676	0.579	1.169

A "regress" paragraph is needed to indicate which variable is the dependent variable. Also, the form of the independent variables are specified. Here we say that they are all interval so that we will not have to use the "design" variable sentence. The enter and remove values in this program are not F levels. Instead, they are P values. The P values for enter must be less than the P for remove. In this example we set the P value for enter at 0.2, so that some variables are highly likely to enter, and the P value for remove at 1, so that no variable is removed. In most cases we recommend that you use the default option. Note the use of f1 instead of f1.0 in the "format" sentence. This shortened version can be used when there are no decimal places.

The program prints the first ten observations so that you can check that it is reading the data you want it to use. It also prints simple summary statistics.

Using the default options results in a stepwise inclusion of variables. At step 0 the constant term is entered, and a goodness-of-fit chi-square is printed. A table of approximate F-to-enter values is printed, and variable "sex" that has the largest numerical value is entered. At step 2 income is entered, and finally at step 3 age enters. The printout includes the coefficients along with their standard errors and each coefficient divided by its standard error, as shown in Table 12.3 for steps 2 and 3.

In our run we noted that the coefficient for income at step 2 was 0.031 but it increased to 0.037 when age entered at the third step. One question, then, is whether a significant age-income interaction or other interaction exists among the independent variables. This issue can be checked by adding the following sentence to the "regress" paragraph:

```
model=age*sex*income.
```

This sentence causes the interaction of age and sex and income, and all subsets of this interaction (age and sex; age and income; sex and income; age, sex, and income), to be considered as possible variables for entry by the stepwise procedure.

The program also prints the number and the percent of correct and incorrect predictions, as was discussed in Section 12.5.

SUMMARY

In this chapter we presented another method of classifying individuals into one of two populations. It is based on the assumption that the logarithm of the odds of belonging to one population is a linear function of the variables used for classification. The result is that the probability of belonging to the population is a multiple logistic function.

We described in some detail how this relatively new method can be implemented by packaged programs, and we made the comparison with the linear discriminant function. Thus logistic regression is appropriate when both categorical and continuous variables are used, while the linear discriminant function is preferable when the multivariate normal model can be assumed. We also described situations in which logistic regression is a useful method of analysis.

BIBLIOGRAPHY

*Breslow, N.E., and Day, N.E. 1980. *Statistical methods in cancer research.* IARC Scientific Publications No. 32. Lyons, France: World Health Organization.

Brittain, E. 1980. *Probability of developing coronary heart disease.* Technical Report 54. Stanford, Calif: Stanford University, Division of Biostatistics.

*Cox, D. R. 1970. *Analysis of binary data.* London: Methuen.

Efron, B. 1975. The efficiency of logistic regression compared to normal discriminant analysis. *Journal of the American Statistical Association* 70:892–898.

Fleiss, J. L. 1981. *Statistical methods for rates and proportions*. New York: Wiley.

Halperin, M.; Blackwelder, W. C.; and Verter, J.I. 1971. Estimation of the multivariate logistic risk function: A comparison of the discriminant function and maximum likelihood approach. *Journal of Chronic Diseases* 24:125–158.

Holford, T. R.; White, C.; and Kelsey, J. L. 1978. Multivariate analyses for matched case-control studies. *American Journal of Epidemiology* 107:245–256.

Lemeshow, S., and Hosmer, D. W. 1982. A review of goodness-of-fit statistics for use in the development of logistic regression models. *American Journal of Epidemiology* 115:92–106.

O'Hara, T. F.; Hosmer, D. W.; Lemeshow, S.; and Hartz, S. C. 1982. A comparison of discriminant function and maximum likelihood estimates of logistic coefficients for categorical data. University of Massachusetts, Amherst, Mass.

Press, S. J., and Wilson, S. 1978. Choosing between logistic regression and discriminant analysis. *Journal of the American Statistical Society* 73:699–705.

Reinhardt, P. S. 1980. *SAS supplemental library user's guide, 1980*. Cary, N.C. SAS Institute, Inc., Box 8000.

Reynolds, H. T. 1977. *Analysis of nominal data*. Beverly Hills: Sage.

Schlesselman, J. J. 1982. *Case-control studies*. New York: Oxford University Press.

Woolson, R. F., and Lachenbruch, P. A. 1982. Regression analysis of matched case-control data. *American Journal of Epidemiology* 115:444–452.

PROBLEMS

12.1   If the probability of an individual getting a hit in baseball is 0.20, then the odds of getting a hit are 0.25. Check to determine that the previous statement is true. Would you prefer to be told that your chances are one in five of a hit or that for every four hitless times at bat you can expect to get one hit?

12.2   Using the formula odds $= P /(1 - P)$, fill in the accompanying table.

Odds	P
0.25	0.20
0.5	
1.0	0.5
1.5	
2.0	
2.5	
3.0	0.75
5.0	

12.3 The accompanying table presents the number of individuals by smoking and disease status. What are the odds that a smoker will get disease A? That a nonsmoker will get disease A? What is the odds ratio?

	Disease A		
Smoking	Yes	No	Total
Yes	80	120	200
No	20	280	300
Total	100	400	500

12.4 Perform a logistic regression analysis with the same variables and data used in the example in Section 11.12 and compare the results.

12.5 Perform a logistic regression analysis on the data described in Problem 11.2. Compare the two analyses.

12.6 Repeat Problem 11.5, using stepwise logistic regression. Compare the results.

12.7 Repeat Problem 11.7, using stepwise logistic regression. Compare the results.

12.8 Repeat Problem 11.10, using stepwise logistic regression. Compare the results.

13 PRINCIPAL COMPONENTS ANALYSIS

13.1 WHAT WILL YOU LEARN FROM THIS CHAPTER?

From this chapter you will learn:

▶ How to recognize when the use of principal components will make your data analysis simpler (13.2, 13.3).

▶ About the basic concepts of restructuring data by using principal components (13.4).

▶ How to decide if it was worthwhile doing the restructuring (13.5).

▶ How to decide on an appropriate computer program (13.6).

13.2 WHEN IS PRINCIPAL COMPONENTS ANALYSIS USED?

Principal components analysis is performed in order to simplify the description of a set of interrelated variables. In principal components analysis the variables are treated equally; i.e., they are not divided into dependent and independent variables, as in regression analysis.

The technique can be summarized as a method of transforming the original variables into new, uncorrelated variables. The new variables are called the *principal components*. Each principal component is a linear combination of the original variables. One measure of the amount of information conveyed by each principal component is its variance. For this reason the principal components are arranged in order of decreasing variance. Thus the most informative principal component is the first, and the least informative is the last (a variable with zero variance does not distinguish between the members of the population).

An investigator may wish to reduce the dimensionality of the problem, i.e., reduce the number of variables without losing much of the information. This objective can be achieved by choosing to analyze only the first few principal components. The principal components not analyzed convey only a small amount of information since their variances are small. This technique is attractive for another reason, namely, that the principal components are not intercorrelated. Thus instead of analyzing a large number of original variables with complex interrelationships, the investigator can analyze a small number of uncorrelated principal components.

The selected principal components may also be used to test for their normality. *If the principal components are not normally distributed, then neither are the original variables.* Another use of the principal components is to search for outliers. A histogram of each of the principal components can identify those individuals with very large or very small values; these values are candidates for outliers or blunders.

In regression analysis it is sometimes useful to obtain the first few principal components corresponding to the X variables and then perform the regression on the selected components. This tactic is useful for overcoming the problem of multicollinearity since the principal components are uncorrelated (see Chatterjee and Price 1977). Principal components analysis can also be viewed as a step toward factor analysis (see Chapter 14).

Principal components analysis is considered to be an exploratory technique that may be useful in gaining a better understanding of the interrelationships among the variables. This idea will be discussed further in Section 13.5.

The original application of principal components analysis was in the field of educational testing. Hotelling (1933) developed this technique and showed that there are two major components to responses on entry-examination tests: verbal and quantitative ability. Principal components and factor analysis are also used extensively in psychological applications in an attempt to discover underlying structure. In addition, principal components analysis has been used in biological and medical applications (see Seal 1964; Morrison 1976).

TABLE 13.1. Sample Statistics for Hypothetical Data Set

	Variable	
Statistic	NORM1	NORM2
N	100	100
Mean	101.63	50.71
Standard deviation	10.47	7.44
Variance	109.63	55.44

13.3 DATA EXAMPLE

The depression data set will be used later in the chapter to illustrate the technique. However, to simplify the exposition of the basic concepts, we generated a hypothetical data set using the BMDP package. These hypothetical data consist of 100 random pairs of observations, X_1 and X_2. The population distribution of X_1 is normal with mean 100 and variance 100. For X_2 the distribution is normal with mean 50 and variance 50. The population correlation between X_1 and X_2 is $1/\sqrt{2} = 0.707$. Figure 13.1 shows a scatter diagram of the 100 random pairs of points. The two variables are denoted in this graph by NORM1 and NORM2. The sample statistics are shown in Table 13.1. The sample correlation is $r = 0.757$.

The advantage of this data set is that it consists of only two variables, so it is easily graphed. In addition, it satisfies the usual normality assumption made in statistical theory.

13.4 BASIC CONCEPTS OF PRINCIPAL COMPONENTS ANALYSIS

Again, to simplify the exposition of the basic concepts, we present first the case of two variables X_1 and X_2. Later we discuss the general case of P variables.

Suppose that we have a random sample of N observations on X_1 and X_2.

FIGURE 13.1. Scatter Diagram of NORM1 Versus NORM2

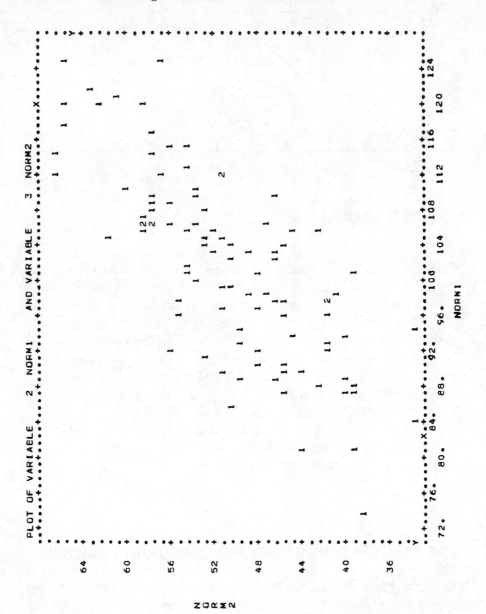

For ease of interpretation we subtract the sample mean from each observation, thus obtaining

$$x_1 = X_1 - \overline{X}_1$$

and
$$x_2 = X_2 - \overline{X}_2$$

Note that this technique makes the means of x_1 and x_2 equal to zero but does not alter the sample variances S_1^2 and S_2^2 or the correlation r.

The basic idea is to create two new variables, C_1 and C_2, called the *principal components*. These new variables are linear functions of x_1 and x_2 and can therefore be written as

$$C_1 = a_{11}x_1 + a_{12}x_2$$

and
$$C_2 = a_{21}x_1 + a_{22}x_2$$

We note that for any set of values of the coefficients a_{11}, a_{12}, a_{21}, a_{22}, we can introduce the N observed x_1 and x_2 and obtain N values of C_1 and C_2. The means and variances of the N values of C_1 and C_2 are

$$\text{mean } C_1 = \text{mean } C_2 = 0$$

$$\text{Var } C_1 = a_{11}^2 S_1^2 + a_{12}^2 S_2^2 + 2a_{11}a_{12}rS_1S_2$$

and
$$\text{Var } C_2 = a_{21}^2 S_1^2 + a_{22}^2 S_2^2 + 2a_{21}a_{22}rS_1S_2$$

where $S_i^2 = \text{Var } X_i$.

The *coefficients* are chosen to satisfy three requirements:

1. The Var C_1 is as large as possible.
2. The N values of C_1 and C_2 are uncorrelated.
3. $a_{11}^2 + a_{12}^2 = a_{21}^2 + a_{22}^2 = 1$.

The mathematical solution for the coefficients was derived by Hotelling (1933). The solution is illustrated graphically in Figure 13.2. Principal components analysis amounts to rotating the original x_1 and x_2 axes to new C_1 and C_2 axes. The angle of rotation is determined uniquely by the requirements just stated. For a given point x_1, x_2 (see Figure 13.2) the values of C_1 and C_2 are found by drawing perpendicular lines to the new C_1 and C_2 axes. The N values of C_1 thus obtained will have the largest variance according to requirement 1. The N values of C_1 and C_2 will have a zero correlation.

FIGURE 13.2. Illustration of Two Principal Components C_1 and C_2

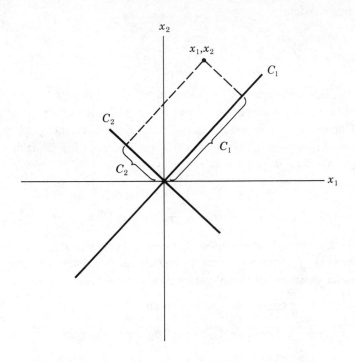

In our hypothetical data example, the two principal components are

$$C_1 = 0.841x_1 + 0.539x_2$$

$$C_2 = -0.539x_1 + 0.841x_2$$

where $\qquad\qquad x_1 = \text{NORM1} - \text{mean(NORM1)}$

and $\qquad\qquad x_2 = \text{NORM2} - \text{mean(NORM2)}$

Note that

$$0.841^2 + 0.539^2 = 1$$

and $\qquad\qquad (-0.539)^2 + 0.841^2 = 1$

as required by requirement 3 above. Also note that—for the *two*-variable case only—$a_{11} = a_{22}$ and $a_{12} = -a_{21}$.

Figure 13.3 illustrates the x_1 and x_2 axes (after subtracting the means)

FIGURE 13.3. Plot of Principal Components for Bivariate Hypothetical Data

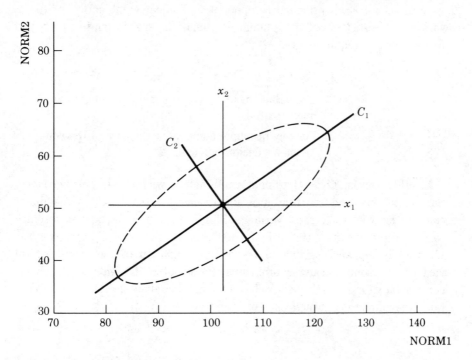

and the rotated C_1 and C_2 principal component axes. Also drawn in the graph is an ellipse of concentration of the original bivariate normal distribution.

The variance of C_1 is 147.44, and the variance of C_2 is 17.59. These two variances are commonly known as the first and second *eigenvalues*, respectively (synonyms used for eigenvalue are characteristic root, latent root, and proper value). Note that the sum of these two variances is 165.03. This quantity is equal to the sum of the original two variances (109.63 + 55.40 = 165.03). This result will always be the case; i.e., the total variance is preserved under rotation of the principal components. Note also that the lengths of the axes of the ellipse of concentration (Figure 13.3) are proportional to the sample standard deviations of C_1 and C_2, respectively. These standard deviations are the square roots of the eigenvalues. It is therefore easily seen from Figure 13.3 that C_1 has a larger variance than C_2. In fact, C_1 has a larger variance than either of the original variables x_1 and x_2.

These basic ideas are easily extended to the case of P variables $x_1, x_2, \ldots, x_P$. Each principal component is a linear combination of the x variables. Coefficients of these linear combinations are chosen to satisfy the following three requirements:

1. Var $C_1 \geq$ Var $C_2 \geq \cdots \geq$ Var C_P.

2. The values of any two principal components are uncorrelated.

3. For any principal component the sum of the squares of the coefficients is one.

In other words, C_1 is the linear combination with the largest variance. Subject to the condition that it is uncorrelated with C_1, C_2 is the linear combination with the largest variance. Similarly, C_3 has the largest variance subject to the condition that it is uncorrelated with C_1 and C_2; etc. The Var C_i are the *eigenvalues*. These P variances add up to the original total variance. In some packaged programs the set of coefficients of the linear combination for the ith principal component is called the ith *eigenvector* (also known as the characteristic or latent vector).

13.5 INTERPRETATION

In this section we discuss how many components should be retained for further analysis, and we present the analysis for standardized x variables. Application to the depression data set is given as an example.

Number of Components Retained

As mentioned earlier, one of the objectives of principal components analysis is *reduction of dimensionality*. Since the principal components are arranged in decreasing order of variance, we may select the first few as representatives of the original set of variables. The number of components selected may be determined by examining the proportion of total variance explained by each component. The cumulative proportion of total variance indicates to the investigator just how much information is retained by selecting a specified number of components.

In the hypothetical example the total variance is 165.03. The variance of the first component is 147.44, which is 147.44/165.03 = 0.893, or 89.3%, of the total variance. It can be argued that this amount is a sufficient percentage of the total variation, and therefore the first principal component is a reasonable representative of the two original variables NORM1 and NORM2 (see Morrison 1976 or Eastment and Krzanowski 1982 for further discussion).

To interpret the meaning of the first principal component, we recall that it was expressed as

$$C_1 = 0.841x_1 + 0.539x_2$$

The coefficient 0.841 can be transformed into a correlation between x_1 and C_1. In general, the correlation between the ith principal component and the jth x variable is

$$r_{ij} = \frac{a_{ij}\sqrt{\text{Var } C_i}}{\sqrt{\text{Var } x_j}}$$

where a_{ij} is the coefficient of x_j for the ith principal component. For example, the correlation between C_1 and x_1 is

$$r_{11} = \frac{0.84\sqrt{147.44}}{\sqrt{109.63}} = 0.975$$

and the correlation between C_1 and x_2 is

$$r_{12} = \frac{0.539\sqrt{147.44}}{\sqrt{55.40}} = 0.880$$

Note that both of these correlations are fairly high and positive. As can be seen from Figure 13.3, when either x_1 or x_2 increases, so will C_1. This result occurs often in principal components analysis whereby the first component is positively correlated with all of the original variables.

Using Standardized Variables

Investigators frequently prefer to *standardize* the x variables prior to performing the principal components analysis. Standardization is achieved by dividing each variable by its sample standard deviation. This analysis is then equivalent to analyzing the correlation matrix instead of the covariance matrix. When we derive the principal components from the correlation matrix, the interpretation becomes easier in two ways:

1. The total variance is simply the number of variables P, and the proportion explained by each principal component is the corresponding eigenvalue divided by P.

2. The correlation between the ith principal component C_i and the jth variable x_j is

$$r_{ij} = a_{ij}\sqrt{\text{Var } C_i}$$

Therefore for a given C_i we can compare the a_{ij} to quantify the relative degree of dependence of C_i on each of the standardized variables.

In our hypothetical example the correlation matrix is as follows:

x_1/S_1	x_2/S_2
1.000	0.757
0.757	1.000

Here S_1 and S_2 are the standard deviations of the first and second variables. Analyzing this matrix results in the following two principal components:

$$C_1 = 0.707\,\frac{\text{NORM1}}{S_1} + 0.707\,\frac{\text{NORM2}}{S_2}$$

which explains $1.757/2 \times 100 = 87.8\%$ of the total variance, and

$$C_2 = 0.707\frac{\text{NORM1}}{S_1} - 0.707\,\frac{\text{NORM2}}{S_2}$$

explaining $0.243/2 \times 100 = 12.2\%$ of the total variance. So the first principal component is equally correlated with the two standardized variables. The second principal component is also equally correlated with the two standardized variables, but in the opposite direction.

The case of two standardized variables is illustrated in Figure 13.4. Forcing both standard deviations to be one causes the vast majority of the data to

FIGURE 13.4. Principal Components for Two Standardized Variables

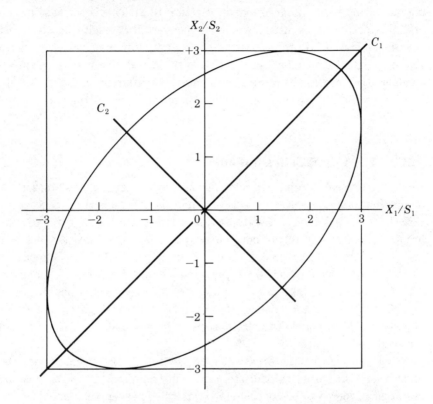

be contained in the square shown in the figure (e.g., 99.7% of normal data must fall within plus or minus three standard deviations of the mean). Because of the symmetry of this square, the first principal component will be in the direction of the 45° line for the case of two variates.

In terms of the original variables NORM1 and NORM2, the principal components based on the correlation matrix are

$$C_1 = 0.707 \frac{\text{NORM1}}{S_1} + 0.707 \frac{\text{NORM2}}{S_2} = 0.707 \frac{\text{NORM1}}{10.47} + 0.707 \frac{\text{NORM2}}{7.44}$$
$$= 0.0675\text{NORM1} + 0.0953\text{NORM2}$$

Similarly,

$$C_2 = +0.0675\text{NORM1} - 0.0953\text{NORM2}$$

Note that these results are very different from the principal components obtained from the covariance matrix (Section 13.4). This is the case in general. In fact, there is no easy way to convert the results based on the covariance matrix into those based on the correlation matrix, or vice versa. The majority of researchers prefer to use the correlation matrix because it compensates for the units of measurement of the different variables. But if it is used, then all interpretations must be made in terms of the standardized variables.

Analysis of Depression Data Set

Next, we present a principal components analysis of a real data set, the depression data. We select for this example the 20 items that make up the CESD scale. Each item is a statement to which the response categories are ordinal. The answer rarely or none of the time (less than 1 day) is coded as 0, some or a little of the time (1–2 days) as 1, occasionally or a moderate amount of the time (3–4 days) as 2, and most or all of the time (5–7 days) as 3. The values of the response categories are reversed for the positive-affect items (see Table 3.2 for a listing of the items) so that a high score indicates likelihood of depression. The CESD score is simply a sum of the scores for these 20 items.

We emphasize that these variables do not satisfy the assumptions often made in statistics of a multivariate normal distribution. In fact, they cannot even be considered to be continuous variables. However, they are typical of what is found in real life applications.

In this example we used BMDP4R with the correlation matrix in order to be consistent with the factor analysis we present in Chapter 14. The eigenvalues (variances of the principal components) are plotted in Figure 13.5b. Since the correlation matrix is used, the total variance is the number of variables, 20. By dividing each eigenvalue by 20 and multiplying by 100, we obtain the percentage of total variance explained by each principal component. Adding these percentages successively produces the cumulative percentages plotted in 13.5a.

These eigenvalues and cumulative percentages are found in the output of standard packaged programs. They enable the user to determine whether and how many principal components should be used. If the variables were uncorrelated to begin with, then each principal component, based on the correlation matrix, would explain the *same* percentage of the total variance,

FIGURE 13.5. Eigenvalues and Cumulative Percentages of Total Variance for
Depression Data

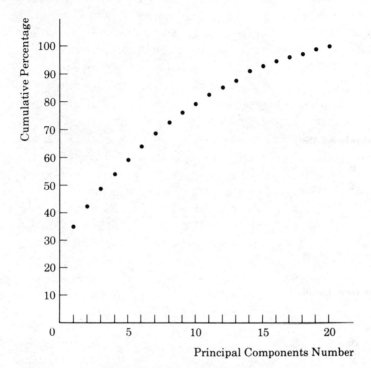

a. Cumulative Percentage

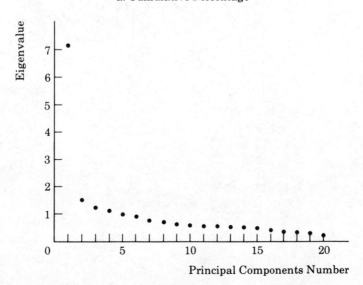

b. Eigenvalue

namely $100/P$. If this were the case, a principal components analysis would be unnecessary. Typically, the first principal component explains a much larger percentage than the remaining components, as shown in Figure 13.5.

Ideally, we wish to obtain a small number of principal components, say two or three, which explain a large percentage of the total variance, say 80% or more. In this example, as is the case in many applications, this ideal is not achieved. We therefore must compromise by choosing a few principal components as possible to explain a reasonable percentage of the total variance. A rule of thumb adopted by many investigators is to select only the principal components explaining at least $100/P$ percent of the total variance (at least 5% in our example). This rule applies whether the covariance or the correlation matrix is used. In our example we would select the first five principal components if we followed this rule. These five components explain 59% of the total variance, as seen in Figure 13.5. Note, however, that the next two components each explain nearly 5%. Some investigators would therefore select the first seven components, which explain 69% of the total variance.

It should be explained that the eigenvalues are estimated variances of the principal components and are therefore subject to large sample variations. Arbitrary cutoff points should thus not be taken too seriously. Ideally, the investigator should make the selection of the number of principal components on the basis of some underlying theory.

Once the number of principal components is selected, the investigator should examine the coefficients defining each of them in order to assign an interpretation to the components. As was discussed earlier, a high coefficient of a principal component on a given variable is an indication of high correlation between that variable and the principal component (see the formulas given earlier). Principal components are interpreted in the context of the variables with high coefficients.

For the depression data example, Table 13.2 shows the coefficients for the first five components. For each principal component the variables with a correlation greater than 0.5 with that component are underlined. For the sake of illustration this value was taken as a cutoff point. Recall that the correlation r_{ij} is $a_{ij}\sqrt{\mathrm{Var}\,C_i}$, and therefore a coefficient a_{ij} is underlined if it exceeds $0.5/\sqrt{\mathrm{Var}\,C_i}$ (see Table 13.2).

As the table shows, many variables are highly correlated (greater than 0.5) with the first component. Note also that the correlations of all the variables with the first component are positive (recall that the scaling of the response

TABLE 13.2. Principal Components Analysis for Standardized CESD Scale Items
(Depression Data Set)

Item	Principal Component				
	1	2	3	4	5
Negative Affect					
1. I felt that I could not shake off the blues even with the help of my family or friends.	0.2774	0.1450	0.0577	−0.0027	0.0883
2. I felt depressed.	0.3132	−0.0271	0.0316	0.2478	0.0244
3. I felt lonely.	0.2678	0.1547	0.0346	0.2472	−0.2183
4. I had crying spells.	0.2436	0.3194	0.1769	−0.0716	−0.1729
5. I felt sad.	0.2868	0.0497	0.1384	0.2794	−0.0411
6. I felt fearful.	0.2206	−0.0534	0.2242	0.1823	−0.3399
7. I thought my life had been a failure.	0.2844	0.1644	−0.0190	−0.0761	−0.0870
Positive Affect					
8. I felt that I was as good as other people.	0.1081	0.3045	0.1103	−0.5567	−0.0976
9. I felt hopeful about the future.	0.1758	0.1690	−0.3962	−0.0146	0.5355
10. I was happy.	0.2766	0.0454	−0.0835	0.0084	0.3651
11. I enjoyed life.	0.2433	0.1048	−0.1314	0.0414	0.2419
Somatic and Retarded Activity					
12. I was bothered by things that usually don't bother me.	0.1790	−0.2300	0.1634	0.1451	0.0368
13. I did not feel like eating; my appetite was poor.	0.1259	−0.2126	0.2645	−0.5400	0.0953
14. I felt that everything was an effort.	0.1803	−0.4015	−0.1014	−0.2461	−0.0847
15. My sleep was restless.	0.2004	−0.2098	0.2703	0.0312	0.0834
16. I could not "get going."	0.1924	−0.4174	−0.1850	−0.0467	−0.0399
17. I had trouble keeping my mind on what I was doing.	0.2097	−0.3905	−0.0860	−0.0684	−0.0499
18. I talked less than usual.	0.1717	−0.0153	0.2019	−0.0629	0.2752
Interpersonal					
19. People were unfriendly.	0.1315	−0.0569	−0.6326	−0.0232	−0.3349
20. I felt that people disliked me.	0.2357	0.2283	−0.1932	−0.2404	−0.2909
Eigenvalues or Var C_i	7.055	1.486	1.231	1.066	1.013
Cumulative proportion explained	0.353	0.427	0.489	0.542	0.593
$0.5/\sqrt{\text{Var}C_i}$	0.188	0.410	0.451	0.484	0.497

for items 8–11 was reversed so that a high score indicates likelihood of depression). The first principal component can therefore be viewed as a weighted average of most of the items. A high value of C_1 is an indication that the respondent had many of the symptoms of depression.

On the other hand, the only item with more than 0.5 absolute correlation with C_2 is item 16, although items 17 and 14 have absolute correlations close to 0.5. The second principal component, therefore, can be interpreted as a measure of lethargy or energy. A low value of C_2 is an indication of a lethargic state and a high value of C_2 is an indication of a high level of energy. By construction, C_1 and C_2 are uncorrelated.

Similarly, C_3 measures the respondent's feeling toward how others perceive him or her; a low value is an indication that the respondent believes that people are unfriendly. Similar interpretations can be made for the other two principal components in terms of the items corresponding to the underlined coefficients.

The numerical values of C_i for $i = 1$ to 5, or 1 to 7, for each individual can be used in subsequent analyses. For example, we could use the value of C_1 instead of the CESD score as a dependent variable in a regression analysis such as that given in Problem 8.1. Had the first component explained a higher proportion of the variance, this procedure might have been better than simply using a sum of the scores.

The depression data example illustrates a situation in which the results are *not* clear-cut. This conclusion may be reached from observing Figure 13.5, where we saw that it is difficult to decide how many components to use. It is not possible to explain a very high proportion of the total variance with a small number of principal components. Also, the interpretation of the components in Table 13.2 is not straightforward. This is frequently the case in real life situations. Occasionally, situations do come up in which the results are clear-cut. For examples of such situations, see Seal (1964), Cooley and Lohnes (1971), or Harris (1975).

In biological examples the first component is often called the size component, and subsequent components are called the shape components. This terminology is especially appropriate when the data being analyzed are length and girth measurements of animals.

Test of hypotheses regarding principal components are discussed in Lawley (1956), Cooley and Lohnes (1971), Jackson and Hearne (1973), and Morrison (1976). These tests have not been implemented in the computer programs since the data are often not multivariate normal, and most users view principal components analysis as an exploratory technique.

13.6 DISCUSSION OF COMPUTER PROGRAMS

Each of the three major packages has at least one program for performing a principal components analysis. For each package a factor analysis program may also be used for this purpose, although other programs may be easier to use.

For the BMDP package, BMDP4R is straightforward and will give all of the desired output with a minimum of specifications. It does require, however, that a variable be named as the dependent variable since the program is intended primarily as a regression program. Any variable other than $X_1, X_2, \ldots, X_P$ may be used for this purpose. The program gives the user the option of employing either the covariance or the correlation matrix to obtain the principal components. The program then proceeds to obtain the regression equation of the dependent variable on the principal components successively. It should be noted that performing the regression on the principal components is a method of analyzing data with a high degree of multicollinearity (see Chatterjee and Price 1977).

With SAS the easier procedure is PRINCOMP. It also offers the user the option of choosing either the correlation or the covariance matrix.

With the SPSS–X package the FACTOR program can be used to perform a principal components analysis. The correlation matrix is the only option available. The user should specify EXTRACTION = PA1 or PC and also CRITERIA = FACTORS(P), where P is the total number of variables.

Table 13.3 gives a summary of the types of output available from each of the three packaged programs.

Now we will consider an example. In Table 13.2 the principal components coefficients were listed for standardized CESD scale items. However, we may argue that since these 20 variables all have the same unit of measurement, it is not necessary to standardize them. But the 20 items do not all have the same standard deviation, so the results of a principal components analysis will not be the same for the standardized and the unstandardized variables. One advantage of using the unstandardized principal components scores is

TABLE 13.3. Summary of Computer Output for Principal Components Analysis

Output	BMDP	SAS	SPSS–X
Means	BMDP4R*	FACTOR, PRINCOMP	FACTOR
Variances or standard deviations	BMDP4R	FACTOR, PRINCOMP	FACTOR
Covariance matrix	BMDP4R	FACTOR, PRINCOMP	
Correlation matrix	BMDP4R	FACTOR, PRINCOMP	FACTOR
Coefficients from covariance matrix	BMDP4R	FACTOR, PRINCOMP	
Coefficients from correlation matrix	BMDP4R	FACTOR, PRINCOMP	FACTOR
Eigenvalues	BMDP4R	FACTOR, PRINCOMP	FACTOR
Cumulative proportion of variance explained	BMDP4R	FACTOR, PRINCOMP	

*BMDP4M (factor analysis) may also be used for principal components; note the example in this section.

that if we want to use the same coefficients to compute principal components for a new sample, we can use the raw items directly from the new sample.

A question a researcher might ask is whether it is better to use the first principal component as a measure of depression or simply to sum up the item scores, as is usually done. Suppose you wish to use the variables sex, age, income, and health to predict the level of depression. Can you obtain a higher multiple correlation coefficient by predicting CESD (the sum) or the first principal component of the unstandardized scores in this sample?

One method of performing the computations to answer this question is to calculate the first principal component scores and save them. Then a regression program can be run twice, first with the first principal component scores as a dependent variable and then with CESD as the dependent variable. Comparison of the results can indicate which correlation is higher.

For our depression data example we can also use BMDP4M for such a comparison. When BMDP4M is used to produce the principal components, the components are labeled as factors. The first part of the control statements include the following two paragraphs:

```
/factor     method = pca.
            form = cova.
            constant = 0.
/rotate     method = none.
```

The first sentence specifies that we wish a principal components analysis (as explained in the next chapter, this program is capable of performing other analyses). The "form = cova" sentence specifies that we want the covariance matrix since the data are not standardized. The "constant = 0" sentence forces the program to print all the principal components. Finally, the "method = none" sentence in the rotate paragraph prevents the program from "rotating" the principal components, a concept explained in the next chapter on factor analysis.

If the save option is used with the default option, a new file is set up that includes the original data and the just-computed principal components added to the end of the data file. For a microprocessor version of the BMDP programs the following statements are made:

```
/save           file = 'pc.save',
                code = pc,
                new,
```

Here "pc.save" is an arbitrary name given a file, and pc is an arbitrary code. Then when the regression analysis is run, the input paragraph has to include the following:

```
/input          file = 'pc.save',
                code = pc,
```

On a mainframe computer a unit = number (where the number depends on the particular installation used) would also be added. The input paragraph must also include the unit sentence if a mainframe computer is used.

When PRINCOMP in SAS is used, the statement OUT = PC produces the same results as those above for a file named PC. In SPSS–X the SAVE subcommand can be used with the default option.

When BMDP2R is run with the standard options, using the first principal component as the dependent variable, the independent variables entered are health, age, and income, in that order. Higher levels of depression, as measured by the first principal component, are associated with being in poor health, younger, and poorer. The multiple correlation is $R = 0.348$. When CESD is used as the dependent variable, the same variables enter in the same order, and $R = 0.357$. Thus the results are very similar, and the simple addition of the scores (i.e., CESD) gives essentially the same results as using the first principal component for this one-regression comparison.

Finally, note that the principal component scores may be used as independent variables in a regression or discriminant function analysis to predict some other dependent variable or outcome.

SUMMARY

In the two previous chapters we presented methods of selecting variables for regression and discriminant function analyses. These methods include stepwise and subset procedures. In those analyses a dependent variable is present, implicitly or explicitly. In this chapter we presented another method for summarizing the data. It differs from variable selection procedures in two ways: (1) no dependent variable exists, and (2) variables are not eliminated but rather summary variables—i.e., principal components—are computed from all of the original variables.

The major ideas underlying the method of principal components analysis were presented in this chapter. We also discussed how to decide on the number of principal components retained and how to use them in subsequent analyses. Further, methods for attaching interpretations or "names" to the selected principal components were given.

BIBLIOGRAPHY

Chatterjee, S., and Price, B. 1977. *Regression analysis by example*. New York: Wiley.

*Cooley, W. W., and Lohnes, P. R. 1971. *Multivariate data analysis*. New York: Wiley.

Dunn, O. J., and Clark, V. A. 1974. *Applied statistics: Analysis of variance and regression*. New York: Wiley.

Eastment, H. T., and Krzanowski, W. J. 1982. Cross-validory choice of the number of components from a principal component analysis. *Technometrics* 24:73–77.

Harris, R. J. 1975. *A primer of multivariate statistics*. New York: Academic Press.

*Hotelling, H. 1933. Analysis of a complex of statistical variables into principal components. *Journal of Educational Psychology* 24:417–441.

Jackson, J. E., and Hearne, F. T. 1973. Relationships among coefficients of vectors in principal components. *Technometrics* 15:601–610.

*Lawley, D. N. 1956. Tests of significance for the latent roots of covariance and correlation matrices. *Biometrika* 43:128–136.

Morrison, D. F. 1976. *Multivariate statistical methods*. New York: McGraw-Hill.

Seal, H. 1964. *Multivariate statistical analysis for biologists*. New York: Wiley.

*Tatsuoka, M. M. 1971. *Multivariate analysis: Techniques for educational and psychological research*. New York: Wiley.

PROBLEMS

13.1 For the depression data set, perform a principal components analysis on the last seven variables (DRINK through CHRONILL); see Table 3.2. Interpret the results.

13.2 Continuation of Problem 13.1: Perform a regression analysis of CASES on the last seven variables as well as on the principal components. What does the regression represent? Interpret the results.

13.3 For the data generated in Problem 7.7, perform a principal components analysis on $X_1, X_2, \ldots, X_9$. Compare the results with what is known about the population.

13.4 Continuation of Problem 13.3: Perform the regression of Y on the principal components. Compare the results with the multiple regression of Y on X1 to X9.

13.5 Perform a principal components analysis on the data in Table 8.1 (not including the variable P/E). Interpret the components. Then perform a regression analysis with P/E as the dependent variable, using the relevant principal components. Compare the results with those in Chapter 8.

14 FACTOR ANALYSIS

14.1 WHAT WILL YOU LEARN FROM THIS CHAPTER?

From this chapter you will learn a useful extension of principal components analysis called factor analysis that will enable you to obtain more distinct new variables. The methods given in this chapter for describing the interrelationships among variables and obtaining new variables have been deliberately limited to a small subset of the methods available in the literature in order to make the explanation more understandable. In particular, you will learn:

▸ When to use factor analysis (14.2, 14.3, and 14.9).
▸ About the basic factor model (14.4).
▸ How to obtain the initial factors (14.5, 14.6).
▸ How to rotate the initial factors to obtain more distinct factors (14.7).
▸ How to obtain factor scores to use in future analyses (14.8).
▸ About which options are available on the various computer packages (14.10).

14.2 WHEN IS FACTOR ANALYSIS USED?

Factor analysis is similar to principal components analysis in that it is a technique for examining the interrelationships among a set of variables. Both of these techniques differ from regression analysis in that we do not have a *dependent* variable to be explained by a set of *independent* variables. However, principal components analysis and factor analysis also differ from each other. In principal components analysis the major objective is to select a number of components that explain as much of the total variance as possible. The values of the principal components for a given individual are relatively simple to compute and interpret. On the other hand, the *factors* obtained in factor analysis are selected mainly to explain the interrelationships among the original variables. Ideally, the number of factors expected is known in advance. The major emphasis is placed on obtaining easily understandable factors that convey the essential information contained in the original set of variables.

Areas of application of factor analysis are the same as those mentioned in Section 13.2 for principal components analysis. Chiefly, applications have come from the social sciences, particularly psychometrics. A certain degree of resistance to using factor analysis in other disciplines has been prevalent; perhaps because of the heuristic nature of the technique and the special jargon employed.

In this chapter no attempt will be made to present a comprehensive treatment of the subject. Rather, we adopt a simple geometric approach to explain only the most important concepts and options available in standard packaged programs. Interested readers should refer to the Bibliography for texts that give more comprehensive treatments of factor analysis.

14.3 DATA EXAMPLE

As in Chapter 13, we generated a hypothetical data set to help us present the fundamental concepts. In addition, the same depression data set used in Chapter 13 will be subjected to a factor analysis here. This example should serve to illustrate the differences between principal components and factor analysis. It will also provide a real life application.

The hypothetical data set consists of 100 data points on five variables, X_1, $X_2, \ldots, X_5$. The data were generated from a multivariate normal distribution with zero means. (The statements used to generate these data are given in Section 14.10.) The sample means and standard deviations are shown in Table 14.1. The correlation matrix is presented in Table 14.2.

TABLE 14.1. Means and Standard Deviations of 100 Hypothetical Data Points

Variable	Mean	Standard Deviation
X_1	0.163	1.047
X_2	0.142	1.489
X_3	0.098	0.966
X_4	−0.039	2.185
X_5	−0.013	2.319

TABLE 14.2. Correlation Matrix of 100 Hypothetical Data Points

	X_1	X_2	X_3	X_4	X_5
X_1	1.000				
X_2	0.757	1.000			
X_3	0.047	0.054	1.000		
X_4	0.115	0.176	0.531	1.000	
X_5	0.279	0.322	0.521	0.942	1.000

We note, on examining the correlation matrix, that there exists a high correlation between X_4 and X_5, a moderately high correlation between X_1 and X_2, and a moderate correlation between X_3 and each of X_4 and X_5. The remaining correlations are fairly low.

14.4 BASIC CONCEPTS OF FACTOR ANALYSIS

In factor analysis we begin with a set of variables X_1, X_2,..., X_P. These variables are usually standardized by the computer program so that their variances are each equal to 1 and their covariances are correlation coefficients. In the remainder of this chapter we therefore assume that each x_i is a *standardized variable*, i.e., $x_i = (X_i - \overline{X}_i)/S_i$. In the jargon of factor analysis the x_i's are called the original or *response variables*.

The object of factor analysis is to represent each of these variables as a linear combination of a smaller set of *common factors* plus a factor unique to each of the response variables. We express this representation as

$$x_1 = l_{11}F_1 + l_{12}F_2 + \cdots + l_{1m}F_m + e_1$$

$$x_2 = l_{21}F_1 + l_{22}F_2 + \cdots + l_{2m}F_m + e_2$$

$$\vdots$$

$$x_P = l_{P1}F_1 + l_{P2}F_2 + \cdots + l_{Pm}F_m + e_P$$

where the following assumptions are made:

1. m is the number of common factors (typically this number is much smaller than P).
2. $F_1, F_2, \ldots, F_m$ are the *common factors*. These factors are assumed to have zero means and unit variances.
3. l_{ij} is the coefficient of F_j in the linear combination describing x_i. This term is called the *loading* of the ith variable on the jth common factor.
4. $e_1, e_2, \ldots, e_P$ are *unique factors*, each relating to one of the original variables.

The above equations and assumptions constitute the so-called *factor model*. Thus each of the response variables is composed of a part due to the common factors and a part due to its own unique factor. The part due to the common factors is assumed to be a linear combination of these factors.

As an example, suppose that x_1, x_2, x_3, x_4, x_5 are the standardized scores of an individual on five tests. If $m = 2$, we assume the following model:

$$x_1 = l_{11}F_1 + l_{12}F_2 + e_1$$

$$x_2 = l_{21}F_1 + l_{22}F_2 + e_2$$

$$\vdots$$

$$x_5 = l_{51}F_1 + l_{52}F_2 + e_5$$

Each of the five scores consists of two parts: a part due to the common factors F_1 and F_2 and a part due to the unique factor for that test. The common factors F_1 and F_2 might be considered the individual's verbal and quantitative abilities, respectively. The unique factors express the individual variation on each test score. The unique factor includes all other effects that keep the common factors from completely defining a particular x_i.

In a sample of N individuals we can express the equations by adding a subscript to each x_i, F_j, and e_i to represent the individual. For the sake of simplicity this subscript was omitted in the model presented here.

The factor model is, in a sense, the mirror image of the principal components model, where each principal component is expressed as a linear combination of the variables. Also, the number of principal components is equal to the number of original variables (although we may not use all of the principal components). On the other hand, in factor analysis we choose the number of factors to be smaller than the number of response variables. Ideally, the number of factors should be known in advance, although this is often not the case. However, as we will discuss later, it is possible to allow the data themselves to determine this number.

The factor model, by breaking each response variable x_i into two parts, also breaks the variance of x_i into two parts. Since x_i is standardized, its variance is 1. This variance of 1 is composed of the following two parts:

1. The *communality*, i.e., the part of the variance that is due to the common factors.
2. The *specificity*, i.e., the part of the variance that is due to the unique factor e_i.

Denoting the *communality* of x_i by h_i^2 and the *specificity* by u_i^2, we can write the variance of x_i as Var $x_i = 1 = h_i^2 + u_i^2$. In words, the variance of x_i equals the communality plus the specificity.

The numerical aspects of factor analysis are concerned with finding estimates of the *factor loadings* (l_{ij}) and the *communalities* (h_i^2). There are many ways available to numerically solve for these quantities. The solution process is called *initial factor extraction*. The next two sections discuss two such extraction methods. Once a set of initial factors is obtained, the next major step in the analysis is to obtain new factors, called the *rotated factors*, in order to improve the interpretation. Methods of rotation are discussed in Section 14.7.

In any factor analysis the number m of common factors is required. As mentioned earlier, this number is, ideally, known prior to the analysis. If it is not known, most investigators use a default option available in standard computer programs whereby the number of factors is the number of eigenvalues greater than 1 (see Chapter 13 for the definition and discussion of eigenvalues). Also, since the numerical results are highly dependent on the chosen number m, many investigators run the analysis with several values in an effort to get further insights into their data.

14.5 INITIAL FACTOR EXTRACTION: PRINCIPAL COMPONENTS ANALYSIS

In this section and the next we discuss two methods for the initial extraction of common factors. We begin with the *principal components analysis method*, which can be found in most of the standard factor analysis programs. It is called PCA in BMDP, PA1 or PC in SPSS–X, and PRINCIPAL (PRIN or P) in SAS. The basic idea is to choose the first m principal components and modify them to fit the factor model defined in the previous section. The reason for choosing the first m principal components, rather than any others, is that they explain the greatest proportion of the variance and are therefore the most important. Note that the principal components are also uncorrelated and thus present an attractive choice as factors.

To satisfy the assumption of unit variances of the factors, we divide each principal component by its standard deviation. That is, we define the jth common factor F_j as $F_j = C_j/(\text{Var } C_j)^{1/2}$, where C_j is the jth principal component.

To express each variable x_i in terms of the F_j's, we first recall the relationship between the variables x_i and the principal components C_j. Specifically,

$$C_1 = a_{11}x_1 + a_{12}x_2 + \cdots + a_{1P}x_P$$

$$C_2 = a_{21}x_1 + a_{22}x_2 + \cdots + a_{2P}x_P$$

$$\vdots$$

$$C_P = a_{P1}x_1 + a_{P2}x_2 + \cdots + a_{PP}x_P$$

It may be shown mathematically that this set of equations can be inverted to express the x_i's as functions of the C_j's. The result is (see Harmon 1976 or Afifi and Azen 1979):

$$x_1 = a_{11}C_1 + a_{21}C_2 + \cdots + a_{P1}C_P$$

$$x_2 = a_{12}C_1 + a_{22}C_2 + \cdots + a_{P2}C_P$$

$$\vdots$$

$$x_P = a_{1P}C_1 + a_{2P}C_2 + \cdots + a_{PP}C_P$$

Note that the rows of the first set of equations become the columns of the second set of equations.

Now since $F_j = C_j/(\text{Var } C_j)^{1/2}$, it follows that $C_j = F_j(\text{Var } C_j)^{1/2}$, and we can then express the ith equation as

$$x_1 = a_{1i}F_1(\text{Var } C_1)^{1/2} + a_{2i}F_2(\text{Var } C_2)^{1/2} + \cdots + a_{Pi}F_P(\text{Var } C_P)^{1/2}$$

This last equation is now modified in two ways:

1. We use the notation $l_{ij} = a_{ji}(\text{Var } C_j)^{1/2}$ for the first m components.
2. We combine the last $P - m$ terms and denote the result by e_i. That is,

$$e_i = a_{m+1,i}F_{m+1}(\text{Var } C_{m+1})^{1/2}$$
$$+ \cdots + a_{Pi}F_P(\text{Var } C_P)^{1/2}$$

With these manipulations we have now expressed each variable x_i as

$$x_i = l_{i1}F_1 + l_{i2}F_2 + \cdots + l_{im}F_m + e_i$$

for $i = 1, 2, \ldots, P$. In other words, we have transformed the principal components model to produce the factor model. For later use, note also that when the original variables are standardized, the factor loading l_{ij} turns out to be the *correlation* between x_i and F_j (see Section 13.5). Furthermore, it can be shown mathematically that the communality of x_i is $h_i^2 = l_{i1}^2 + l_{i2}^2 + \cdots + l_{im}^2$.

For example, in our hypothetical data set the eigenvalues of the correlation matrix (or Var C_i) are

$$\text{Var } C_1 = 2.578$$

$$\text{Var } C_2 = 1.567$$

$$\text{Var } C_3 = 0.571$$

$$\text{Var } C_4 = 0.241$$

$$\text{Var } C_5 = 0.043$$

$$\text{total} = 5.000$$

Note that the sum 5.0 is equal to P, the total number of variables. Based on the rule of thumb of selecting only those principal components corresponding to eigenvalues of 1 or more, we select $m = 2$. Using BMDP4M, we obtain the principal components analysis factor loadings, the l_{ij}'s, shown in Table 14.3. For example, the loading of x_1 on F_1 is $l_{11} = 0.511$ and on F_2 is $l_{12} = 0.782$. Thus the first equation of the factor model is

$$x_1 = 0.511 F_1 + 0.782 F_2 + e_1$$

Table 14.3 also shows the variance explained by each factor. For example, the variance explained by F_1 is 2.578, or 51.6%, of the total variance of 5.

The communality column, h_i^2, in Table 14.3 shows the part of the variance of each variable explained by the common factors. For example,

$$h_1^2 = 0.511^2 + 0.782^2 = 0.873$$

Finally, the specificity u_i^2 is the part of the variance not explained by the common factors. In this example we use standardized variables x_i whose variances are each equal to one. Therefore for each x_i, $u_i^2 = 1 - h_i^2$.

It should be noted that the sum of the communalities is equal to the cumulative part of the total variance explained by the common factors. In this example it is seen that

$$0.873 + 0.875 + 0.586 + 0.898 + 0.913 = 4.145$$

and $$2.578 + 1.567 = 4.145$$

thus verifying the above statement. Similarly, the sum of the specificities is equal to the total variance minus the sum of the communalities.

TABLE 14.3. Initial Factor Analysis Summary for Hypothetical Data Set from Principal Components Extraction Method

Variable	Factor Loadings		Communality	Specificity
	F_1	F_2	h_i^2	u_i^2
x_1	0.511	0.782	0.873	0.127
x_2	0.553	0.754	0.875	0.125
x_3	0.631	−0.433	0.586	0.414
x_4	0.866	−0.386	0.898	0.102
x_5	0.929	−0.225	0.913	0.087
Variance explained	2.578	1.567	$\Sigma h_i^2 = 4.145$	$\Sigma u_i^2 = 0.855$
Percentage	51.6%	31.3%	82.9%	17.1%

A valuable graphical aid to interpreting the factors is the *factor diagram* shown in Figure 14.1. Unlike the usual scatter diagrams where each point represents an individual case and each axis a variable, in this graph each point represents a response variable and each axis a common factor. For example, the point labeled 1 represents the response variable x_1. The coordinates of that point are the loadings of x_1 on F_1 and F_2, respectively. The other four points represent the remaining four variables.

Since the factor loadings are the correlations between the standardized variables and the factors, the range of values shown on the axes of the factor diagram is −1 to +1. It can be seen from Figure 14.1 that x_1 and x_2 load more on F_2 than they do on F_1. Conversely, x_3, x_4, and x_5 load more on F_1 than on F_2. However, these distinctions are not very clear-cut, and the technique of factor rotation will produce clearer results (see Section 14.7).

An examination of the correlation matrix shown in Table 14.2 confirms that x_1 and x_2 form a block of correlated variables. Similarly, x_4 and x_5 form another block, with x_3 also correlated to them. These two major blocks are only weakly correlated with each other. (Note that the correlation matrix for the standardized x's is the same as that for the unstandardized X's.)

FIGURE 14.1. Factor Diagram for Principal Components Extraction Method

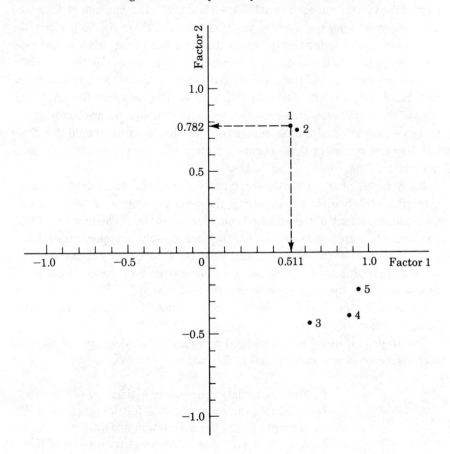

14.6 INITIAL FACTOR EXTRACTION: ITERATED PRINCIPAL COMPONENTS

The second method of extracting initial factors is a modification of the principal components analysis method. It has different names in different packages. It is called the principal factor analysis or PFA method in BMDP, PA2 or PAF in SPSS–X, and PRINIT in SAS.

To understand this method, you should recall that the communality is the part of the variance of each variable associated with the common factors. The principle underlying the *iterated solution* states that we should perform the factor analysis by using the communalities in place of the original variance. This principle entails substituting communality estimates for the 1s representing the variances of the standardized variables along the diagonal of the correlation matrix. With 1s in the diagonal we are factoring the total variance of the variables; with communalities in the diagonal we are factoring the variance associated with the common factors. Thus with communalities along the diagonal we select those common factors that maximize the total communality.

Many factor analysts consider maximizing the total communality a more attractive objective than maximizing the total proportion of the explained variance, as is done in the principal components method. The problem is that communalities are not known before the factor analysis is performed. Some initial estimates of the communalities must be obtained prior to the analysis. Various procedures exist, and we recommend, in the absence of a priori estimates, that the investigator use the default option in the particular program since the resulting factor solution is usually little affected by the initial communality estimates.

The steps performed by a packaged program in carrying out the *iterated factor extraction* are summarized as follows:

1. Find the initial communality estimates.
2. Substitute the communalities for the diagonal elements (1s) in the correlation matrix.
3. Extract m principal components from the modified matrix.
4. Multiply the principal components coefficients by the standard deviation of the respective principal components to obtain factor loadings.
5. Compute new communalities from the computed factor loadings.
6. Replace the communalities in step 2 with these new communalities and repeat steps 3, 4, and 5. This step constitutes an *iteration*.
7. Continue iterating, stopping when the communalities stay essentially the same in the last two iterations.

TABLE 14.4. Initial Factor Analysis Summary for Hypothetical Data Set from Iterated Principal Factor Extraction Method

Variable	Factor Loadings		Communality	Specificity
	F_1	F_2	h_i^2	u_i^2
x_1	0.470	0.734	0.759	0.241
x_2	0.510	0.704	0.756	0.244
x_3	0.481	−0.258	0.298	0.702
x_4	0.888	−0.402	0.949	0.051
x_5	0.956	−0.233	0.968	0.032
Variance explained	2.413	1.317	$\Sigma h_i^2 = 3.730$	$\Sigma u_i^2 = 1.270$
Percentage	48.3%	26.3%	74.6%	25.4%

For our hypothetical data example the results of using this method are shown in Table 14.4. In comparing this table with Table 14.3, we note that the total communality is higher for the principal components method (82.9% versus 74.6%). This result is generally the case since in the iterative method we are factoring the total communality, which is by necessity smaller than the total variance. The individual loadings do not seem to be very different in the two methods; compare Figures 14.1 and 14.2. The only apparent difference in loadings is in the case of the third variable. Note that Figure 14.2, the factor diagram for the iterated method, is constructed in the same way as Figure 14.1.

We point out that the factor loadings extracted by the iterated method depend on the number of factors extracted. For example, the loadings for the first factor would depend on whether we extract two or three common factors. This condition does not exist for the uniterated principal components method.

Regardless of the choice of the extraction method, if the investigator does not have a preconceived number of factors (m) from knowledge of the subject matter, then several methods are available for choosing one numerically. Reviews of these methods are given in Harmon (1976), Gorsuch (1974), and Thorndike (1978). In Section 14.4 the commonly used technique of including any factor whose eigenvalue is greater than or equal to 1 was used. This criterion is based on theoretical rationales developed using true population correlation coefficients. It is commonly thought to yield about one factor

FIGURE 14.2.   Factor Diagram for Iterated Principal Factor Extraction Method

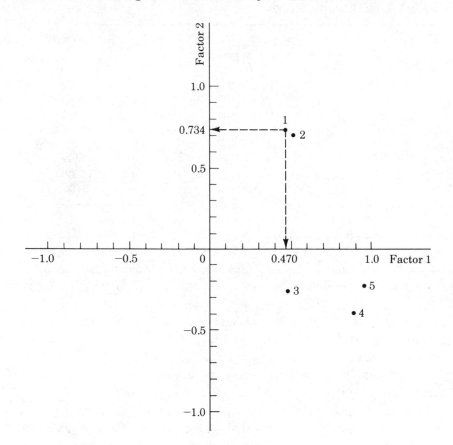

to every three to five variables. It appears to correctly estimate the number of factors when the communalities are high and the number of variables is not too large. As an alternative method, called the *scree method*, some investigators will simply plot the eigenvalues on the vertical axis versus the number of factors on the horizontal axis and look for the place where a change in the slope of the curve connecting successive points occurs. Examining the eigenvalues listed in Section 14.5, we see that compared with the first two eigenvalues (2.578 and 1.567), the values of the remaining eigenvalues are low (0.571, 0.241, and 0.043). This result indicates that $m = 2$ is a reasonable choice. The SAS FACTOR program will print a plot of the eigenvalues if the SCREE option is used.

In terms of initial factor extraction methods the iterated principal factor solution is the method employed most frequently by social scientists, the main users of factor analysis. For theoretical reasons many mathematical statisticians are more comfortable with the principal components method. Many other methods are available that may be preferred in particular situations (see Table 14.7 given later in the chapter). In particular, if the investigator is convinced that the factor model is valid and that the variables have a multivariate normal distribution, then the *maximum likelihood* (ML) *method* should be used. If these assumptions hold, then the ML procedure enables the investigator to perform certain tests of hypotheses or compute confidence intervals (see Lawley and Maxwell 1963).

14.7 FACTOR ROTATIONS

Recall that the main purpose of factor analysis is to derive from the data easily interpretable common factors. The initial factors, however, are often difficult to interpret. For the hypothetical data example we noted that the first factor is essentially an average of all variables. We also noted earlier that the factor interpretations in that example are not clear-cut. This situation is often the case in practice, regardless of the method used to extract the initial factors.

Fortunately, it is possible to find new factors whose loadings are easier to interpret. These new factors, called the *rotated factors*, are selected so that (ideally) some of the loadings are very large (near ±1) and the remaining loadings are very small (near zero). Conversely, we would ideally wish, for any given variable, that it have a high loading on only one factor. If this is the case, it is easy to give each factor an interpretation arising from the variables with which it is highly correlated (high loadings).

Theoretically, factor rotations can be done in an infinite number of ways. If you are interested in detailed descriptions of these methods, refer to some of the books listed in the Bibliography. In this section we highlight only two typical factor rotation techniques. The most commonly used technique is the *varimax rotation*. This method is the default option in most packaged programs and we present it first. The other technique, discussed later in this section, is called *oblique rotation*; we describe one commonly used oblique rotation, the *direct quartimin rotation*.

Varimax Rotation

We have reproduced in Figure 14.3 the principal component factors that were shown in Figure 14.1 so that you can get an intuitive feeling for what factor rotation does. As shown in Figure 14.3, factor rotation consists of finding

FIGURE 14.3.   Varimax Orthogonal Rotation for factor Diagram of Figure 14.1: Rotated Axes Go through Clusters of Variables and are Perpendicular to Each Other

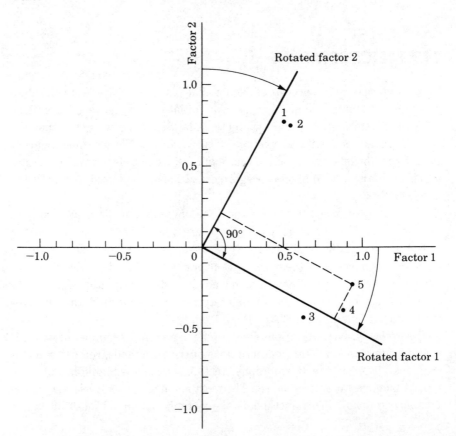

new axes to represent the factors. These new axes are selected so that they go through clusters or subgroups of the points representing the response variables. The *varimax procedure* further restricts the new axes to being orthogonal (perpendicular) to each other. Figure 14.3 shows the results of the rotation. Note that the new axis representing rotated factor 1 goes through the cluster of variables x_3, x_4, and x_5. The axis for rotated factor 2 is close to the cluster of variables x_1 and x_2 but cannot go through them since it is restricted to being orthogonal to factor 1.

The result of this varimax rotation is that variables x_1 and x_2 have high loadings on rotated factor 2 and nearly zero loadings on rotated factor 1. Similarly, x_3, x_4, and x_5 load heavily on rotated factor 1 but not on rotated factor 2. Figure 14.3 clearly shows that the rotated factors are orthogonal since the angle between the axes representing the rotated factors is 90°. In statistical terms the orthogonality of the rotated factors is equivalent to the fact that they are uncorrelated with each other.

Computationally, the *varimax rotation* is achieved by maximizing the sum of the variances of the squared factor loadings within each factor. Further, these factor loadings are adjusted by dividing each of them by the communality of the corresponding variable. This adjustment is known as the *Kaiser normalization* (see Harmon 1967 or Afifi and Azen 1979). This adjustment tends to equalize the impact of variables with varying communalities. If it were not used, the variables with higher communalities would highly influence the final solution.

Any method of rotation may be applied to any initially extracted factors. For example, in Figure 14.4 we show the varimax rotation of the factors initially extracted by the iterated principal factor method (Figure 14.2). Note the similarity of the graphs shown in Figures 14.3 and 14.4. This similarity is further reinforced by examination of the two sets of rotated factor loadings shown in Tables 14.5 and 14.6. In this example both methods of initial extraction produced a rotated factor 1 associated mainly with x_3, x_4, and x_5, and a rotated factor 2 associated with x_1 and x_2. The points in Figures 14.3 and 14.4 are plots of the rotated loadings (Tables 14.5 and 14.6) with respect to the rotated axes.

In comparing Tables 14.5 and 14.6 with Tables 14.3 and 14.4, we note that the *communalities are unchanged* after the varimax rotation. This result is always the case for any orthogonal rotation. Note also that the percentage of the variance explained by the rotated factor 1 is less than that explained by unrotated factor 1. However, the cumulative percentage of the

TABLE 14.5. Varimax Rotated Factors: *Principal Components Extraction*

| Variable | Factor Loadings* | | Communality |
	F_1	F_2	h_i^2
x_1	0.055	0.933	0.873
x_2	0.105	0.929	0.875
x_3	0.763	−0.062	0.586
x_4	0.943	0.095	0.898
x_5	0.918	0.266	0.913
Variance explained	2.328	1.817	4.145
Percentage	46.6%	36.3%	82.9%

* Vertical lines indicate large loadings.

TABLE 14.6. Varimax Rotated Factors: *Iterated Principal Factors Extraction*

| Variable | Factor Loadings* | | Communality |
	F_1	F_2	h_i^2
x_1	0.063	0.869	0.759
x_2	0.112	0.862	0.756
x_3	0.546	0.003	0.298
x_4	0.972	0.070	0.949
x_5	0.951	0.251	0.968
Variance explained	2.164	1.566	3.730
Percentage	43.3%	31.3%	74.6%

* Vertical lines indicate large loadings.

FIGURE 14.4. Varimax Orthogonal Rotation for Factor Diagram of Figure 14.2

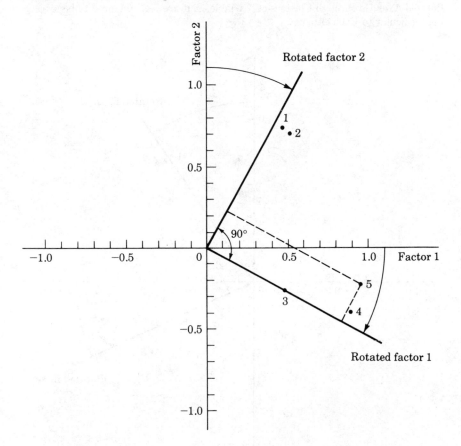

variance explained by all common factors remains the same after orthogonal rotation. Furthermore, the loadings of any rotated factor depend on how many other factors are selected, regardless of the method of initial extraction.

Oblique Rotation

Some factor analysts are willing to relax the restriction of orthogonality of the rotated factors, thus permitting a further degree of flexibility. Nonorthogonal rotations are called *oblique rotations*. The origin of the term

FIGURE 14.5.   Direct Quartimin Oblique Rotation for Factor Diagram of Figure 14.1:
Rotated Axes Go through Clusters of Variables But are Not Required to be
Perpendicular to Each Other

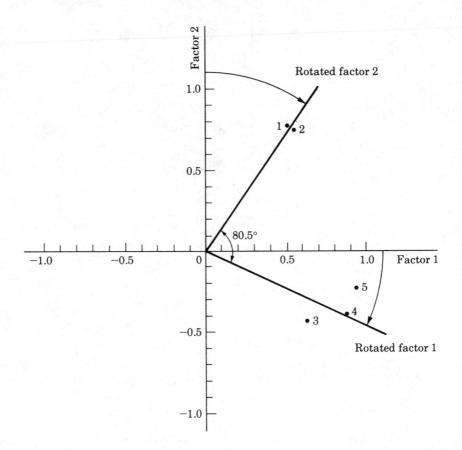

oblique lies in geometry, whereby two crossing lines are called oblique if they
are not perpendicular to each other. Oblique rotated factors are correlated
with each other, and in some applications it may be harmless or even desirable to have correlated common factors (see Mulaik 1972).

A commonly used oblique rotation is called the *direct quartimin procedure*. The results of applying this method to our hypothetical data example
are shown in Figures 14.5 and 14.6. Note that the rotated factors go neatly

FIGURE 14.6. Direct Quartimin Oblique Rotation for Factor Diagram of Figure 14.2

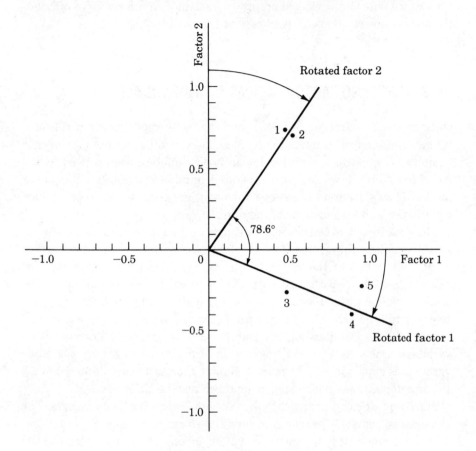

through the centers of the two variable clusters. The angle between the rotated factors is not 90°. In fact, the cosine of the angle between the two factor axes is the sample correlation between them. For example, for the rotated principal components the correlation between rotated factor 1 and rotated factor 2 is 0.165, which is the cosine of 80.5°.

In general, a factor analysis computer program will print the correlations between rotated factors if an oblique rotation is performed. The rotated oblique factor loadings also depend on how many factors are selected. Finally, we note that packaged programs generally offer a choice of orthogonal and oblique rotations, as discussed in Section 14.10.

14.8 ASSIGNING FACTOR SCORES TO INDIVIDUALS

Once the initial extraction of factors and the factor rotations are performed, it may be of interest to obtain the score an individual has for each factor. For example, if two factors, verbal and quantitative abilities, were derived from a set of test scores, it would be desirable to determine equations for computing an individual's score on these two factors from a set of test scores. Such equations are linear functions of the original variables.

Theoretically, it is conceivable to construct factor score equations in an infinite number of ways (see Gorsuch 1974). But perhaps the simplest way is to add the values of the variables loading heavily on a given factor. For example, for the hypothetical data discussed earlier we would obtain the score of rotated factor 1 as $x_3 + x_4 + x_5$ for a given individual. Similarly, the score of rotated factor 2 in that example would be $x_1 + x_2$. In effect, the factor analysis identifies x_3, x_4, and x_5 as a subgroup of intercorrelated variables. One way of combining the information conveyed by the three variables is simply to add them up. A similar approach is used for x_1 and x_2. In some applications such a simple approach may be sufficient.

More sophisticated approaches do exist for computing factor scores. The three major computer packages include the so-called *regression procedure*. This method combines the intercorrelations among the x_i variables and the factor loadings to produce quantities called *factor score coefficients*. These coefficients are used in a linear fashion to combine the values of the *standardized* x_i's into factor scores. For example, in the hypothetical data example using principal component factors rotated by the varimax method (see Table 14.5), the scores for rotated factor 1 are obtained as

$$\text{factor score } 1 = -0.076x_1 - 0.053x_2 + 0.350x_3 + 0.414x_4 + 0.384x_5$$

The large factor score coefficients for x_3, x_4, and x_5 correspond to the large factor *loadings* shown in Table 14.5. Note also that the coefficients of x_1 x_2 are close to zero and those of x_3, x_4, and x_5 are each approximately and 0.4. Factor score 1 can therefore be approximated by $0.4x_3 + 0.4x_4$

$+\ 0.4x_5 = 0.4(x_3 +\ x_4 + x_5)$, which is proportional to the simple additive factor score given in the previous paragraph.

The factor scores can themselves be used as data for additional analyses. The major packages facilitate this technique by offering the user the option of storing the factor scores in a file to be used for subsequent analyses, as was discussed in Section 13.6.

14.9 AN APPLICATION OF FACTOR ANALYSIS TO THE DEPRESSION DATA

In Chapter 13 we presented the results of a principal components analysis for the 20 CESD items in the depression data set. Table 13.2 listed the coefficients for the first five principal components. However, to be consistent with published literature on this subject, we now choose $m = 4$ factors (not 5) and proceed to perform an orthogonal varimax rotation using the principal components as the method of factor extraction. The results for the rotated factors are given in Table 14.7.

In comparing these results with those in Table 13.1, we note that, as expected, rotated factor 1 explains less of the total variance than does the first principal component. The four rotated factors together explain the same proportion of the total variance as do the first four principal components (54.2%). This result occurs because the unrotated factors were the principal components.

In interpreting the factors, we see that factor 1 loads heavily on variables 1 through 7. These items are known as negative-affect items. Factor 2 loads mainly on items 12 through 18, which measure somatic and retarded activity. Factor 3 loads on the two interpersonal items, 19 and 20, as well as some positive-affect items (9 and 11). Factor 4 does not represent a clear pattern. Its highest loading is associated with the positive-affect item 8. The factors can thus be loosely identified as negative affect, somatic and retarded activity, interpersonal relations, and positive affect, respectively.

Overall, these rotated factors are much easier to interpret than the original principal components. This example is an illustration of the usefulness of factor analysis in identifying important interrelations among measured variables. Further results and discussion regarding the application of factor analysis to the CESD scale my be found in Radloff (1977) and Clark et al. (1981).

TABLE 14.7 Varimax Rotation, Principal Component Factors for Standardized CESD · Scale Items (Depression Data Set)

Item	F_1	F_2	F_3	F_4	Communalities h_i^2
Negative Affect					
1. I felt that I could not shake off the blues even with the help of my family or friends.	0.638	0.146	0.268	0.280	0.5784
2. I felt depressed.	0.773	0.296	0.272	-0.003	0.7598
3. I felt lonely.	0.726	0.054	0.275	0.052	0.6082
4. I had crying spells.	0.630	-0.061	0.168	0.430	0.6141
5. I felt sad.	0.797	0.172	0.160	0.016	0.6907
6. I felt fearful.	0.624	0.234	-0.018	0.031	0.4448
7. I thought my life had been a failure.	0.592	0.157	0.359	0.337	0.6173
Positive Affect					
8. I felt that I was as good as other people.	0.093	-0.051	0.109	0.737	0.5655
9. I felt hopeful about the future.	0.238	0.033	0.621	0.105	0.4540
10. I was happy.	0.557	0.253	0.378	0.184	0.5516
11. I enjoyed life.	0.498	0.147	0.407	0.146	0.4569
Somatic and Retarded activity					
12. I was bothered by things that usually don't bother me.	0.449	0.389	-0.049	-0.065	0.3600
13. I did not feel like eating; my appetite was poor.	0.070	0.504	-0.173	0.535	0.5760
14. I felt that everything was an effort.	0.117	0.695	0.180	0.127	0.5459
15. My sleep was restless.	0.491	0.419	-0.123	-0.089	0.4396
16. I could not "get going."	0.196	0.672	0.263	-0.070	0.5646
17. I had trouble keeping my mind on what I was doing.	0.270	0.664	0.192	0.000	0.5508
18. I talked less than usual.	0.409	0.212	-0.026	0.223	0.2628
Interpersonal					
19. People were unfriendly.	-0.015	0.237	0.746	-0.088	0.6202
20. I felt that people disliked me.	0.358	0.091	0.506	0.429	0.5770
Variance explained	4.795	2.381	2.111	1.551	10.838
Percentage	24.0%	11.9%	10.6%	7.8%	54.2%

14.10 DISCUSSION OF COMPUTER PROGRAMS

In this section we present the main features of the three packaged programs, and we illustrate their use with an example.

Features of Packaged Programs

Each of the three packages has a major factor analysis program. The three programs offer a large variety of options, not all the same. Table 14.8 summarizes the most important options offered by each package, including some not discussed in the text here. If you are interested in those extra options, consult the corresponding package manual and the references listed in the bibliography.

Each program can factor-analyze the correlation matrix, while BMDP4M and SAS FACTOR can analyze other matrices. The three packages agree on two methods of initial factor extraction, with each program offering other methods. In estimating the diagonal elements (i.e., communalities), each program offers the option of leaving the diagonal element unaltered or accepting user-supplied estimates. The BMDP4M program and SAS FAC-TOR allow estimation of them by other methods. Each of the programs allows the user to specify the number of factors or specify that a factor should be selected only if its corresponding eigenvalue exceeds a certain value. In addition, SAS FACTOR allows selection of the number of factors to explain a certain cumulative proportion of the total variance if the PROPOR-TION option is used.

For orthogonal rotation each program allows the use of varimax, equimax, and quartimax. The latter two rotation methods are discussed in Gorsuch (1974). For oblique rotations each program has its own methods. In addition, BMDP4M offers two other rotation methods, one orthogonal and one that combines orthogonal and oblique ideas.

Finally, each of the programs can produce factor scores to be used in future analysis and can print, upon request, equations for computing these scores.

Example

To illustrate some typical output for factor analysis, we used the BMDP4M factor analysis program to produce the results shown in Tables 14.3 and

14.5. For data we generated normal random deviates. The data were generated within the computer by the BMDP TRANSFORM paragraph. The following paragraphs were used:

```
/PROBLEM        TITLE IS 'PCA PHONY'.
/INPUT          VAR = 0. CASES = 100.
/VARIABLE       ADD = 5.
                NAMES = NORM1,NORM2,NORM3,NORM4,NORM5.
/TRANSFORM      NORM1 = RNDG(46791).
                NORM2 = RNDG(56705).
                NORM3 = RNDG(66717).
                NORM4 = RNDG(77729).
                NORM5 = RNDG(37255).
                NORM2 = NORM1 + NORM2.
                NORM4 = NORM3 + 2*NORM4 + 0.3*NORM2.
                NORM5 = NORM4 + 0.7*NORM5 + 0.3*NORM1.
/PRINT          FSFC.
/PLOT           INIT = 2.
/END
```

In the INPUT paragraph we are reading data on 100 cases but no variables. In effect, we are reserving space for 100 cases in the computer. In the VARIABLE paragraph we add five variables and give them names.

In the TRANSFORM paragraph we instruct the computer to create five new variables by using the built-in normal random generator. Each random number requires a 5-digit number to start it off that must end in an odd digit. The other three statements modify these created variables to produce variables with the desired correlation structure. Otherwise, the five variables would be mutually independent and no further factor analysis would be necessary. No more new variables have to be added to the VARIABLE paragraph because the same names are used again. The variable is simply redefined.

If the investigator does not generate the data, the only paragraphs that are essential in order to run BMDP4M are the INPUT paragraph and the END paragraph. With the default options the initial factor extraction is done by the method of principal components and a varimax rotation is performed. The investigator is urged to use these default options for the first run in order to save time and to obtain some output that can be examined as an aid in planning other runs.

TABLE 14.8. Summary of Computer Output for Factor Analysis

Output	BMDP	SAS	SPSS–X
Matrix Analyzed			
Correlation	BMDP4M	FACTOR	FACTOR
Covariance	BMDP4M	FACTOR	
Correlation about origin	BMDP4M	FACTOR	
Covariance about origin	BMDP4M	FACTOR	
Method of Initial Extraction			
Principal component	BMDP4M	FACTOR	FACTOR
Iterative principal factor	BMDP4M	FACTOR	FACTOR
Maximum likelihood	BMDP4M	FACTOR	FACTOR
Alpha		FACTOR	FACTOR
Image		FACTOR	FACTOR
Least squares			FACTOR
Kaiser little jiffy	BMDP4M		
Communality Estimates			
Unaltered diagonal	BMDP4M	FACTOR	FACTOR
Multiple correlation squared	BMDP4M	FACTOR	FACTOR
Maximum correlation in row	BMDP4M	FACTOR	
Estimates, user-supplied	BMDP4M	FACTOR	FACTOR
Number of Factors			
Specify number	BMDP4M	FACTOR	FACTOR
Specify minimum eigenvalue	BMDP4M	FACTOR	FACTOR
Cumulative eigenvalues		FACTOR	
Orthogonal Rotations			
Varimax	BMDP4M	FACTOR	FACTOR
Equamax	BMDP4M	FACTOR	FACTOR
Quartimax	BMDP4M	FACTOR	FACTOR
Orthomax with gamma		FACTOR	
Orthogonal with gamma	BMDP4M		
Promax		FACTOR	
Oblique Rotations			
Direct quartimin	BMDP4M		
Promax		FACTOR	
With gamma	BMDP4M		
Orthogonal oblique	BMDP4M		
Procrustes		FACTOR	
Direct oblimin			FACTOR
Factor Scores			
Factor scores for individuals	BMDP4M	FACTOR	FACTOR
Factor score coefficients	BMDP4M	FACTOR	FACTOR
Other Output			
Inverse correlation matrix	BMDP4M	FACTOR	FACTOR
Factor structure matrix	BMDP4M	FACTOR	
Plots of factor loadings	BMDP4M	FACTOR	
Scree		FACTOR	FACTOR
Tests of Hypotheses			
Factors sufficient to explain correlation (if initial factor method is ML)		FACTOR	
Mahalanobis distances divided by df (to detect outliers)	BMDP4M		

The output from the run that is necessary to interpret the factor analysis is described next. The program prints the first five cases so that the investigator can check that it is reading the proper data. The mean, standard deviation, smallest and largest values, and smallest and largest standardized scores are printed for each variable. If the data are normally distributed or symmetric, the absolute value of the smallest and largest standardized scores (each observation minus its mean and then divided by its standard deviation, done separately for each variable) should be approximately equal. Absolute values of standardized data that are greater than 4 are quite unusual if the data are normally distributed. Note that it is assumed that the data are normally distributed if tests of hypotheses are to be made or if the maximum likelihood method is to be used.

Next on the computer printout are the correlation matrix given in Table 14.2, the squared multiple correlation (SMC) of each variable with the remaining variables, and the estimated communalities listed in Table 14.3. If any variable has an extremely high squared multiple correlation with all other variables (greater than 0.9999), this variable can probably be expressed as an almost exact linear function of some of the other variables, and the investigator should consider removing it or some other variable. The variance explained is given as listed in Section 14.5, and the cumulative proportion of the total variance is also listed (variance divided by P and cumulated). The unrotated factor loadings (Table 14.3 and Figure 14.1) are presented, followed by the rotated factor loadings as given in Figure 14.3.

In our example, we asked for factor score coefficients (FSFC) to be printed in the PRINT paragraph. These coefficients were presented for the first factor in Section 14.8 to illustrate that they were not equal to the factor loadings.

In the plot paragraph we requested a plot of the initial or unrotated factors, which was presented in Figure 14.1. The rotated factors, plotted as a default option, were displayed in Figure 14.3.

The two factor scores are printed out for each of the 100 cases. For example, for the first case and the first factor, the factor score is

$$-0.488 = -0.076x_1 - 0.053x_2 + 0.350x_3 + 0.414x_4 + 0.384x_5$$

For the first case and the first standardized variable,

$$x_1 = \frac{0.666 - 0.163}{1.047}$$

where 0.666 is the value of x_1 for the first case, 0.163 is the mean for the first variable, and 1.047 is the standard deviation for the first variable (see Table 14.1). The values of x_2, x_3, x_4, and x_5 are obtained in a similar fashion and then substituted to produce factor scores. These factor scores could be employed in subsequent analyses by using the SAVE paragraph with the default option (see Section 13.6).

The program also prints out three Mahalanobis distances divided by their degrees of freedom. These distances measure the following:

1. The distance from each case to the mean of all the cases.

2. The distance from each factor score to the mean of the factor scores.

3. The differences between these two distances adjusted for the appropriate degrees of freedom.

They are distributed approximately according to a chi-square distribution divided by its degrees of freedom if the original distribution is multivariate normal. The investigator may wish to examine values larger than 12.1 for 1 df, 7.6 for 2 df, 5.9 for 3 df, 5.0 for 4 df, 4.4 for 5 df, 4.0 for 6 df, 3.7 for 7 df, 3.5 for 8 df, and 3.1 for 10 df. These correspond to the 99.95 percentile of the chi-square distribution after dividing by the degrees of freedom. Doing so is useful in detecting multivariate outliers.

In the generated data used in this example, the largest distance from each case to the mean of all cases is 3.926 for case 85. The output indicates that this value should be compared with a chi-square divided by degrees of freedom with 5 degrees of freedom. It happens that case 85 has values of $x_1 = -1.304$, $x_2 = -0.949$, $x_3 = 0.689$, $x_4 = 2.457$, and $x_5 = 4.521$, none of which are the most extreme values when the variables are examined one at a time. The distance for the factor scores is only 1.911 for this case, so it does not result in an unusual factor score.

The simplest way of checking for outliers among the factor scores is to examine the plot of the factor scores. Figure 14.7 is taken directly from the BMDP4M output and illustrates such a plot. If an unusual point has been found, the listing of the factor scores can be used to determine which case it is. Note that the factor score plot in Figure 14.7 is typical of a bivariate normal plot with zero correlation; this data set was generated by using

FIGURE 14.7. Plot of Factor Score 1 Versus Factor Score 2 for the Hypothetical Data Set

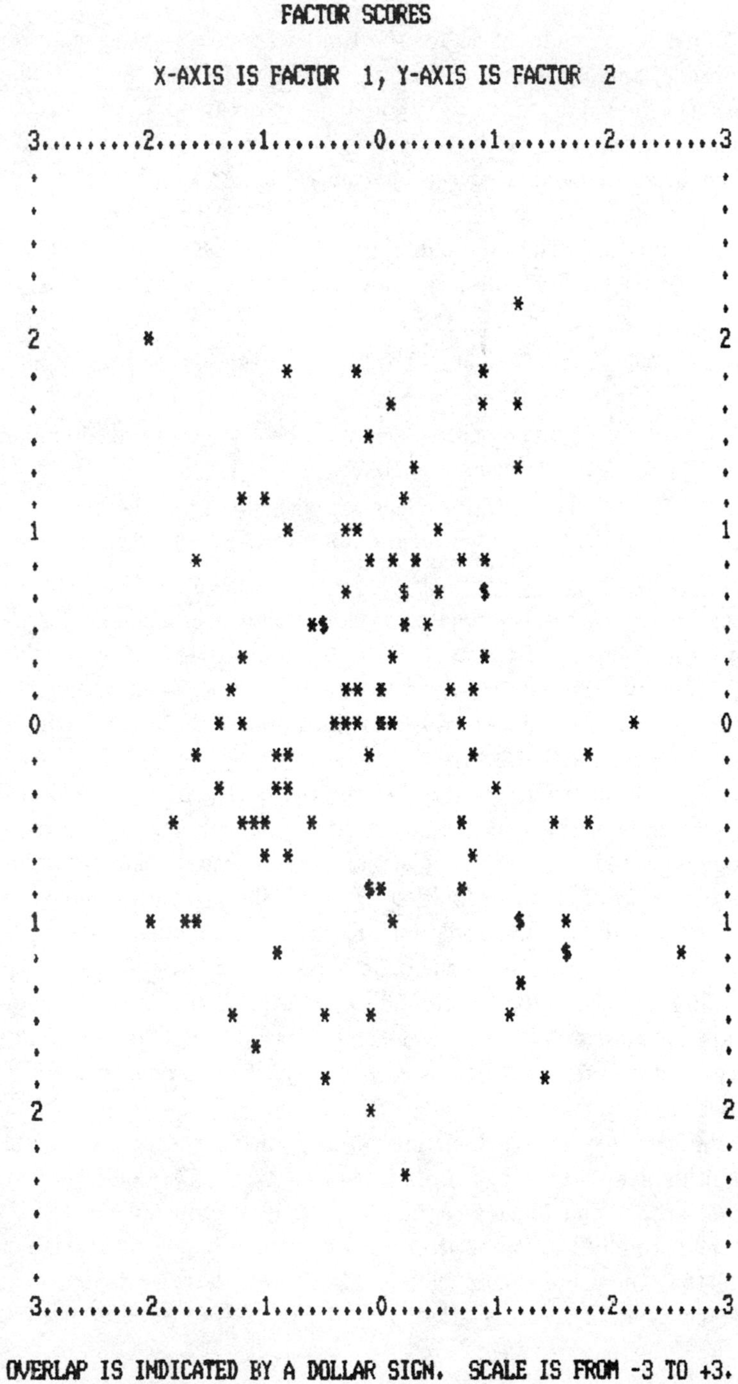

FACTOR SCORES

X-AXIS IS FACTOR 1, Y-AXIS IS FACTOR 2

OVERLAP IS INDICATED BY A DOLLAR SIGN. SCALE IS FROM -3 TO +3.

random normal deviates. Factor 1 scores range in value from −1.98 to 2.60, and factor 2 scores from −2.27 to 2.21.

To obtain Tables 14.4 and 14.6 where the iterated principal factor extraction method is used instead of the default option, the following FACTOR paragraph is added after the TRANSFORM paragraph:

```
/FACTOR          METHOD = PFA.
```

To obtain an oblique rotation such as the direct quartimin procedure, a ROTATE paragraph can be added, as follows:

```
/ROTATE          METHOD = DQUART.
```

SUMMARY

In this chapter we presented the most essential features of factor analysis, a technique heavily used in social science research. The major phases of a factor analysis are factor extraction, factor rotation, and factor score computation. We gave examples of its use for the depression data set.

A fact that should be remembered by investigators is that factor analysis is, in reality, an exploratory technique. Its main usefulness is to give the investigator an impression of the interrelationships present in the data and offer a preliminary method for summarizing them. Factor analysis often suggests certain hypotheses to be further examined in future research. On the one hand, the investigator should not hesitate to replace the results of any factor analysis by scientifically based theories derived at a later time. On the other hand, in the absence of such theory, factor analysis is a convenient and useful tool for searching for relationships among variables.

Since factor analysis is an exploratory technique, there does not exist an optimum way of performing it. We advise the investigator to try various combinations of extraction and rotation of main factors. Subjective judgment is then needed to select those results that seem most appealing, on the basis of the investigator's own knowledge of the underlying subject matter.

BIBLIOGRAPHY

Afifi, A. A., and Azen, S. P. 1979. *Statistical analysis: A computer oriented approach*. 2nd ed. New York: Academic Press.

Clark, V. A.; Aneshensel, C. S.; Frerichs, R. R.; and Morgan, T. M. 1981. Analysis of effects of sex and age in response to items on the CES-D scale. *Psychiatry Research* 5 : 171–181.

*Gorsuch, R. L. 1974. *Factor analysis*. Philadelphia: Saunders.

*Harmon, H. H. 1976. *Modern factor analysis*. Chicago: University of Chicago Press.

Kim, J. O., and Mueller, C. W. 1978*a*. *Introduction to factor analysis*. Beverly Hills: Sage.

————. 1978*b*. *Factor analysis*. Beverly Hills: Sage.

*Lawley, D. N., and Maxwell, A. E. 1963. *Factor analysis as a statistical method*. London: Butterworths.

Mulaik, S. A. 1972. *The foundations of factor analysis*. New York: McGraw-Hill.

Radloff, L. S. 1977. The CES-D scale: A self-report depression scale for research in the general population. *Applied Psychological Measurement* 1 : 385–401.

Thorndike, R. M. 1978. *Correlational procedures for research*. New York: Gardner Press.

PROBLEMS

14.1 The CESD scale items (C1–C20) from the depression data set in Chapter 3 were used to obtain the factor loadings listed in Table 14.7. The initial factor solution was obtained from the principal components method, and a varimax rotation was performed. Analyze this same data set by using an oblique rotation such as the direct quartimin procedure. Compare the results.

14.2 Repeat the analysis of Problem 14.1 and Table 14.7, but use an iterated principal factor solution instead of the principal components method. Compare the results.

14.3 Using the sentences given in the TRANSFORM paragraph in Section 14.10, which were used to generate the hypothetical data set, explain why the correlations given in Table 14.2 were obtained.

14.4 If a BMDP program is used with the same TRANSFORM paragraph (including the same numbers for each random normal deviate generated) given in Section 14.10, then the same data should result. Run BMDP4M with these data but use the maximum likelihood method of extracting factors. Examine the output to determine whether this method leads to different results.

14.5 For the data generated in Problem 7.7, perform four factor analyses, using two different initial extraction methods and both orthogonal and oblique rotations. Interpret the results.

15 | CANONICAL CORRELATION ANALYSIS

15.1 WHAT WILL YOU LEARN FROM THIS CHAPTER?

From this chapter you will learn how to analyze data when you have a set of X variables and more than one Y variable. In particular, you will learn:

▶ When to use canonical correlation analysis (15.2, 15.3).

▶ About the basic concepts used in analyzing linear relationships between two sets of data (15.4).

▶ How to interpret the linear combinations obtained in canonical correlation analysis (15.4).

▶ How to decide how many canonical correlations to obtain (15.4).

▶ About other methods for analyzing this type of problem (15.5).

▶ About the options available in the three statistical packages (15.6).

15.2 WHEN IS CANONICAL CORRELATION ANALYSIS USED?

The technique of *canonical correlation analysis* is best understood by considering it as an extension of multiple regression and correlation analysis. In multiple regression analysis we find the best linear combination of P variables, $X_1, X_2, \ldots, X_P$, to predict one variable Y. The multiple correlation coefficient is the simple correlation between Y and its predicted value $\hat{Y}$. In multiple regression and correlation analysis our concern was therefore to examine the relationship between the X variables and the single Y variable.

In canonical correlation analysis we examine the linear relationships between a set of X variables and a set of more than one Y variable. So the only difference is that we now have more than one Y variable. The technique consists of finding several linear combinations of the X variables and the same number of linear combinations of the Y variables in such a way that these linear combinations best express the correlations between the two sets. Those linear combinations are called the *canonical variables*, and the correlations between corresponding pairs of canonical variables are called *canonical correlations*.

In a common application of this technique the Y's are interpreted as outcome or dependent variables, while the X's represent independent or predictive variables. The Y variables may be harder to measure than the X variables, as in the calibration situation discussed in Section 6.10.

Canonical correlation analysis applies to situations in which regression techniques are appropriate and where there exists more than one dependent variable. Another useful application is for testing independence between the sets of Y and X variables. This application will be discussed further in Section 15.4.

An example of an application is given by Waugh (1942); he studied the relationship between characteristics of certain varieties of wheat and characteristics of the resulting flour. Waugh was able to conclude that desirable wheat was high in texture, density, and protein content, and low on damaged kernels and foreign materials. Similarly, good flour should have high crude protein content and low scores on wheat per barrel of flour and ash in flour. Canonical correlation has also been used in psychology by Meredith (1964) to calibrate two sets of intelligence tests given to the same individuals. In addition, it has been used to relate linear combinations of personality scales to linear combinations of achievement tests (Tatsuoka 1971). Hopkins (1969) discusses several health-related applications of canonical correlation analysis, including, for example, a relationship between illness and housing quality.

Canonical correlation analysis is one of the less commonly used multivariate techniques. Its limited use may be due, in part, to the difficulty often encountered in trying to interpret the results. Also, prior to the advent of computers the calculations seemed forbidding.

15.3 DATA EXAMPLE

The depression data set presented in previous chapters is used again here to illustrate canonical correlation analysis. We select two dependent variables, CESD and health. The variable CESD is the sum of the scores on the 20 depression scale items; thus a high score indicates likelihood of depression. Likewise, "health" is a rating scale going from 1 to 4, where 4 signifies poor health and 1 signifies excellent health. The set of independent variables includes "sex," transformed so that 0 = male and 1 = female; "age" in years; "education," from 1 = less than high school up to 7 = doctorate; and "income" in thousands of dollars per year.

The summary statistics for the data are given in Tables 15.1 and 15.2. Note that the average score on the depression scale (CESD) is 8.9 in a possible range of 0 to 60. The average on the health variables is 1.8, indicating an average perceived health level falling between excellent and good. The average educational level of 3.5 shows that an average person has finished high school and perhaps attended some college.

Examination of the correlation matrix in Table 15.2 shows that neither CESD nor health is highly correlated with any of the independent variables. In fact, the highest correlation in this matrix is between education and income. Also, CESD is negatively correlated with age, education, and income (the younger, less educated, and lower-income person tends to be more depressed). The positive correlation between CESD and sex shows that females tend to be more depressed than males. Persons who perceived their health as good are more apt to be high on income and education and low on age.

TABLE 15.1. Means and Standard Deviations for Depression Data Set

Variable	Mean	Standard Deviation
CESD	8.88	8.82
Health	1.77	0.84
Sex	0.62	0.49
Age	44.41	18.09
Education	3.48	1.31
Income	20.57	15.29

TABLE 15.2. Correlation Matrix for Depression Data Set

	CESD	Health	Sex	Age	Education	Income
CESD	1	0.212	0.124	−0.164	−0.101	−0.158
Health		1	0.098	0.308	−0.270	−0.183
Sex			1	0.044	−0.106	−0.180
Age				1	−0.208	−0.192
Education					1	0.492
Income						1

In the following sections we will examine the relationship between the dependent variables (perceived health and depression) and the set of independent variables.

15.4 BASIC CONCEPTS OF CANONICAL CORRELATION

Suppose we wish to study the relationship between a set of variables $x_1, x_2, \ldots, x_P$ and another set $y_1, y_2, \ldots, y_Q$. The x variables can be viewed as the independent or predictor variables, while the y's are considered dependent or outcome variables. We assume that in any given sample the mean of each variable has been subtracted from the original data so that the sample means of all x and y variables are zero. In this section we discuss how the degree of association between the two sets of variables is assessed, and we present some related tests of hypotheses.

First Canonical Correlation

The basic idea of canonical correlation analysis begins with finding one linear combination of the y's, say

$$U_1 = a_1 y_1 + a_2 y_2 + \cdots + a_Q y_Q$$

and one linear combination of the x's, say

$$V_1 = b_1 x_1 + b_2 x_2 + \cdots + b_P x_P$$

For any particular choice of the coefficients, the a's and the b's, we can compute values of U_1 and V_1 for each individual in the sample. From the N

TABLE 15.3. Canonical Correlation Coefficients for First Correlation (Depression Data Set)

Coefficients	Standardized Coefficients
$b_1 = 0.051$ (sex) $a_1 = -0.055$ (CESD)	$b_1 = 0.025$ (sex) $a_1 = -0.490$ (CESD)
$b_2 = 0.048$ (age) $a_2 = 1.17$ (health)	$b_2 = 0.871$ (age) $a_2 = +0.982$ (health)
$b_3 = -0.29$ (education)	$b_3 = -0.383$ (education)
$b_4 = +0.005$ (income)	$b_4 = 0.082$ (income)

individuals in the sample we can then compute the simple correlation between the N pairs of U_1 and V_1 values in the usual manner. The resulting correlation depends on the choice of the a's and the b's.

In canonical correlation analysis we select values of a and b coefficients so as to *maximize* the correlation between U_1 and V_1. With this particular choice the resulting linear combination U_1 is called the *first canonical variable* of the y's and V_1 is called the *first canonical variable* of the x's. Note that both U_1 and V_1 have a mean of zero. The resulting correlation between U_1 and V_1 is called the *first canonical correlation*.

The first canonical correlation is thus the highest possible correlation between a linear combination of the x's and a linear combination of the y's. In this sense it is the maximum linear correlation between the set of x variables and the set of y variables. The first canonical correlation is analogous to the multiple correlation coefficient between a single Y variable and the set of X variables. The difference is that in canonical correlation analysis we have several y variables and we must find a linear combination of them also.

The BMDP6M program computed the coefficients (a's and b's) as shown in Table 15.3. Note that the first set of coefficients is used to compute the values of the canonical variables U_1 and V_1.

Table 15.4 shows the process used to compute the canonical correlation. For each individual we compute V_1 from the b coefficients and the individual's X variable values after subtracting the means. We do the same for U_1. These computations are shown for the first three individuals. The correlation coefficient is then computed from the 294 values of U_1 and V_1. Note that the variances of U_1 and V_1 are each equal to 1.

The standardized coefficients are also shown in Table 15.3, and they are to be used with the standardized variables. The standardized coefficients can be

TABLE 15.4. Computation of Correlation Between U_1 and V_1

Individual	Sex $b_1(X_1 - \bar{X}_1)$	Age $+ \ b_2(X_2 - \bar{X}_2)$	Education $+ \ b_3(X_3 - \bar{X}_3)$	Income $+ \ b_4(X_4 - \bar{X}_4) \rightarrow$
1	$0.051(1 - 0.62)$	$+ \ 0.048(68 - 44.4)$	$- \ 0.29(2 - 3.48)$	$+ \ 0.0054(4 - 20.57)$
2	$0.051(0 - 0.62)$	$+ \ 0.048(58 - 44.4)$	$- \ 0.29(4 - 3.48)$	$+ \ 0.0054(15 - 20.57)$
3	$0.051(1 - 0.62)$	$+ \ 0.048(45 - 44.4)$	$- \ 0.29(3 - 3.48)$	$+ \ 0.0054(28 - 20.57)$
.				
.				
.				
294				

$\rightarrow V_1$	U_1	CESD $\leftarrow \ a_1(Y_1 - \bar{Y}_1)$	Health $a_2(Y_2 - \bar{Y}_2)$
$= 1.49$	$0.76 =$	$-0.055(0 - 8.88)$	$+ \ 1.17(2 - 1.77)$
$= 0.44$	$-0.64 =$	$-0.055(4 - 8.88)$	$+ \ 1.17(1 - 1.77)$
$= 0.23$	$0.54 =$	$-0.055(4 - 8.88)$	$+ \ 1.17(2 - 1.77)$

Correlation between U_1 and V_1 = canonical correlation = 0.405

obtained by multiplying each standardized coefficient by the standard deviation of the corresponding variable. For example, the unstandardized coefficient of y_1 (CESD) is $a_1 = -0.0555$, and from Table 15.1 the standard deviation of y_1 is 8.82. Therefore the standardized coefficient of y_1 is -0.490.

In this example the resulting canonical correlation is 0.405. This value represents the highest possible correlation between any linear combination of the independent variables and any linear combination of the dependent variables. In particular, it is larger than any simple correlation between an x variable and a y variable (see Table 15.2). One method for interpreting the linear combination is by examining the standardized coefficients. For the x variables the canonical variable is determined largely by age and education. Thus a person who is relatively old and relatively uneducated would score high on canonical variable V_1. The canonical variable based on the y's gives a large positive weight to the perceived health variables and a negative weight to CESD. Thus a person with a high health value (perceived poor health) and a low depression score would score high on canonical variable U_1. In contrast, a young person with relatively high education would score low on V_1, and a person in good perceived health but relatively depressed would score low on U_1. Sometimes, because of high intercorrelations between two variables in the same set, one variable may result in another having a small coefficient (see Levine 1977) and thus make the interpretation difficult. No very high correlations within a set existed in the present example.

In summary, we conclude that older but uneducated people tend to be relatively undepressed although they perceive their health as relatively poor. Because the first canonical correlation is the largest possible, this impression is the strongest conclusion we can make from this analysis of the data. However, there may be other important conclusions to be drawn from the data, which will be discussed next.

Other Canonical Correlations

Additional interpretation of the relationship between the x's and the y's is obtained by deriving other sets of canonical variables and their corresponding canonical correlations. Specifically, we derive a second canonical variable V_2 (linear combination of the x's) and a corresponding canonical variable U_2 (linear combination of the y's). The coefficients for these linear combinations are chosen so that the following conditions are met:

1. V_2 is uncorrelated with V_1 and U_1.
2. U_2 is uncorrelated with V_1 and U_1.
3. Subject to conditions 1 and 2, U_2 and V_2 have the maximum possible correlation.

The correlation between U_2 and V_2 is called the *second canonical correlation* and will necessarily be less than or equal to the first canonical correlation.

In our example the second set of canonical variables expressed in terms of the standardized coefficients is

$$V_2 = 0.396(\text{sex}) - 0.443(\text{age}) - 0.448(\text{education}) - 0.555(\text{income})$$

and $U_2 = 0.899(\text{CESD}) + 0.288(\text{health})$

Note that U_2 gives a high positive weight to CESD and a low positive weight to health. In contrast, V_2 gives approximately the same moderate weight to all four variables, with sex having the only positive coefficient. A large value of V_2 is associated with young, poor, uneducated females. A large value of U_2 is associated with a high value of CESD (depressed) and to a lesser degree with a high value of health (poor perceived health). The value of the second canonical correlation is 0.266.

In general, this process can be continued to obtain other sets of canonical variables U_3, V_3; U_4, V_4; etc. The maximum number of canonical correlations and their corresponding sets of canonical variables is equal to the minimum of P (the number of x variables) and Q (the number of y variables). In our data example $P = 4$ and $Q = 2$, so the maximum number of canonical correlations is 2.

Tests of Hypotheses

Most packaged computer programs print the coefficients for all of the canonical variables, the values of the canonical correlations, and the values of the canonical variables for each individual in the sample. Some programs also print the computed values of Bartlett's chi-square test statistic. This test is an approximate test of the null hypothesis that the k smallest population canonical correlations are zero. A large chi-square is an indication that not all of those k correlations are zero. This test was derived with the assumption that the x's and the y's are jointly distributed according to a multivariate normal distribution (see Bartlett 1941 and Lawley 1959).

In our data example the Bartlett chi-square for the hypothesis that both canonical correlations are zero was computed by BMDP6M to be 73.04 with 8 degrees of freedom. Since the P value is less than 0.00001, we conclude that at least one canonical correlation is nonzero and proceed to test the hypothesis that the smallest one is zero. The test statistic chi-square equals 21.17 with 3 degrees of freedom. The P value is 0.001, and we conclude that both canonical correlations are significantly different from zero.

In data sets with more variables Bartlett's test can be a useful guide for selecting the number of significant canonical correlations. The test results are examined to determine at which step the remaining canonical correlations can be considered zero.

15.5 OTHER TOPICS RELATED TO CANONICAL CORRELATION

In this section we discuss some useful optional output available from packaged programs. We also present the relationship between canonical correlation and discriminant analysis.

Plots of Canonical Variable Scores

A useful option available in some programs is a plot of the canonical variable scores U_i versus V_i. In Figure 15.1 we show a scatter diagram of U_1 versus V_1 for the depression data. The first individual from Table 15.4 is indicated on the graph. The degree of scatter gives the impression of a somewhat weak but significant canonical correlation (0.405). For multivariate normal data the graph would approximate an ellipse of concentration. Such a plot can be useful in highlighting unusual cases in the sample as possible outliers or blunders. For example, the individual with the lowest value on U_1 is case number 289. This individual is a 19-year-old female with some high school education and with $28,000-per-year income. These scores produce a value of $V_1 = -0.73$. Also, this woman perceives her health as excellent (1) and is very depressed (CESD = 47), resulting in $U_1 = -3.02$. This individual, then, represents an extreme case in that she is uneducated and young but has a good income. In spite of the fact that she perceives her health as excellent, she is extremely depressed. Although this case gives an unusual combination, it is not necessarily a blunder.

Another Interpretation of Canonical Variables

Another useful optional output is the set of correlations between the canonical variables and the original variables used in deriving them. This output provides a way of interpreting the canonical variables when some of the variables within either the set of independent or the set of dependent variables are highly intercorrelated with each other (see Cooley and Lohnes 1971 or Levine 1977). For the depression data example these correlations are as shown in Table 15.5. These correlations are sometimes called *canonical variable loadings*.

FIGURE 15.1.   Plot of 294 Pairs of Values of the Canonical Variables U_1 and V_1 for the Depression Data Set (Canonical Correlation = 0.405)

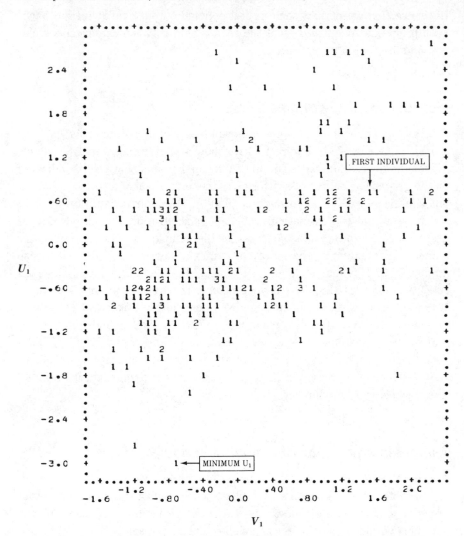

TABLE 15.5. Correlations Between Canonical Variables and Corresponding Variables (Depression Data Set)

	U_1	U_2
CESD	−0.281	0.960
Health	0.878	0.478

	V_1	V_2
Sex	0.089	0.525
Age	0.936	−0.225
Education	−0.532	−0.636
Income	−0.254	−0.737

These correlations may be interpreted in the same way as the loadings in factor analysis. Thus U_1 loads heavily on health and U_2 on CESD. Also V_1 loads heavily on age and education and V_2 on all variables except age. These interpretations are usually similar to those obtained by examining the standardized canonical variable coefficients. These canonical variable loadings may be plotted and rotated in the same manner as factor loadings are.

Examining Correlations Using Principal Components

Another method of examining the relationships between a set of X and Y variables is as follows:

1. Obtain the principal components of $y_1, y_2, \ldots, y_Q$ and denote them by $D_1, D_2, \ldots, D_Q$.
2. Obtain the principal components of $x_1, x_2, \ldots, x_P$ and denote them by $C_1, C_2, \ldots, C_P$.
3. Choose the first m of the principal components of each set, as discussed in Chapter 13.
4. Compute the correlations between C_1 and D_1; C_2 and D_2;...; C_m and D_m.

The correlations computed in step 4 are, in general, smaller than those in the canonical correlations. However, the principal components may be of interest in and of themselves. The principal components explain the maximum variance *within* a set of variables, whereas the canonical variables maximize the correlation *between* two sets of variables.

Since C_1 and D_2, for example, may have a nonzero correlation, it may be useful to compute such correlations in addition to those in step 4. Another variation might be to select a different number of principal components in the two sets of variables and examine their cross-correlations.

Application to Discriminant Analysis

In Section 11.11 we discussed classification of an individual into one of $k \geq 2$ populations on the basis of measurements $X_1, X_2, \ldots, X_P$. To classify an individual, we compute each of the k classification functions and assign the individual to the population for which the computed function is largest. This procedure is the predictive aspect of classification. For descriptive purposes we also mentioned that packaged programs derive canonical variables (called canonical discriminant functions by SPSS–X). Recall that these variables were derived in such a way as to exhibit the maximum possible differences among the groups. We now present these functions in the light of canonical correlation analysis.

We begin by defining a set of new variables called $Y_1, Y_2, \ldots, Y_{k-1}$. These are dummy or indicator variables that show which group each member of the sample came from. Note that as discussed in Chapter 10, we need $k - 1$ dummy variables to describe k groups. For example, suppose that there are $k = 4$ groups. Then the dummy variables Y_1, Y_2, and Y_3 are formed as follows:

Group	Y_1	Y_2	Y_3
1	1	0	0
2	0	1	0
3	0	0	1
4	0	0	0

Thus an individual coming from group 1 would be assigned a value of 1 on Y_1, 0 on Y_2, and 0 on Y_3, etc.

If we make $Q = k - 1$, then we have a sample with two sets of variables,

$Y_1, Y_2, \ldots, Y_Q$ and $X_1, X_2, \ldots, X_P$. We now perform a canonical correlation analysis of these variables. This analysis will result in a set of U_i variables and a set of V_i variables. As explained earlier, the number of these pairs of variables is the smaller of P and Q. Thus this number is the smaller of P and $k - 1$.

The V_i variables are the canonical discriminant functions we were looking for in Chapter 11. (The coefficients are proportional to those given in the discriminant function outputs.) Thus V_1 is the linear combination of the X variables with the maximum correlation with U_1. In this sense V_1 maximizes the correlation with the dummy variables representing the groups, and therefore it exhibits the maximum difference among the groups. Similarly, V_2 exhibits this maximum difference with the condition that V_2 is uncorrelated with V_1; etc.

Once derived, the canonical discriminant variables V_i should be examined as discussed in this chapter in an effort to give them a meaningful interpretation. Such an interpretation can be based on the magnitude of the standardized coefficients given in the outputs. The variables with the largest standardized coefficients can help give "names" to the canonical discriminant functions. To further assist the user in this interpretation, the SPSS–X DISCRIMINANT procedure offers the option of rotating the canonical discriminant functions by the varimax method (explained in Chapter 14 for factor analysis).

15.6 DISCUSSION OF COMPUTER PROGRAMS

The BMDP and SAS packages each contain a canonical correlation program. The various options for these programs are summarized in Table 15.6. The SPSS–X MANOVA program will also perform canonical correlation analysis.

An example of using MANOVA from SPSS–X to perform a canonical correlation analysis is given in the user's manual. This program is somewhat more difficult to interpret than the older SPSS CANCORR program (see Chapter 3 Bibliography for the SPSS manual). We therefore include in Table 15.6 the features of the SPSS CANCORR procedure.

For the data example described in Section 15.3 we want to relate reported health and depression levels to several typical demographic variables. The

TABLE 15.6. Summary of Computer Output for Canonical Correlation

Output	BMDP	SAS	SPSS
Means of original data	BMDP6M	CANCORR	CANCORR
Standard deviation of data or variances	BMDP6M	CANCORR	CANCORR
Correlation matrix	BMDP6M	CANCORR	CANCORR
Covariance matrix	BMDP6M	CANCORR	
Canonical correlation	BMDP6M	CANCORR	CANCORR
Canonical coefficients	BMDP6M	CANCORR	CANCORR
Standard canonical coefficients	BMDP6M	CANCORR	CANCORR
Canonical variable loadings	BMDP6M	CANCORR	
Canonical variable scores	BMDP6M		
Means of canonical variable scores		CANCORR	
Plots of canonical variable scores	BMDP6M		
Bartlett's test and P value	BMDP6M		CANCORR
Wilk's lambda		CANCORR	CANCORR
F statistic		CANCORR	
Save canonical scores	BMDP6M	CANCORR	

BMDP6M program was used to obtain the results reported in this chapter. The input and variable paragraphs are the usual ones. Because our data are divided into two sets of variables, we have to indicate which variables are in each set. This indication is given in the canonical paragraph. The "first= CESD,health" sentence defines which variables are the dependent variables. The second sentence defines the independent variables. Unlike the regression programs, this program does not need enter or remove sentences because the program will automatically compute up to a minimum of P or Q (the number of independent and dependent variables). The results of the Bartlett's test are used to decide how many significant correlations exist between the two sets of variables.

We ask that the correlation matrix be printed so that we can see all the simple correlations among the variables. We also ask for the coefficients of the canonical variables (a's and b's) and for the canonical variable loadings (the correlation between each variable and the canonical variable score; see Section 15.5).

Finally, we ask to see the plots of the first and second canonical variables (U_1 versus V_1 and U_2 versus V_2). These plots enable us to search for outliers and to interpret the correlation between the canonical variables. For example, if the relationship between U_1 and V_1 is not linear, then this plot enables us to assess the nonlinearity.

The paragraphs for our example are as follows:

```
/problem      title is `canonical analysis`.
/input        file is depress.
              variables are 6.
              cases are 294.
              format is free.
/variable     names are (2)sex,(3)age,(5)educat,
              (7)income,(29)cesd,(32)health.
/canonical    first =cesd,health.
              second=sex,age,educat,income.
/print        matr=corr,load,coef.
/plot         xvar=cnvrs1,cnvrs2.
              yvar=cnvrf1,cnvrf2.
/end
```

The program lists the first five cases so that we can check that it used the correct data; then it gives us simple summary statistics for each variable. It then prints the correlation matrix (see Table 15.2), and we readily see that although most of the variables are somewhat correlated, no very high correlations exist. Thus one variable cannot be considered a direct substitute for another. The program also prints the squared multiple correlation of each variable with all other variables in its set, which provides another method of checking multicollinearity. If a very high multiple correlation exists, then one strategy would be to discard a variable since it is not needed.

Next, the program prints out the canonical correlations between the two sets of canonical variables (0.405 and 0.266); these values are listed in order of magnitude. The square of these correlations is also printed and labeled "eigenvalue." Bartlett's test is also presented, as discussed in the last part of Section 15.4.

The unstandardized and standardized coefficients are printed for both sets of variables for the two canonical correlations. Also, the correlations of the canonical variables with the original variables are given in a separate table. In the case of the first canonical correlation a comparison of (1) the standardized coefficients and (2) the correlations of the canonical variables with the original variables shows no striking differences, although the sign is different for income and the two measures do not have a constant ratio from variable to variable, as shown below:

	First Canonical Correlation	
Variable	Standardized Coefficients	Correlations
CESD	−0.490	−0.281
Health	0.982	0.878
Sex	0.025	0.089
Age	0.871	0.936
Education	−0.383	−0.532
Income	0.082	−0.254

The correlation of age and health with their canonical variable is very high, indicating the close association between the outcome of those variables and the first canonical variable.

The plot of U_1 versus V_1 does not result in a nonlinear-appearing scatter diagram, nor does it look like a bivariate normal distribution (ellipse in shape), as can be seen in Figure 15.1. It may be that the skewness present in the CESD distribution has resulted in a somewhat skewed pattern for U_1 even though health has a greater overall effect on the first canonical variable. If this pattern were more extreme, it might be worthwhile to consider transformations on some of the variables such as CESD.

The SAS CANCORR procedure is very straightforward to run. You simply call the procedure and specify the following:

```
CANCORR DATA=DEPRESS ALL
VAR CESD HEALTH;
WITH SEX AGE EDUCAT INCOME;
```

Here the name of the data set is DEPRESS. The ALL statement produces all the optional output. The first set of variables represents the dependent variables, and the second set is the independent variables.

The SAS CANCORR procedure will also perform the canonical correlation analysis starting with the partial correlations. Thus the linear effects of a variable or variables can be removed prior to obtaining the canonical correlation.

For the SPSS CANCORR program the instructions are also straightforward. The pertinent statements are simply as follows:

```
CANCORR          VARIABLES=CESD HEALTH
                 WITH SEX AGE EDUCAT INCOME/
STATISTICS       ALL
```

These statements instruct the computer to perform the same analysis obtained from the other two packages, as discussed above.

SUMMARY

In this chapter we presented the basic concepts of canonical correlation analysis, an extension of multiple regression and correlation analysis. The extension is that the dependent variable is replaced by two or more dependent variables. If Q, the number of dependent variables, equals 1, then canonical correlation reduces to multiple regression analysis.

In general, the resulting canonical correlations quantify the strength of the association between the dependent and independent sets of variables. The derived canonical variables show which combinations of the original variables best exhibit this association. The canonical variables can be interpreted in a manner similar to the interpretation of principal components or factors.

In this chapter we also showed the relationship between canonical correlation analysis and discriminant function analysis. In addition, we described some optional computer output related to canonical correlation.

BIBLIOGRAPHY

*Bartlett, M. S. 1941. The statistical significance of canonical correlations. *Biometrika* 32 : 29–38.

*Cooley, W. W., and Lohnes, P. R. 1971. *Multivariate data analysis*. New York: Wiley.

Hopkins, C. E. 1969. Statistical analysis by canonical correlation: A computer application. *Health Services Research* (Winter); 4: 304–312.

*Hotelling, H. 1936. Relations between two sets of variables. *Biometrika* 28 : 321–377.

*Lawley, D. N. 1959. Tests of significance in canonical analysis. *Biometrika* 46 : 59–66.

Levine, M. S. 1977. *Canonical analysis and factor comparison*. Sage University Paper. Beverly Hills: Sage.

Meredith, W. 1964. Canonical correlation with fallible data. *Psychometrika* 29 : 55–65.

*Morrison, D. F. 1976. *Multivariate statistical methods*. New York: McGraw-Hill.

*Tatsuoka, M. M. 1971. *Multivariate analysis: Techniques for educational and psychological research*. New York: Wiley.

Thorndike, R. M. 1978. *Correlational procedures for research*. New York: Gardner Press.

Waugh, F. V. 1942. Regressions between sets of variables. *Econometrica* 10 : 290–310.

PROBLEMS

15.1 For the depression data set, perform a canonical correlation analysis between the following:

> ▶ Set 1: AGE, MARITAL (married versus other), EDUCAT (high school or less versus other), EMPLOY (full-time versus other), and INCOME.

> ▶ SET 2: The last seven variables.

Perform separate analyses for men and women. Interpret the results.

15.2 For the data set given in Appendix B, do a canonical correlation analysis on height, weight, FVC, and FEV1 for fathers versus the same variables for mothers. Interpret.

15.3 For the chemical company data given in Table 8.1, perform a canonical correlation analysis using P/E and EPS5 as dependent variables and the remaining variables as independent variables. Write the interpretations of the significant canonical correlations in terms of their variables.

<table>
<tr><td>

16

</td><td>

CLUSTER ANALYSIS

</td></tr>
</table>

16.1 *WHAT WILL YOU LEARN FROM THIS CHAPTER?*

From this chapter you will learn:

▶ What cluster analysis is (16.2).

▶ About distance measures used in cluster analysis (16.4).

▶ When hierarchical clustering is appropriate (16.5).

▶ When K-means clustering is appropriate (16.5).

▶ About simple graphical descriptions of data patterns that are useful in interpreting clusters (16.3,16.4,16.6).

▶ How to choose the appropriate computer program (16.7).

16.2 *WHEN IS CLUSTER ANALYSIS USED?*

Cluster analysis is a technique for grouping individuals or objects into *unknown* groups. It differs from other methods of classification, such as discriminant analysis, in that in cluster analysis the number and characteristics of the groups are to be derived from the data and are not usually known prior to the analysis.

In biology, cluster analysis has been used for decades in the area of

FIGURE 16.1. Example of Taxonomic Classification

A.   Modern Humans

KINGDOM: Animalia (animals)
   PHYLUM: Chordata (chordates)
      SUBPHYLUM: Vertebrata (vertebrates)
         CLASS: Mammalia (mammals)
            ORDER: Primates (primates)
               FAMILY: Hominidae (humans and close relatives)
                  GENUS: Homo (modern humans and precursors)
                     SPECIES: sapiens (modern humans)

B.   Domestic Rose

KINGDOM: Plantae (plants)
   PHYLUM: Tracheophyta (vascular plants)
      SUBPHYLUM: Pteropsida (ferns and seed plants)
         CLASS: Dicotyledoneae (dicots)
            ORDER: Rosales (saxifrages, psittosporums, sweet gum,
                                 plane trees, roses, and relatives)
               FAMILY: Rosaceae (cherry, plum, hawthorn, roses,
                                 and relatives)
                  GENUS: Rosa (roses)
                     SPECIES: galliea (domestic roses)

taxonomy. In taxonomy, living things are classified into arbitrary groups on
the basis of their characteristics. The classification proceeds from the most
general to the most specific, in steps. For example, classifications for domestic roses and for modern humans are illustrated in Figure 16.1 (see Wilson et
al. 1973). The most general classification kingdom, followed by phylum,
subphylum, etc. The use of cluster analysis in taxonomy is explained by
Sneath and Sokal (1973).

Cluster analysis has been used in medicine to assign patients to specific
diagnostic categories on the basis of their presenting symptoms and signs. In
particular, cluster analysis has been used in classifying types of depression
(see, e.g., Andreasen and Grove 1982). It has also been used in anthropology
to classify stone tools, shards, or fossil remains by the civilization that
produced them. Consumers can be clustered on the basis of their choice of

purchases in marketing research. In short, it it possible to find applications of cluster analysis in virtually any field of research.

We point out that cluster analysis is highly empirical. Different methods can lead to very different groupings, both in number and in content. Furthermore, since the groups are not known a priori, it is usually difficult to judge whether the results make sense in the context of the problem being studied. We note also that programs exist for clustering variables. However, we will discuss in detail only clustering cases or observations.

16.3 DATA EXAMPLES

A hypothetical data set was created to illustrate several of the concepts discussed in this chapter. Figure 16.2 shows a plot of five points for the two variables X_1 and X_2. This small data set will simplify the presentation since the analysis can be preformed by hand.

Another data set we will use includes financial performance data from the January 1981 issue of *Forbes*. The variables used are those defined in

FIGURE 16.2.   Plot of Hypothetical Cluster Data Points

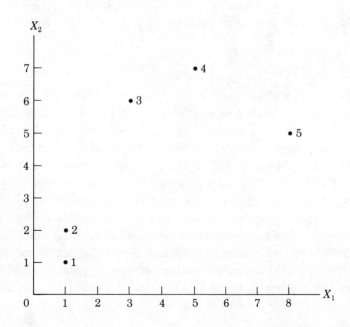

Section 8.3. Table 16.1 shows the data for 25 companies from three industries: chemical companies (the first 14 of the 31 discussed in Section 8.3), health care companies, and supermarket companies. The column labeled "Type" in Table 16.1 lists the abbreviations Chem, Heal, and Groc for these three industries. In Section 16.6 we will use two clustering techniques to group these companies and then check the agreement with their industrial type. These three industries were selected because they represent different stages of growth, different product lines, different management philosophies, different labor and capital requirements, etc. Among the chemical companies all of the large diversified firms were selected. From the major supermarket chains, the top six rated for return on equity were included. In the health care industry four of the five companies included were those connected with hospital management; the remaining company involves hospital supplies and equipment.

16.4 BASIC CONCEPTS: INITIAL ANALYSES AND DISTANCE MEASURES

In this section we present some preliminary graphical techniques for clustering. Then we discuss distance measures that will be useful in later sections.

Scatter Diagrams

Prior to using any of the analytical clustering procedures (see Section 16.5), most investigators begin with simple graphical displays of their data. In the case of two variables a scatter diagram can be very helpful in displaying some of the main characteristics of the underlying clusters. In the hypothetical data example shown in Figure 16.2, the points closest to each other are points 1 and 2. This observation may lead us to consider these two points as one cluster. Another cluster might contain points 3 and 4, with point 5 perhaps constituting a third cluster. On the other hand, some investigators may consider points 3, 4, and 5 as the second cluster. This example illustrates the indeterminacy of cluster analysis, since even the number of clusters is usually unknown. Note that the concept of closeness was implicitly used in defining the clusters. Later in this section we expand on this concept by presenting several definitions of distance.

TABLE 16.1. Financial Performance Data for Diversified Chemical, Health, and Supermarket Companies

Type	Symbol	Observation Number	ROR5	D/E	SALESGR5	EPS5	NPM1	P/E	PAYOUTR1
Chem	dia	1	13.0	0.7	20.2	15.5	7.2	9	0.426398
Chem	dow	2	13.0	0.7	17.2	12.7	7.3	8	0.380693
Chem	stf	3	13.0	0.4	14.5	15.1	7.9	8	0.406780
Chem	dd	4	12.2	0.2	12.9	11.1	5.4	9	0.568182
Chem	uk	5	10.0	0.4	13.6	8.0	6.7	5	0.324544
Chem	psm	6	9.8	0.5	12.1	14.5	3.8	6	0.508083
Chem	gra	7	9.9	0.5	10.2	7.0	4.8	10	0.378913
Chem	hpc	8	10.3	0.3	11.4	8.7	4.5	9	0.481928
Chem	mtc	9	9.5	0.4	13.5	5.9	3.5	11	0.573248
Chem	acy	10	9.9	0.4	12.1	4.2	4.6	9	0.490798
Chem	cz	11	7.9	0.4	10.8	16.0	3.4	7	0.489130
Chem	ald	12	7.3	0.6	15.4	4.9	5.1	7	0.272277
Chem	rom	13	7.8	0.4	11.0	3.0	5.6	7	0.315646
Chem	rci	14	6.5	0.4	18.7	-3.1	1.3	10	0.384000
Heal	hum	15	9.2	2.7	39.8	34.4	5.8	21	0.390879
Heal	hca	16	8.9	0.9	27.8	23.5	6.7	22	0.161290
Heal	nme	17	8.4	1.2	38.7	24.6	4.9	19	0.303030
Heal	ami	18	9.0	1.1	22.1	21.9	6.0	19	0.303318
Heal	ahs	19	12.9	0.3	16.0	16.2	5.7	14	0.287500
Groc	lks	20	15.2	0.7	15.3	11.6	1.5	8	0.598930
Groc	win	21	18.4	0.2	15.0	11.6	1.6	9	0.578313
Groc	sgl	22	9.9	1.6	9.6	24.3	1.0	6	0.194946
Groc	slc	23	9.9	1.1	17.9	15.3	1.6	8	0.321070
Groc	kr	24	10.2	0.5	12.6	18.0	0.9	6	0.453731
Groc	sa	25	9.2	1.0	11.6	4.5	0.8	7	0.594966
Means			10.45	0.70	16.79	13.18	4.30	10.16	0.408
Standard deviation			2.65	0.54	7.91	8.37	2.25	4.89	0.124

Data abstracted from *Forbes* 127, no. 1 (January 5, 1981).

If the number of variables is small, it is possible to examine scatter diagrams of each pair of variables and search for possible clusters. But this technique may become unwieldy if the number of variables exceeds four, particularly if the number of points is large.

Profile Diagram

A helpful technique for a moderate number of variables is a *profile diagram*. To plot a profile of an individual case in the sample, the investigator customarily first standardizes the data by subtracting the mean and dividing by the standard deviation for each variable. However, this step is omitted by some researchers, especially if the units of measurement of the variables are comparable. In the financial data example the units are not the same, so standardization seems helpful. The standardized financial data for the 25 companies are shown in Table 16.2. A profile diagram, as shown in Figure 16.3, lists the variables along the horizontal axis and the standardized value scale along the vertical axis. Each point on the graph indicates the value of the corresponding variable. The profile for the first company in the sample has been graphed in Figure 16.3. The points are connected in order to facilitate the visual interpretation. We see that this company hovers around the mean, being at most 1.3 standard deviations away on any variable.

A preliminary clustering procedure is to graph the profiles of all cases on the same diagram. To illustrate this procedure, we plotted the profiles of seven companies (15 through 21) in Figure 16.4. To avoid unnecessary clutter, we reversed the sign of the values of ROR5 and PAYOUTR1. Using the original data on these two variables would have caused the lines connecting the points to cross each other excessively, thus making it difficult to identify single company profiles.

Examining Figure 16.4, we note the following:

1. Companies 20 and 21 are very similar.
2. Companies 16 and 18 are similar.
3. Companies 15 and 17 are similar.
4. Company 19 stands out alone.

Thus it is possible to identify the above four clusters. It is also conceivable to identify three clusters: (15, 16, 17, 18), (20, 21), and (19). These clusters are

TABLE 16.2 Standardized Financial Performance Data for Diversified Chemical, Health, and Supermarket Companies

Type	Symbol	Observation Number	ROR5	D/E	SALESGR5	EPS5	NPM1	P/E	PAYOUTR1
						Standardized Input Data			
Chem	dia	1	0.963	-0.007	0.431	0.277	1.289	-0.237	0.151
Chem	dow	2	0.963	-0.007	0.052	-0.057	1.334	-0.442	-0.193
Chem	stf	3	0.963	-0.559	-0.290	0.230	1.601	-0.442	-0.007
Chem	dd	4	0.661	-0.927	-0.492	-0.248	0.488	-0.237	1.291
Chem	uk	5	-0.171	-0.559	-0.403	-0.618	1.067	-1.056	-0.668
Chem	psm	6	-0.246	-0.375	-0.593	0.158	-0.224	-0.851	0.807
Chem	gra	7	-0.209	-0.375	-0.833	-0.737	0.221	-0.033	-0.231
Chem	hpc	8	-0.057	-0.743	-0.681	-0.534	0.087	-0.237	0.597
Chem	mtc	9	-0.360	-0.559	-0.416	-0.869	-0.358	0.172	1.331
Chem	acy	10	-0.209	-0.559	-0.593	-1.072	0.132	-0.237	0.668
Chem	cz	11	-0.964	-0.589	-0.757	0.337	-0.402	-0.647	0.655
Chem	ald	12	-1.191	-0.191	-0.176	-0.988	0.354	-0.647	-1.089
Chem	rom	13	-1.002	-0.559	-0.732	-1.215	0.577	-0.647	-0.740
Chem	rei	14	-1.494	-0.559	0.241	-1.943	-1.337	-0.033	-0.190
Heal	hum	15	-0.473	3.672	2.908	2.534	0.666	2.218	-0.135
Heal	hca	16	-0.587	0.361	1.366	1.233	1.067	2.422	-1.981
Heal	nme	17	-0.775	0.913	2.769	1.364	0.265	1.809	-0.841
Heal	ami	18	-0.549	0.729	0.671	1.042	0.755	1.809	-0.839
Heal	ahs	19	0.925	-0.743	-0.100	0.361	0.621	0.786	-0.966
Groc	lks	20	1.794	-0.007	-0.189	-0.188	-1.248	-0.442	1.538
Groc	win	21	3.004	-0.927	-0.226	-0.188	-1.204	-0.237	1.372
Groc	sgl	22	-0.209	1.649	-0.909	1.328	-1.471	-0.851	-1.710
Groc	slc	23	-0.209	0.729	0.140	0.254	-1.204	-0.442	-0.696
Groc	kr	24	-0.095	-0.375	-0.530	0.576	-1.515	-0.851	0.370
Groc	sa	25	-0.473	0.545	-0.656	-1.036	-1.560	-0.647	1.506

Note: * See Table 16.1 for the original data source.

FIGURE 16.3. Profile Diagram of a Chemical Company (dia) Using Standardized Financial Performance Data

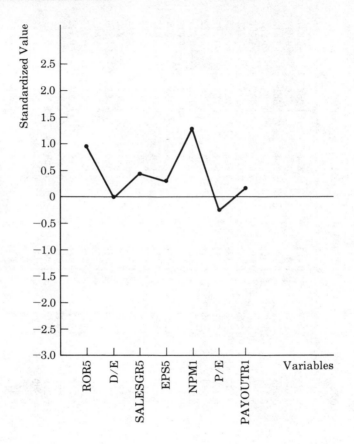

consistent with the types of the companies, especially noting that company 19 deals with hospital supplies.

Although this technique's effectiveness is not affected by the number of variables, it fails when the number of observations is large. In Figure 16.4 the impression is clear because we plotted only 7 companies. Plotting all 25 companies would have produced too cluttered a picture.

Distance Measures

For a large data set analytical methods such as those described in the next section are necessary. All of these methods require defining some measure of

FIGURE 16.4. Profile Plot of Health and Supermarket Companies with Standardized Financial Performance Data

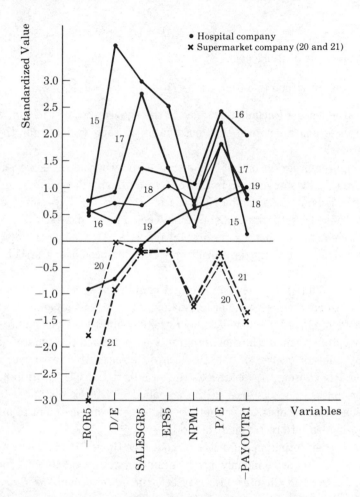

closeness or *similarity* of two observations. The converse of similarity is *distance*. Before defining distance measures, though, we warn the investigator that many of the analytical techniques are particularly sensitive to outliers. Some preliminary checking for outliers and blunders is therefore advisable. This check may be facilitated by the graphical methods just described.

The most commonly used distance is the *Euclidian distance*. In two dimensions, suppose that two points have coordinates (X_{11}, X_{21}) and

(X_{12}, X_{22}), respectively. Then the Euclidian distance between the two points is defined as

$$\text{distance} = \sqrt{(X_{11} - X_{12})^2 + (X_{21} - X_{22})^2}$$

For example, the distance between points 4 and 5 in Figure 16.2 is

$$\text{distance (points 4,5)} = \sqrt{(5-7)^2 + (8-5)^2} = \sqrt{13} = 3.61$$

For P variables the Euclidian distance is the square root of the sum of the squared differences between the coordinates of each variable for the two observations.

In computer program output the distances between all possible pairs of points are usually summarized in the form of a matrix. For example, for our hypothetical data, the Euclidian distances between the five points are as given in Table 16.3. Since the distance between a point and itself is zero, the diagonal elements of this matrix are always zero. Also, since the distances are symmetric, many programs print only the distances above or below the diagonal.

Since the square root operation does not change the order of how close the points are to each other, some programs use the sum of the *squared differences* instead of the Euclidian distance (i.e., they don't take the square root). Another option available in some programs is to replace the squared differences by another *power* of the absolute differences. For example, if the power of 1 is chosen, the distance is the sum of the absolute differences of the coordinates. The distance is the so-called city-block distance. In two dimensions it is the distance you must walk to get from one point to another in a city divided into rectangular blocks.

Several other definitions of distance exist (see Gower 1971). Here we will give only one more commonly used definition, the Mahalanobis distance discussed earlier in Chapter 11. In effect, the *Mahalanobis distance* is a generalization of the idea of standardization. The squared Euclidian distance based on standardized variables is the sum of the squared differences, each divided by the appropriate variance. When the variables are correlated, a distance can be defined to take this correlation into account. The Mahalanobis distance does just that. For two variables X_1 and X_2, suppose that the sample variances are S_1^2 and S_2^2, respectively, and that the correlation is r. The squared Euclidian distance based on the original values is

$$(\text{Euclidian distance})^2 = (X_{11} - X_{22})^2 + (X_{21} - X_{22})^2$$

TABLE 16.3. Euclidian Distance Between Five Hypothetical Points

	1	*2*	*3*	*4*	*5*
1	0	1.00	5.39	7.21	8.06
2	1.00	0	4.47	6.40	7.62
3	5.39	4.47	0	2.24	5.10
4	7.21	6.40	2.24	0	3.61
5	8.08	7.62	5.10	3.61	0

The same quantity based on standardized variables is

$$(\text{standardized Euclidian distance})^2 = \frac{(X_{11} - X_{12})^2}{S_1^2} + \frac{(X_{21} - X_{22})^2}{S_2^2}$$

If $r = 0$, then the last quantity is also the Mahalanobis distance. If $r \neq 0$, the the Mahalanobis distance is

$$D^2 = \frac{1}{1 - r^2}\left[\frac{(X_{11} - X_{12})^2}{S_1^2} + \frac{(X_{21} - X_{22})^2}{S_2^2} - \frac{2r(X_{11} - X_{12})(X_{21} - X_{22})}{S_1 S_2}\right]$$

For more than two variables the Mahalanobis distance is easily defined in terms of vectors and matrices (see, e.g., Afifi and Azen 1979). Some computer programs, such as BMDPKM, offer the Mahalanobis distance as an option, with estimates of the sample variances and correlations obtained from the data. It is noted that, strictly speaking, the within-group sample covariance matrix should be used. However, before the investigator finds clusters, no such estimates exist, and the total group covariance matrix is the only one available. When the latter is used, the computed Mahalanobis distance is then somewhat different from that presented in Chapter 11.

In most situations different distance measures will give different distance matrices, in turn leading to different clusters. When the variables have different units, it is advisable to standardize the data before computing the distances. This procedure is particularly helpful when the range of one variable is much larger than the others. Furthermore, when high positive or negative correlations exist among the variables, it may be helpful to consider computing the Mahalanobis distance.

In the next section we discuss two of the more commonly used analytical cluster techniques. These techniques make use of the distance functions just defined.

16.5 ANALYTICAL CLUSTERING TECHNIQUES

The commonly used methods of clustering fall into two general categories: hierarchical and nonhierarchical. First, we discuss the hierarchical techniques.

Hierarchical Clustering

Hierarchical methods can be either agglomerative or divisive. In the *agglomerative methods* we begin with N clusters; i.e., each observation constitutes its own cluster. In successive steps we combine the two closest clusters, thus reducing the number of clusters by one in each step. In the final step all observations are grouped into one cluster. In *divisive methods* we begin with one cluster containing all of the observations. In successive steps we split off the cases that are most dissimilar to the remaining ones. Most of the commonly used programs are of the agglomerative type, and we therefore do not discuss divisive methods further.

An example of the agglomerative methods is the SAS CLUSTER procedure, which allows three methods, differing in how the distance between clusters is computed: *centroid*, *Ward's methods*, and *average linkage* of squared Euclidian distances (see the manual). In the centroid method the distance between two clusters is defined as the distance between the group centroids (the centroid is the point whose coordinates are the means of all the observations in the cluster). If a cluster has one observation, then the centroid is the observation itself. The process proceeds by combining groups according to the distance between their centroids, the groups with the shortest distance being combined first. The centroid method is also used in BMDP2M.

The centroid method is illustrated in Figure 16.5 for our hypothetical data. Initially, the closest two centroids (points) of the five hypothetical observations plotted in Figure 16.2 are points 1 and 2, so they are combined first and their centroid is obtained in step 1. In step 2, centroids (points) 3 and 4 are combined (and their centroid is obtained), since they are the closest now that points 1 and 2 have been replaced by their centroid. At step 3 the centroid of points 3 and 4 and centroid (point) 5 are combined, and the centroid is obtained. Finally, at the last step the centroid of points 1 and 2 and the centroid of points 3, 4, and 5 are combined to form a single group.

Both BMDP and SAS offer optional distances based on standardized vari-

FIGURE 16.5. Hierarchical Cluster Analysis Using Unstandardized Hypothetical Data Set

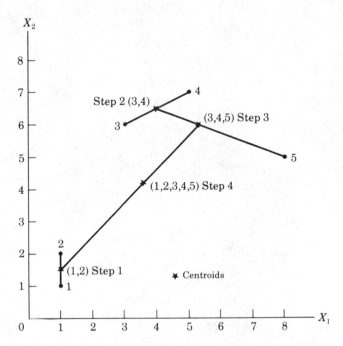

ables. Figure 16.6 illustrates the clustering steps based on the standardized hypothetical data, using BMDP2M. The results are identical to the previous results, although this is not the case in general. In addition, BMDP2M allows the optional distance of the sum of powers of differences (note that this option includes a power of 1, which would represent the city-block distance).

With two variables and a large number of data points, the representation of the steps in a graph similar to Figure 16.5 can get too cluttered to interpret. Also, if the number of variables is more than two, such a graph is not feasible. A clever device called the *dendrogram* or *tree graph* has therefore been incorporated into packaged computer programs to summarize the clustering at successive steps. The dendrogram for the hypothetical data set is illustrated in Figure 16.7. The horizontal axis lists the observations in a particular order. In this example the natural order is convenient. The vertical axis shows the successive steps. At step 1 points 1 and 2 are combined. Similarly, points 3 and 4 are combined in step 2; point 5 is combined with

FIGURE 16.6. Hierarchical Cluster Analysis Using Standardized Hypothetical Data Set

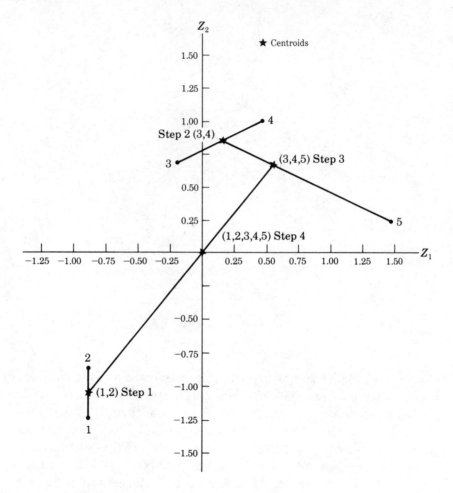

cluster (3,4) in step 3; and finally clusters (1,2) and (3,4,5) are combined. In each step two clusters are combined.

A tree graph is a default output of BMDP2M, and it can be obtained from SAS by calling the TREE procedure, using the output from the CLUSTER program. In TREE the number of clusters is printed on the vertical axis instead of the step number shown in Figure 16.7, while in BMDP2M the distance between the clusters just combined is printed. The order of the

FIGURE 16.7.   Dendrogram for Hierarchical Cluster Analysis of Hypothetical Data Set

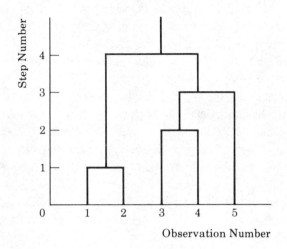

observations on the horizontal axis is helpful in indicating which observations are sufficiently similar to be combined in the early steps. The investigator should be aware that it is difficult to produce connected straight lines on a computer printer. For this reason the dendrograms produced by standard packaged programs are not as easy to read as the one shown in Figure 16.7.

The BMDP2M program offers also the option of using the nearest-neighbor distance estimator. This estimator measures the distance between two clusters as the distance between the two nearest points, one from each cluster, rather than between the two centroids. The program then proceeds as before, using the nearest-neighbor distance in place of the centroid distance. This procedure is further generalized to the so-called kth nearest-neighbor distance (see the BMDP manual).

As mentioned earlier, SAS offers two alternatives to the centroid option: Ward's method and average linkage. In Ward's method clusters resulting in the minimum within-cluster sums of squares are obtained (see Ward 1963 or Everitt 1980). In the average linkage method the distance between clusters is defined as the average distance between all pairs of observations, one from each cluster. A discussion of the performance of these methods is given in the SAS manual. Further discussion can be found in Anderberg (1973) or Everitt (1980).

Hierarchical procedures do have the advantage of being fast—faster, for example, than examining all possible combinations of observations. They are also appealing in a taxonomic application. Such procedures can be misleading, however, in certain situations. For example, an undesirable early combination persists throughout the analysis and may lead to artificial results. The investigator may wish to perform the analysis several times after deleting certain suspect observations.

For large sample sizes the printed dendrograms become very large and unwieldy to read. One statistician noted that they were more like wallpaper than comprehensible results for large N. Note, for example, that BMDP2M will run up to 500 cases with 5 variables and still print the distance between each pair of cases, as well as the dendrogram. Though computationally possible, the resulting output is difficult to carry, much less to interpret.

An important problem is how to select the number of clusters. No standard objective procedure exists for making the selection. The distances between clusters at successive steps may serve as a guide. The investigator can stop when this distance exceeds a specified value or when the successive differences in distances between steps make a sudden jump. Also, the underlying situation may suggest a natural number of clusters. If such a number is known, a particularly appropriate technique is the K-means clustering techniques of MacQueen (1967).

K-Means Clustering

The K-means clustering is a popular nonhierarchical clustering technique. For a specified number of clusters K the basic algorithm proceeds in the following steps:

1. Divide the data into K initial clusters. The members of these clusters may be specified by the user or may be selected by the program, according to an arbitrary procedure.
2. Calculate the means or centroids of each of the K clusters.
3. For a given case, calculate its distance to each centroid. If the case is closest to the centroid of its own cluster, leave it in that cluster; otherwise, reassign it to the cluster whose centroid is closest to it.

4. Repeat step 3 for each case.
5. Repeat steps 2, 3, and 4 until no cases are reassigned.

Individual programs implement the basic algorithm in different ways. The default option of BMDPKM begins by considering all of the data as one cluster. For the hypothetical data set this step is illustrated in Figure 16.8a. The program then searches for the variable with the highest variance, in this case X_1. The original cluster is now split into two clusters, using the midrange of X_1 as the dividing point, as shown in Figure 16.8b. If the data are standardized, then each variable has a variance of one. In that case the variable with the smallest range is selected to make the split. The program, in general, proceeds in this manner by further splitting the clusters until the specified number K is achieved. That is, it successively finds that particular variable and the cluster producing the largest variance and splits that cluster accordingly, until K clusters are obtained. At this stage step 1 of the basic algorithm is completed and the program proceeds with the other steps.

For the hypothetical data example with $K = 2$, it is seen that every case already belongs to the cluster whose centroid is closest to it (Figure 16.8c). For example, point 3 is closer to the centroid of cluster (1,2,3) than it is to the centroid of cluster (4,5). Therefore it is not reassigned. Similar results hold for the other cases. Thus the algorithm stops, with the two clusters selected being cluster (1,2,3) and cluster (4,5).

The SAS procedure FASTCLUS is recommended especially for large data sets. The user specifies the maximum number of clusters allowed, and the program starts by first selecting cluster "seeds," which are used as initial guesses of the means of the clusters. The first observation with no missing values in the data set is selected as the first seed. The next complete observation that is separated from the first seed by an optional, specified minimum distance becomes the second seed (the default minimum distance is zero). The program then proceeds to assign each observation to the cluster with the nearest seed. The user can decide whether to update the cluster seeds by cluster means each time an observation is assigned, using the DRIFT option, or only after all observations are assigned. Limiting seed replacement results in the program using less computer time. Note that the initial, and possibly the final, results depend on the order of the observations in the data set. Specifying the initial cluster seeds can lessen this dependence. Finally, we note that BMDPKM also permits the user to specify seed points.

FIGURE 16.8. Successive Steps in K-Means Cluster Analysis for $K = 2$, Using BMDPKM for Hypothetical Data Set with Default Options

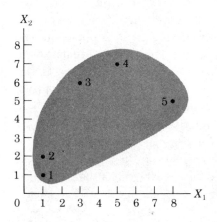

a. Start with All Points
   in One Cluster

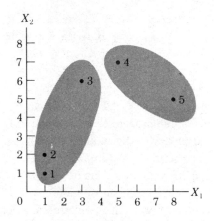

b. Cluster Is Split into
   Two Clusters at Midrange
   of X_1 (Variable with
   Largest Variance)

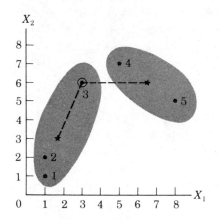

c. Point 3 Is Closer to
   Centroid of Cluster (1,2,3)
   and Stays Assigned to
   Cluster (1,2,3)

d. Every Point Is Now Closest
   to Centroid of Its Own
   Cluster; Stop

For either program judicious choice of the seeds or leaders can improve the results.

As mentioned earlier, the selection of the number of clusters is a troublesome problem. If no natural choice of K is available, it is useful to try various values and compare the results. One aid to this examination is a plot of the points in the clusters, labeled by the cluster they belong to. If the number of variables is more than two, some two-dimensional plots are available from certain programs, such as BMDPKM. Another possibility is to plot the first two principal components, labeled by cluster membership. A second aid is to examine the F ratio for testing the hypothesis that the cluster means are equal. This test may be done for each variable separately, as in BMDPKM. An alternative possibility is to supply the data with each cluster as a group as input to a K group discriminant analysis program. Such a program would then produce a test statistic for the equality of means of all variables simultaneously, as was discussed in Chapter 11. Then for each variable, separately or as a group, comparison of the P values for various values of K may be a helpful indication of the value of K to be selected. Note, however, that these P values are valid only for comparative purposes and not as significance levels in a hypothesis-testing sense, even if the normality assumption is justified. The individual F statistics for each variable indicate the relative importance of the variables in determining cluster membership (again, they are not valid for hypothesis testing).

Since cluster analysis is an empirical technique, it may be advisable to try several approaches in a given situation. In addition to the hierarchical and K-means approaches discussed above, several other methods are available in the literature cited in the Bibliography. It should be noted that the K-means approach is gaining acceptability in the literature over the hierarchical approach. In any case, unless the underlying clusters are clearly separated, different methods can produce widely different results. Even with the same program, different options can produce quite different results.

16.6 CLUSTER ANALYSIS FOR FINANCIAL DATA SET

In this section we apply some of the standard procedures to the financial performance data set shown in Table 16.1. In all of our runs the data are first standardized as shown in Table 16.2. Recall that in cluster analysis the total sample is considered as a single sample. Thus the information on type of company is not used to derive the clusters. However, this information will be used to interpret the results of the various analyses.

Hierarchical Clustering

The dendrogram resulting from the hierarchical BMDP2M program is shown in Figure 16.9. Default options including the centroid method with Euclidian distance were used with the standardized data. The horizontal axis lists the observation numbers in a particular order, which prevents the lines in the dendrogram from crossing each other. One result of this arrangement is that certain subgroups appearing near each other on the horizontal axis constitute clusters at various steps. Note that the distance is shown on the right vertical axis. These distances are measured between the *centers* of the two clusters just joined. On the left vertical axis the number of clusters is listed.

In the Figure companies 1, 2, and 3 form a single cluster, with the grouping being completed when there are 22 clusters. Similarly, at the opposite end 15, 17, 18, and 16 (all health care companies) form a single cluster at the step in which there are two clusters. Company 22 stays by itself until there are only three clusters.

The distance axis indicates how disparate the two clusters just joined are. The distances are progressively increasing. As a general rule, large increases in the sequence should be a signal for examining those particular steps. A large distance indicates that at least two very different observations exist, one from each of the two clusters just combined. Note, for example, the large increase in distance occurring when we combine the last two clusters into a single cluster.

It is instructive to look at the industry groups and ask why the cluster analysis, as in Figure 16.9, for example, did not completely differentiate them. It is clear that the clustering is quite effective. First, 13 of the chemical companies, all except no. 14 (rci), are clustered together with only one nonchemical firm, no. 19 (ahs), when the number of clusters is eight. This result is impressive when one considers that these are large diversified companies with varied emphasis, ranging from industrial chemicals to textiles to oil and gas production.

At the level of nine clusters three of the four hospital management firms, nos. 16, 17, and 18 (hca, nme, and ami), have also been clustered together, and the other, no. 15 (hum), is added to that cluster before it is aggregated

FIGURE 16.9. BMDP2M Dendrogram of Standardized Financial Performance Data Set

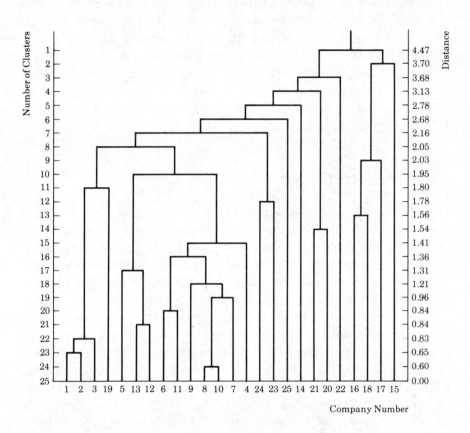

Company Number

with any nonhospital management firms. A look at the data in Tables 16.1 and 16.2 shows that no. 15 (hum) has a clearly different D/E value than the others, suggesting a more highly leveraged operation and probably a different management style. The misfit in this health group is no. 19 (ahs), clustered with the chemical firms instead of with the hospital management firms. Further examination shows that, in fact, no. 19 is a large, established supplies and equipment firm, probably more akin to drug firms than to the fast-growing hospital management business, and so it could be expected to share some financial characteristics with the chemical firms.

The grocery firms do not cluster tightly. In scanning the data of Tables 16.1 and 16.2 we note that they vary substantially on most variables. In particular, no. 22 (sgl) is highly leveraged (high D/E), and no. 21 (win) has low leverage (low D/E) relative to the others. Further, if we examine other characteristics not included in this data set, important disparities show up. Three of the six, nos. 21, 24, and 25 (win, kr, sa), are three of the four largest United States grocery supermarket chains. Two others, nos. 20 and 22 (kls, sgl), have a diversified mix of grocery, drug, department, and other stores. The remaining firm, no. 24 (slc), concentrates on convenience stores (7-Eleven) and has a substantial franchising operation. Thus the six, while all grocery-related, are quite different from each other.

Various K-Means Clusters

Since these companies do not present a natural application of hierarchical clustering such as taxonomy, the K-means procedure may be more appropriate. The natural value of K is three since there are three types of companies. The first four columns of Table 16.4 show the results of one run of the SAS FASTCLUS procedure and three runs of BMDPKM. The numbers in each column indicate the cluster to which each company is assigned. In columns (1) and (2) the default options of the two programs were used with $K = 3$. In column (3) two values of K, 2 and 3, were specified in the same run. Note that the results are different from those in column (2) although the same program was used. The difference is due to the fact that in column (3) the initial three groups are selected by splitting one of the two clusters obtained for $K = 2$, whereas in column (2) the initial splitting is done before any reassignment is carried out (see Section 16.5). Different results still were obtained when the Mahalanobis distance option was used, as shown in column (4).

One method of summarizing the first four columns is shown in column (5). For example, all four runs agreed on assigning companies 1 through 4 to cluster 1. Similarly, every company assigned unanimously to a cluster is so indicated. Other companies were assigned to a cluster if three out of four runs agreed on it. For example, company 8 was assigned to cluster 1 and company 18 was assigned to cluster 2. Seven of the companies could not be assigned to a unique cluster. Each received two votes for cluster 1 and two votes for cluster 3.

It seems, therefore, that the value of K should perhaps be 4, not 3. With $K = 4$ and the BMDPKM program the results are as shown in column (6).

TABLE 16.4. Companies Clustered Together from K-Means Standardized Cluster Analysis (Financial Performance Data Set)

Type of Company	(1) Fast Clus* K = 3 Default	(2) BMDPKM K = 3 Default	(3) BMDPKM K = 2,3 Default	(4) BMDPKM K = 3 Mahalanobis	(5) Summary Of 4 Runs, K = 3	(6) BMDPKM K = 4 Mahalanobis
1 Chem	1	1	1	1	1	1
2 Chem	1	1	1	1	1	1
3 Chem	1	1	1	1	1	1
4 Chem	1	1	1	1	1	4
5 Chem	3	1	3	1	1,3	3
6 Chem	1	3	3	1	1,3	4
7 Chem	3	1	3	1	1,3	4
8 Chem	1	1	3	1	1,3	4
9 Chem	1	3	3	1	1,3	4
10 Chem	1	1	3	1	1	3
11 Chem	3	3	3	1	3	4
12 Chem	3	1	3	1	1,3	4
13 Chem	3	1	3	1	1,3	4
14 Chem	3	3	3	1	3	4
15 Heal	2	2	2	2	2	2
16 Heal	2	2	2	2	2	2
17 Heal	2	2	2	2	2	2
18 Heal	2	2	2	2	2	2
19 Heal	1	1	1	3	1	1
20 Groc	1	3	1	1	1	1
21 Groc	3	3	3	1	1	1
22 Groc	3	3	3	3	3	3
23 Groc	3	3	3	3	3	3
24 Groc	3	3	3	3	3	3
25 Groc	1	3	3	1	1,3	4

* Changing the order of companies 1 and 11 did not change the results.

This run seems basically to group the companies labeled 1 and 3 in column (5) into a new cluster, 4. Otherwise, with the exception of companies 8, 10, and 14, all other companies remain in their original clusters. Cluster 1 now consists of companies 1, 2, 3, 4 (chemicals), 19 (health), and 20, 21 (grocery). Cluster 2 consists exclusively of health companies. Cluster 3 consists of three grocery companies and two chemicals. Finally, cluster 4 consists exclusively of chemical companies. It thus seems, with some exceptions, that the companies within an industry have similar financial data. For example, an exception is company 19, a company that deals with hospital supplies. This exception was evident in the hierarchical results as well as in K-means, where company 19 joined the chemical companies and not the other health companies.

Profile Plots of Means

Some of the output of BMDPKM may be useful in further interpretations of the clusters. A plot of each observation labeled by its cluster is given as part of the default output. The variable means for each cluster are printed and graphed to provide cluster profiles, which aids in understanding the characteristics of each cluster. Figure 16.10 shows the cluster profiles based on the means of the standardized variables ($K = 4$). Note that the values of ROR5 and PAYOUTR1 are plotted with the signs reversed in order to keep the lines from crossing too much. It is immediately apparent that cluster 2 is very different from the remaining three clusters. Cluster 1 is also quite different from clusters 3 and 4. The latter two clusters are most different on EPS5. Thus the four health companies comprising cluster 2 seem to clearly average higher SALESGR5 and P/E. This cluster mean profile is a particularly appropriate display of the K-means output, because the objective of the analysis is to make these means as widely separated as possible and because there are usually a small number of clusters to be plotted.

Use of F Tests

For quantification of the relative importance of each variable the univariate F ratio for testing the equality of each variable's means in the K clusters is given in the BMDPKM output. A large F value is an indication that the corresponding variable is useful in separating the clusters. For example, the largest F value is for P/E, indicating that the clusters are most different

FIGURE 16.10. Profile of Cluster Means for Four Clusters (Financial Performance Data Set)

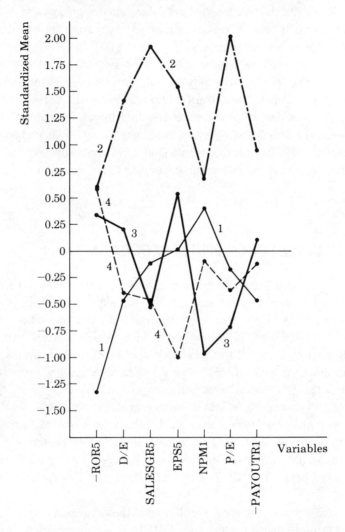

from each other in terms of this variable. Conversely, the F ratio for PAY-OUTR1 is very small, indicating that this variable does not play an important role in defining the clusters for the particular industry groups and companies analyzed. Note that these F ratios are used for comparing the variables only, not for hypothesis-testing purposes.

The above example illustrates how these techniques are used in an exploratory manner. The results serve to point out some possible natural groupings as a first step in generating scientific hypotheses for the purpose of further analyses. A great deal of judgment is required for an interpretation of the results since the outputs of different runs may be different and sometimes even contradictory. Indeed, this result points out the desirability of making several runs on the same data in an attempt to reach some consensus.

16.7 DISCUSSION OF COMPUTER PROGRAMS

The SAS package offers four procedures in its statistics manual: CLUSTER, which does hierarchical clustering of observations using three clustering algorithms; FASTCLUS, which uses the K-means method and is especially well-suited for large sample sizes; VARCLUS, which clusters variables (a topic not covered in detail in this book); and TREE, which draws dendrograms using output from CLUSTER or VARCLUS. Additional procedures are available in the *SAS Supplemental Library User's Guide*.

The BMDP package offers four cluster analysis programs: P1M, which does cluster analysis of variables; P2M, which uses one of four distance criteria to perform hierarchical clustering; PKM, which performs K-means clusters; and P3M , which constructs block clusters for categorical data (also not discussed here). The SPSS–X package does not offer a cluster analysis program.

Table 16.5 summarizes some highlights of the programs employed in this chapter. If the city block is used, then BMDP2M is appropriate for hierarchical analysis. This analysis is achieved by using pth-power distance option with $p = 1$. For K-means clustering BMDPKM offers a variety of optional printed output that may be useful in assessing the results. On the other hand, FASTCLUS is designed to be economical when used on large data sets. For very large data sets it offers options that could cut the cost even further.

TABLE 16.5. Summary of Computer Output for Cluster Analysis

Options	BMDP	SAS
Hierarchical Cluster Analysis		
Not standardized	BMDP2M	CLUSTER
Standardization	BMDP2M	CLUSTER*
Distance		
Euclidian	BMDP2M	CLUSTER
Sum of pth power of absolute distances	BMDP2M	
Average linkage on squared Euclidean distance		CLUSTER
Distances suitable for counted data	BMDP2M	
Ward's method		CLUSTER
Distance Measured Between		
Centroid of clusters	BMDP2M	CLUSTER
Nearest neighbor of clusters	BMDP2M	
Principal Output		
Input data	BMDP2M	CLUSTER
Distances between combined clusters	BMDP2M	CLUSTER+
Dendrogram	BMDP2M	TREE
Shaded distance matrix	BMDP2M	
Initial distances between observations	BMDP2M	
Number of observations per cluster		TREE
K-Means Cluster Analysis		
Transformation of variables	BMDPKM	FASTCLUS
Standardization	BMDPKM‡	FASTCLUS*
Distance		
Euclidean	BMDPKM	FASTCLUS
Mahalanobis (covariance)	BMDPKM‡	
Standardized (unit variance)	BMDPKM‡	FASTCLUS
Distance Measured From		
Leader or seed points	BMDPKM	FASTCLUS
Centroid or center	BMDPKM	FASTCLUS
Printed Output		
Input data	BMDPKM	PRINT
Distances between cluster centers	BMDPKM	FASTCLUS
Distances between points and centers	BMDPKM	FASTCLUS
Cluster means	BMDPKM	FASTCLUS
Cluster standard deviations	BMDPKM	FASTCLUS
F ratios	BMDPKM	FASTCLUS
Cluster mean profiles	BMDPKM	
Plot of cluster membership	BMDPKM	FASTCLUS
Assign cluster membership for further analysis	BMDPKM	FASTCLUS

Note: The SPSS–X package does not offer a cluster analysis program.
* Use STANDARD procedure to obtain standardization before analysis.
+ Several distance measures are available.
‡ These operations can be done either for the total sample or within clusters.

Since cluster analysis is not a standardized technique, the investigator may wish to try other programs in addition to those described above. The results should not be expected to be in agreement with each other. The following are examples of other programs available in the United Kingdom and North America:

▶ MIDAS (CLUSTER): D. J. Fox and K. E. Guire, *Documentation for MIDAS*. 3rd ed. Ann Arbor: Statistical Research Laboratory, University of Michigan, 1976.

▶ OSIRIS (CLUSTER): Survey Research Center Computer Support Group. *OSIRIS IV User's Manual*. 7th ed. Ann Arbor: Institute for Social Research, University of Michigan, 1981.

▶ CLUSTAN: Dr. D. Wishart, c/o Department of Computational Science, University of St. Andrews, North Haugh, St. Andrews KY16 9SX, Scotland.

▶ GENSTAT: Programs Secretary, Statistical Department, Rothamsted Experimental Station, Harpenden, AL5 2JQ, England.

It may also be appropriate to combine two analyses into one. For example, a principal components analysis may be performed on the entire data set and a small number of components selected. A cluster analysis can then be run on the values of the selected principal components. Similarly, factor analysis could be performed first, followed by a cluster analysis of the factor scores. This procedure has the advantage of reducing the original number of variables, making the interpretation possibly easier, especially if the meaningful factors have been selected.

As noted earlier, cluster analysis can be used to cluster variables instead of cases. For example, BMDP1M is designed to perform such an analysis. If the default option is used, the absolute value of the correlation between two variables measures the "distance" between them: The higher the absolute value of the correlation, the closer the two variables are. The program then proceeds to cluster variables in a hierarchical fashion. The VARCLUS procedure in SAS offers several options. It attempts to divide a set of variables into nonoverlapping clusters. A discussion of its use to reduce the number of variables employed in analysis is given in the SAS manual.

The following statements are used to run FASTCLUS on the five hypothetical observations:

```
DATA;
INPUT X1 1.0 X2 1.0;
CARDS;
11
12
36
57
85;
PROC FASTCLUS LIST MAXCLUSTER=2;
PROC PRINT;
PROC MEANS;
```

The first set of statements defines this simple data set of two variables and five observations. Both X1 and X2 are listed in a single column that follow on cards. The FASTCLUS procedure is used with only the LIST and MAXCLUSTER option. Two clusters are specified by MAXCLUSTER=2. The LIST option results in a list of the ID of all the observations along with the number of the cluster to which the observation is assigned and the distance between the observation and the final cluster seed. A large number of statistics can be printed by using the various available printout options. The PROC PRINT statement results in the values of the original data being printed, and PROC MEANS results in the means, standard deviations, and other univariate statistics being printed for X1 and X2.

The FASTCLUS procedure given in Table 16.4 for the financial data was run by using the default options and by using cards. The following control statements were supplied:

```
DATA;
INPUT @ 3 ROR5 3.1 D-E 2.1 SALESGR5 3.1
EPS5 4.1 NPN1 3.1 EPS1 3.2
DIV1 3.2 P E 2.0;
PAYOUTR1=DIV1/EPS1;
DROP DIV1 EPS1;
IF _N_ < 15 THEN TYPE = 'CHEM.';
ELSE IF _N_ <20 THEN TYPE='HEAL.';
```

```
ELSE TYPE='GROC';
CARDS;

(DATA CARDS)

PROC STANDARD MEAN=0 STD=1 OUT=STAN1;
PROC PRINT;
PROC FASTCLUS DATA=STAN1 LIST MAXCLUSTER=3;
```

The @ 3 tells the computer to go to column 3 and start reading the first variable from that column. This pointer control simplifies the use of a subset of the variables. The format of each variable is given following the variables. For example, ROR5 is a three-digit variable with a single number following the decimal point. It is the first variable used on our cards. The first two columns are simply ID numbers going from 1 to 25. The variable PAYOUTR1 is computed by the program as a ratio of dividends to earnings per share during the last year.

The companies are labeled by type (chem, heal, or groc), depending on their order. The first 14 companies are chemical companies, so the statement IF _N_ < 15 followed by the labeling 'CHEM' is used; etc.

The original data are transformed into standardized data since these variables do not have the same units (PROC STANDARD), and a printout of the standardized data is requested (PROC PRINT). Finally, the FASTCLUS procedure is called to be run on the standardized data. Three clusters are requested by using the MAXCLUSTER=3 statement. Additional output can be obtained by following the instructions given in the manual. The final assignments are shown in Table 16.4, column (1), and were discussed earlier in the text.

For a run of a similar program using BMDPKM, the usual INPUT and VARIABLE paragraphs are followed by a CLUSTER paragraph, as follows:

```
/CLUSTER    STAND=WVAR.
            NUMBER=3.
```

The first sentence requests that the data be standardized, using the within-cluster variances. This statement results in the standardization changing at each stage of the clustering. If the sentence STAND=VAR. had been used, then each variable would have been divided by its own standard deviation, using all cases as a single sample. The default option is no standardization. The second sentence requests three clusters. Columns (2), (3), (4), and (6) of Table 16.4 show the outputs of various runs of BMDPKM as discussed in Section 16.6.

SUMMARY

In this chapter we presented a topic still in its evolutionary form. Unlike subjects discussed in previous chapters, cluster analysis has not yet gained a standard methodology. Nonetheless, a number of techniques have been developed for dividing a multivariate sample, the composition of which is not known in advance, into several groups.

In view of the state of the subject, we opted for presenting some of the techniques that have been incorporated into the standard packaged programs. These include a hierarchical and nonhierarchical technique. We also explained the use of profile plots for small data sets. Since the results of any clustering procedure are often not definitive, it is advisable to perform more than one analysis and attempt to collate the results.

BIBLIOGRAPHY

Afifi, A. A., and Azen, S. P. 1979. *Statistical analysis: A computer oriented approach*. 2nd ed. New York: Academic Press.

Anderberg, M. R. 1973. *Cluster analysis for applications*. New York: Academic Press.

Andreasen, N. C., and Grove, W. M. 1982. The classification of depression: Traditional versus mathematical approaches. *American Journal of Psychiatry* 139 : 45–52.

Cormack, R. M. 1971. A review of classification. *Journal of the Royal Statistical Society, Series A* 134 : 321–367.

Everitt, B. 1980. *Cluster analysis*. London: Heinemann Educational Books.

Gordon, A. D. 1981. *Classification*. London: Chapman and Hall.

Gower, J. C. 1971. A general coefficient of similarity and some of its properties. *Biometrics* 27 : 857–874.

Hand, D. J. 1981. *Discrimination and classification*. New York: Wiley.

Hartigan, J. A. 1975. *Clustering algorithms*. New York: Wiley.

Johnson, R. A., and Wichern, D. W. 1982. *Applied multivariate statistical analysis*. Englewood Cliffs, N.J.: Prentice-Hall.

Johnson, S. C. 1967. Hierarchical clustering schemes. *Psychometrika* 32 : 241–244.

MacQueen, J. B. 1967. Some methods for classification and analysis of multivariate observations. *Proceedings of the Fifth Berkeley Symposium on Mathematical Statistical Problems* 1 : 281–297.

Sneath, P. H., and Sokal, R. R. 1973. *Numerical taxonomy*. San Francisco: Freeman.

Ton, J. T., and Gonzalez, R. C. 1974. *Pattern recognition principles*. Reading, Mass.: Addison-Wesley.

Ward, J. H. 1963. Hierarchical grouping to optimize an objective function. *Journal of the American Statistical Association* 58 : 236–244.

Wilson, E. O.; Eisner, T.; Briggs, W. R.; Dickerson, R. E.; Metzenberg, R. L.; O'Brien, R. D.; Susman, M.; and Boggs, W. E. 1973. *Life on earth*. Sunderland, Mass.: Sinauer Associates.

PROBLEMS

16.1   For the depression data set, use the last seven variables to perform a cluster analysis producing two groups. Compare the distribution of CESD and cases in the groups. Compare also the distribution of sex in the groups. Try two different programs with different options. Comment on your results.

16.2   For the situation described in Problem 7.7, modify the data for X1, X2,..., X9 as follows:

▶ For the first 25 cases, add 10 to X1, X2, X3.

▶ For the next 25 cases, add 10 to X4, X5, X6.

▶ For the next 25 cases, add 10 to X7, X8.

▶ For the last 25 cases, add 10 to X9.

Now perform a cluster analysis to produce four clusters. Use two different programs with different options. Compare the derived clusters with the above groups.

16.3   Perform a cluster analysis on the chemical company data in Table 8.1, using the K-means method for $K = 2, 3, 4$.

16.4   For the accompanying small hypothetical data set, plot the data by using methods given in this chapter, and perform both hierarchical and K-means clustering with $K = 2$.

Cases	X1	X2
1	11	10
2	8	10
3	9	11
4	5	4
5	3	4
6	8	5
7	11	11
8	10	12

16.5 Describe how you would expect guards, forwards, and centers in basketball to cluster on the basis of size or other variables. Which variables should be measured?

APPENDIX A
STATISTICAL TABLES

TABLE A.1. Cumulative Normal Distribution

Example: Area to the left of $z = -2.04$ is 0.0207.

z	0.00	0.01	0.02	0.03	0.04	0.05	0.06	0.07	0.08	0.09
−3.8	0.0001	0.0001	0.0001	0.0001	0.0001	0.0001	0.0001	0.0001	0.0001	0.0001
−3.7	0.0001	0.0001	0.0001	0.0001	0.0001	0.0001	0.0001	0.0001	0.0001	0.0001
−3.6	0.0002	0.0002	0.0001	0.0001	0.0001	0.0001	0.0001	0.0001	0.0001	0.0001
−3.5	0.0002	0.0002	0.0002	0.0002	0.0002	0.0002	0.0002	0.0002	0.0002	0.0002
−3.4	0.0003	0.0003	0.0003	0.0003	0.0003	0.0003	0.0003	0.0003	0.0003	0.0002
−3.3	0.0005	0.0005	0.0005	0.0004	0.0004	0.0004	0.0004	0.0004	0.0004	0.0003
−3.2	0.0007	0.0007	0.0006	0.0006	0.0006	0.0006	0.0006	0.0005	0.0005	0.0005
−3.1	0.0010	0.0009	0.0009	0.0009	0.0008	0.0008	0.0008	0.0008	0.0007	0.0007
−3.0	0.0014	0.0013	0.0013	0.0012	0.0012	0.0011	0.0011	0.0011	0.0010	0.0010
−2.9	0.0019	0.0018	0.0018	0.0017	0.0016	0.0016	0.0015	0.0015	0.0014	0.0014
−2.8	0.0026	0.0025	0.0024	0.0023	0.0023	0.0022	0.0021	0.0021	0.0020	0.0019
−2.7	0.0035	0.0034	0.0033	0.0032	0.0031	0.0030	0.0029	0.0028	0.0027	0.0026
−2.6	0.0047	0.0045	0.0044	0.0043	0.0041	0.0040	0.0039	0.0038	0.0037	0.0036
−2.5	0.0062	0.0060	0.0059	0.0057	0.0055	0.0054	0.0052	0.0051	0.0049	0.0048
−2.4	0.0082	0.0080	0.0078	0.0076	0.0073	0.0071	0.0069	0.0068	0.0068	0.0064
−2.3	0.0107	0.0104	0.0102	0.0099	0.0096	0.0094	0.0091	0.0089	0.0087	0.0084
−2.2	0.0139	0.0136	0.0132	0.0129	0.0125	0.0122	0.0119	0.0116	0.0113	0.0110
−2.1	0.0179	0.0174	0.0170	0.0166	0.0162	0.0158	0.0154	0.0150	0.0146	0.0143
−2.0	0.0228	0.0222	0.0217	0.0212	0.0207	0.0202	0.0197	0.0192	0.0188	0.0183
−1.9	0.0287	0.0281	0.0274	0.0268	0.0262	0.0256	0.0250	0.0244	0.0239	0.0233
−1.8	0.0359	0.0351	0.0344	0.0336	0.0329	0.0322	0.0314	0.0307	0.0301	0.0294
−1.7	0.0446	0.0436	0.0427	0.0418	0.0409	0.0401	0.0392	0.0384	0.0375	0.0367
−1.6	0.0548	0.0537	0.0526	0.0516	0.0505	0.0495	0.0485	0.0475	0.0465	0.0455
−1.5	0.0668	0.0655	0.0643	0.0630	0.0618	0.0606	0.0594	0.0582	0.0571	0.0559

TABLE A.1. (*Continued*)

z	0.00	0.01	0.02	0.03	0.04	0.05	0.06	0.07	0.08	0.09
-1.4	0.0808	0.0793	0.0778	0.0764	0.0749	0.0735	0.0721	0.0708	0.0694	0.0681
-1.3	0.0968	0.0951	0.0934	0.0918	0.0901	0.0885	0.0869	0.0853	0.0838	0.0823
-1.2	0.1151	0.1131	0.1112	0.1093	0.1075	0.1057	0.1038	0.1020	0.1003	0.0985
-1.1	0.1357	0.1335	0.1314	0.1292	0.1271	0.1251	0.1230	0.1210	0.1190	0.1170
-1.0	0.1587	0.1562	0.1539	0.1515	0.1492	0.1469	0.1446	0.1423	0.1401	0.1379
-0.9	0.1841	0.1814	0.1788	0.1762	0.1736	0.1711	0.1685	0.1660	0.1635	0.1611
-0.8	0.2119	0.2090	0.2061	0.2033	0.2005	0.1977	0.1949	0.1922	0.1894	0.1867
-0.7	0.2420	0.2389	0.2358	0.2327	0.2297	0.2266	0.2236	0.2206	0.2177	0.2148
-0.6	0.2743	0.2709	0.2676	0.2643	0.2611	0.2578	0.2546	0.2514	0.2483	0.2451
-0.5	0.3085	0.3050	0.3015	0.2981	0.2946	0.2912	0.2877	0.2843	0.2810	0.2776
-0.4	0.3446	0.3409	0.3372	0.3336	0.3300	0.3264	0.3228	0.3192	0.3156	0.3121
-0.3	0.3821	0.3783	0.3745	0.3707	0.3669	0.3632	0.3594	0.3557	0.3520	0.3483
-0.2	0.4207	0.4168	0.4129	0.4090	0.4052	0.4013	0.3974	0.3936	0.3897	0.3859
-0.1	0.4602	0.4562	0.4522	0.4483	0.4443	0.4404	0.4364	0.4325	0.4286	0.4247
-0.0	0.5000	0.4960	0.4920	0.4880	0.4840	0.4801	0.4761	0.4721	0.4681	0.4641
0.0	0.5000	0.5040	0.5080	0.5120	0.5160	0.5199	0.5239	0.5279	0.5319	0.5359
0.1	0.5398	0.5438	0.5478	0.5517	0.5557	0.5596	0.5636	0.5675	0.5714	0.5753
0.2	0.5793	0.5832	0.5871	0.5910	0.5948	0.5987	0.6026	0.6064	0.6103	0.6141
0.3	0.6179	0.6217	0.6255	0.6293	0.6331	0.6368	0.6406	0.6443	0.6480	0.6517
0.4	0.6554	0.6591	0.6628	0.6664	0.6700	0.6736	0.6772	0.6808	0.6844	0.6879
0.5	0.6915	0.6950	0.6985	0.7019	0.7054	0.7088	0.7123	0.7157	0.7190	0.7224
0.6	0.7257	0.7291	0.7324	0.7357	0.7389	0.7422	0.7454	0.7486	0.7517	0.7549
0.7	0.7580	0.7611	0.7642	0.7673	0.7703	0.7734	0.7764	0.7794	0.7823	0.7852
0.8	0.7881	0.7910	0.7939	0.7967	0.7995	0.8023	0.8051	0.8078	0.8106	0.8133
0.9	0.8159	0.8186	0.8212	0.8238	0.8264	0.8289	0.8315	0.8340	0.8365	0.8389
1.0	0.8413	0.8438	0.8461	0.8485	0.8508	0.8531	0.8554	0.8577	0.8599	0.8621
1.1	0.8643	0.8665	0.8686	0.8708	0.8729	0.8749	0.8770	0.8790	0.8810	0.8830
1.2	0.8849	0.8869	0.8888	0.8907	0.8925	0.8943	0.8962	0.8980	0.8997	0.9015
1.3	0.9032	0.9049	0.9066	0.9082	0.9099	0.9115	0.9131	0.9147	0.9162	0.9177
1.4	0.9192	0.9207	0.9222	0.9236	0.9251	0.9265	0.9279	0.9292	0.9306	0.9319
1.5	0.9332	0.9345	0.9357	0.9370	0.9382	0.9394	0.9406	0.9418	0.9429	0.9441
1.6	0.9452	0.9463	0.9474	0.9484	0.9495	0.9505	0.9515	0.9525	0.9535	0.9545
1.7	0.9554	0.9564	0.9573	0.9582	0.9591	0.9599	0.9608	0.9616	0.9625	0.9633
1.8	0.9641	0.9649	0.9656	0.9664	0.9671	0.9678	0.9686	0.9693	0.9699	0.9706
1.9	0.9713	0.9719	0.9726	0.9732	0.9738	0.9744	0.9750	0.9756	0.9761	0.9767
2.0	0.9772	0.9778	0.9783	0.9788	0.9793	0.9798	0.9803	0.9808	0.9812	0.9817
2.1	0.9821	0.9826	0.9830	0.9834	0.9838	0.9842	0.9846	0.9850	0.9854	0.9857
2.2	0.9861	0.9864	0.9868	0.9871	0.9875	0.9878	0.9881	0.9884	0.9887	0.9890
2.3	0.9893	0.9896	0.9898	0.9901	0.9904	0.9906	0.9909	0.9911	0.9913	0.9916
2.4	0.9918	0.9920	0.9922	0.9924	0.9927	0.9929	0.9931	0.9932	0.9934	0.9936
2.5	0.9938	0.9940	0.9941	0.9943	0.9945	0.9946	0.9948	0.9949	0.9951	0.9952
2.6	0.9953	0.9955	0.9956	0.9957	0.9959	0.9960	0.9961	0.9962	0.9963	0.9964
2.7	0.9965	0.9966	0.9967	0.9968	0.9969	0.9970	0.9971	0.9972	0.9973	0.9974

TABLE A.1. (*Concluded*)

z	0.00	0.01	0.02	0.03	0.04	0.05	0.06	0.07	0.08	0.09
2.8	0.9974	0.9975	0.9976	0.9977	0.9977	0.9978	0.9979	0.9979	0.9980	0.9981
2.9	0.9981	0.9982	0.9982	0.9983	0.9984	0.9984	0.9985	0.9985	0.9986	0.9986
3.0	0.9986	0.9987	0.9987	0.9988	0.9988	0.9989	0.9989	0.9989	0.9990	0.9990
3.1	0.9990	0.9991	0.9991	0.9991	0.9992	0.9992	0.9992	0.9992	0.9993	0.9993
3.2	0.9993	0.9993	0.9994	0.9994	0.9994	0.9994	0.9994	0.9995	0.9995	0.9995
3.3	0.9995	0.9995	0.9995	0.9996	0.9996	0.9996	0.9996	0.9996	0.9996	0.9997
3.4	0.9997	0.9997	0.9997	0.9997	0.9997	0.9997	0.9997	0.9997	0.9997	0.9998
3.5	0.9998	0.9998	0.9998	0.9998	0.9998	0.9998	0.9998	0.9998	0.9998	0.9998
3.6	0.9998	0.9998	0.9999	0.9999	0.9999	0.9999	0.9999	0.9999	0.9999	0.9999
3.7	0.9999	0.9999	0.9999	0.9999	0.9999	0.9999	0.9999	0.9999	0.9999	0.9999
3.8	0.9999	0.9999	0.9999	0.9999	0.9999	0.9999	0.9999	0.9999	0.9999	0.9999
3.9	1.0000									

TABLE A.2. Percentiles of the Student's t Distribution

Example: The $q = 95$th percentile for $\nu = 10$ degrees of freedom is 1.812.

q / ν	60	75	90	95	97.5	99	99.5	99.95
1	.325	1.000	3.078	6.314	12.706	31.821	63.657	636.619
2	.289	.816	1.886	2.920	4.303	6.965	9.925	31.598
3	.277	.765	1.638	2.353	3.182	4.541	5.841	12.941
4	.271	.741	1.533	2.132	2.776	3.747	4.604	8.610
5	.267	.727	1.476	2.015	2.571	3.365	4.032	6.859
6	.265	.718	1.440	1.943	2.447	3.143	3.707	5.959
7	.263	.711	1.415	1.895	2.365	2.998	3.499	5.405
8	.262	.706	1.397	1.860	2.306	2.896	3.355	5.041
9	.261	.703	1.383	1.833	2.262	2.821	3.250	4.781
10	.260	.700	1.372	1.812	2.228	2.764	3.169	4.587
11	.260	.697	1.363	1.796	2.201	2.718	3.106	4.437
12	.259	.695	1.356	1.782	2.179	2.681	3.055	4.318
13	.259	.694	1.350	1.771	2.160	2.650	3.012	4.221
14	.258	.692	1.345	1.761	2.145	2.624	2.977	4.140
15	.258	.691	1.341	1.753	2.131	2.602	2.947	4.073
16	.258	.690	1.337	1.746	2.120	2.583	2.921	4.015
17	.257	.689	1.333	1.740	2.110	2.567	2.898	3.965
18	.257	.688	1.330	1.734	2.101	2.552	2.878	3.922
19	.257	.688	1.328	1.729	2.093	2.539	2.861	3.883
20	.257	.687	1.325	1.725	2.086	2.528	2.845	3.850
21	.257	.686	1.323	1.721	2.080	2.518	2.831	3.819
22	.256	.686	1.321	1.717	2.074	2.508	2.819	3.792
23	.256	.685	1.319	1.714	2.069	2.500	2.807	3.767
24	.256	.685	1.318	1.711	2.064	2.492	2.797	3.745
25	.256	.684	1.316	1.708	2.060	2.485	2.787	3.725
26	.256	.684	1.315	1.706	2.056	2.479	2.779	3.707
27	.256	.684	1.314	1.703	2.052	2.473	2.771	3.690
28	.256	.683	1.313	1.701	2.048	2.467	2.763	3.674
29	.256	.683	1.311	1.699	2.045	2.462	2.756	3.659
30	.256	.683	1.310	1.697	2.042	2.457	2.750	3.646
40	.255	.681	1.303	1.684	2.021	2.423	2.704	3.551
60	.254	.679	1.296	1.671	2.000	2.390	2.660	3.460
120	.254	.677	1.289	1.658	1.980	2.358	2.617	3.373
∞	.253	.674	1.282	1.645	1.960	2.326	2.576	3.291

Reproduced with kind permission of the publisher and authors from Afifi and Azen, *Statistical Analysis: A Computer Oriented Approach*, 2nd ed., Academic Press, New York, 1979.

TABLE A.3. Percentiles of the Chi-square Distribution

Example: The 90th percentile for d.f. = 15 is 22.307.

% d.f.	0.5	1	2.5	5	10	20	30	40	50
1	0.0001	0.0002	0.001	0.004	0.016	0.064	0.148	0.275	0.455
2	0.010	0.020	0.051	0.103	0.211	0.446	0.713	1.022	1.386
3	0.072	0.115	0.216	0.352	0.584	1.005	1.424	1.869	2.366
4	0.207	0.297	0.484	0.711	1.064	1.649	2.195	2.753	3.357
5	0.412	0.554	0.831	1.145	1.610	2.343	3.000	3.655	4.351
6	0.676	0.872	1.237	1.635	2.204	3.070	3.828	4.570	5.348
7	0.989	1.239	1.690	2.167	2.833	3.822	4.671	5.493	6.346
8	1.344	1.646	2.180	2.733	3.490	4.594	5.527	6.423	7.344
9	1.735	2.088	2.700	3.325	4.168	5.380	6.393	7.357	8.343
10	2.156	2.558	3.247	3.940	4.865	6.179	7.267	8.295	9.342
11	2.603	3.053	3.816	4.575	5.578	6.989	8.148	9.237	10.341
12	3.074	3.571	4.404	5.226	6.304	7.807	9.034	10.182	11.340
13	3.565	4.107	5.009	5.892	7.042	8.634	9.926	11.129	12.340
14	4.075	4.660	5.629	6.571	7.790	9.467	10.821	12.078	13.339
15	4.601	5.229	6.262	7.261	8.547	10.307	11.721	13.030	14.339
16	5.142	5.812	6.908	7.962	9.312	11.152	12.624	13.983	15.338
17	5.697	6.408	7.564	8.672	10.085	12.002	13.531	14.937	16.338
18	6.265	7.015	8.231	9.390	10.865	12.857	14.440	15.893	17.338
19	6.844	7.633	8.907	10.117	11.651	13.716	15.352	16.850	18.338
20	7.434	8.260	9.591	10.851	12.443	14.578	16.266	17.809	19.337
21	8.034	8.897	10.283	11.591	13.240	15.445	17.182	18.768	20.337
22	8.643	9.542	10.982	12.338	14.041	16.314	18.101	19.729	21.337
23	9.260	10.196	11.689	13.091	14.848	17.187	19.021	20.690	22.337
24	9.886	10.856	12.401	13.848	15.659	18.062	19.943	21.752	23.337
25	10.520	11.524	13.120	14.611	16.473	18.940	20.867	22.616	24.337
26	11.160	12.198	13.844	15.379	17.292	19.820	21.792	23.579	25.336
27	11.808	12.879	14.573	16.151	18.114	20.703	22.719	24.544	26.336
28	12.461	13.565	15.308	16.928	18.939	21.588	23.647	25.509	27.336
29	13.121	14.256	16.047	17.708	19.768	22.475	24.577	26.475	28.336
30	13.787	14.953	16.791	18.493	20.599	23.364	25.508	27.442	29.336
35	17.192	18.509	20.569	22.465	24.797	27.836	30.178	32.282	34.336
40	20.707	22.164	24.433	26.509	29.051	32.345	34.872	37.134	39.335
45	24.311	25.901	28.366	30.612	33.350	36.884	39.585	41.995	44.335
50	27.991	29.707	32.357	34.764	37.689	41.449	44.313	46.864	49.335
60	35.534	37.485	40.482	43.188	46.459	50.641	53.809	56.620	59.335
70	43.275	45.442	48.758	51.739	55.329	59.898	63.346	66.396	69.334
80	51.172	53.540	57.153	60.391	64.278	69.207	72.915	76.188	79.334

TABLE A.3. (*Continued*)

% d.f.	0.5	1	2.5	5	10	20	30	40	50
90	59.196	61.754	65.647	69.126	73.291	78.558	82.511	85.993	89.334
100	67.328	70.065	74.222	77.929	82.358	87.945	92.129	95.808	99.334
120	83.852	86.923	91.573	95.705	100.624	106.806	111.419	115.465	119.334
140	100.655	104.034	109.137	113.659	119.029	125.758	130.766	135.149	139.334
160	117.679	121.346	126.870	131.756	137.546	144.783	150.158	154.856	159.334
180	134.884	138.820	144.741	149.969	156.153	163.868	169.588	174.580	179.334
200	152.241	156.432	162.728	168.279	174.835	183.003	189.049	194.319	199.344

% d.f.	60	70	80	90	95	97.5	99	99.5	99.95
1	0.708	1.074	1.642	2.706	3.841	5.024	6.635	7.879	12.116
2	1.833	2.408	3.219	4.605	5.991	7.378	9.210	10.597	15.202
3	2.946	3.665	4.642	6.251	7.815	9.348	11.345	12.838	17.730
4	4.045	4.878	5.989	7.779	9.488	11.143	13.277	14.860	19.997
5	5.132	6.064	7.289	9.236	11.070	12.833	15.086	16.750	22.105
6	6.211	7.231	8.558	10.645	12.592	14.449	16.812	18.548	24.103
7	7.283	8.383	9.803	12.017	14.067	16.013	18.475	20.278	26.018
8	8.351	9.524	11.030	13.362	15.507	17.535	20.090	21.955	27.868
9	9.414	10.656	12.242	14.684	16.919	19.023	21.666	23.589	29.666
10	10.473	11.781	13.442	15.987	18.307	20.483	23.209	25.188	31.420
11	11.530	12.899	14.631	17.275	19.675	21.920	24.725	26.757	33.137
12	12.584	14.011	15.812	18.549	21.026	23.337	26.217	28.300	34.821
13	13.636	15.119	16.985	19.812	22.362	24.736	27.688	29.819	36.478
14	14.685	16.222	18.151	21.064	23.685	26.119	29.141	31.319	38.109
15	15.733	17.322	19.311	22.307	24.996	27.488	30.578	32.801	39.719
16	16.780	18.418	20.465	23.542	26.296	28.845	32.000	34.267	41.308
17	17.824	19.511	21.615	24.769	27.587	30.191	33.409	35.718	42.879
18	18.868	20.601	22.760	25.989	28.869	31.526	34.805	37.156	44.434
19	19.910	21.689	23.900	27.204	30.144	32.852	36.191	38.582	45.973
20	20.951	22.775	25.038	28.412	31.410	34.170	37.566	39.997	47.498
21	21.991	23.858	26.171	29.615	32.671	35.479	38.932	41.401	49.011
22	23.031	24.939	27.301	30.813	33.924	36.781	40.289	42.796	50.511
23	24.069	26.018	28.429	32.007	35.172	38.076	41.638	44.181	52.000
24	25.106	27.096	29.553	33.196	36.415	39.364	42.980	45.559	53.479
25	26.143	28.172	30.675	34.382	37.652	40.646	44.314	46.928	54.947
26	27.179	29.246	31.795	35.563	38.885	41.923	45.642	48.290	56.407
27	28.214	30.319	32.912	36.741	40.113	43.195	46.963	49.645	57.858
28	29.249	31.391	34.027	37.916	41.337	44.461	48.278	50.993	59.300
29	30.283	32.461	35.139	39.087	42.557	45.722	49.588	52.336	60.735
30	31.316	33.530	36.250	40.256	43.773	46.979	50.892	53.672	62.162
35	36.475	38.859	41.778	46.059	49.802	53.203	57.342	60.275	69.199
40	41.622	44.165	47.269	51.805	56.758	59.342	63.691	66.766	76.095
45	46.761	49.452	52.729	57.505	61.656	65.410	69.957	73.166	82.876
50	51.892	54.723	58.164	63.167	67.505	71.420	76.154	79.490	89.561
60	62.135	65.227	68.972	74.397	79.082	83.298	88.379	91.952	102.695
70	72.358	75.689	79.715	85.527	90.531	95.023	100.425	104.215	115.578

TABLE A.3. (*Concluded*)

% d.f.	60	70	80	90	95	97.5	99	99.5	99.95
80	82.566	86.120	90.405	96.578	101.879	106.629	112.329	116.321	128.261
90	92.761	96.524	101.054	107.565	113.145	118.136	124.116	128.299	140.782
100	102.946	106.906	111.667	118.498	124.342	129.561	135.807	140.169	153.167
120	123.289	127.616	132.806	140.233	146.567	152.211	158.950	163.648	177.603
140	143.604	148.269	153.854	161.827	168.613	174.648	181.840	186.847	201.683
160	163.898	168.876	174.828	183.311	190.516	196.915	204.530	209.824	225.481
180	184.173	189.446	195.743	204.704	212.304	219.044	227.056	232.620	249.048
200	204.434	209.985	216.609	226.021	233.994	241.058	249.445	255.264	272.423

TABLE A.4. Percentiles of the F Distribution

a. Percentiles for $P = 0.25$, 0.15, and 0.10, with 1 and v_2 Degrees of Freedom.

Example: The 75th percentile ($P = 0.25$) with 1 and $v_2 = 10$ df is 1.4915.

v_2	$v_1 = 1$		
	$P = 0.25$	$P = 0.15$	$P = 0.10$
1	5.8285	17.3497	39.864
2	2.5714	5.2072	8.5263
3	2.0239	3.7030	5.5383
4	1.8074	3.1620	4.5448
5	1.6925	2.8878	4.0604
6	1.6214	2.7231	3.7760
7	1.5732	2.6134	3.5894
8	1.5384	2.5352	3.4579
9	1.5121	2.4766	3.3603
10	1.4915	2.4312	3.2850
11	1.4749	2.3949	3.2252
12	1.4613	2.3653	3.1765
13	1.4500	2.3407	3.1362
14	1.4403	2.3198	3.1022
15	1.4321	2.3020	3.0732
16	1.4249	2.2865	3.0481
17	1.4186	2.2730	3.0262
18	1.4130	2.2611	3.0070
19	1.4081	2.2506	2.9899
20	1.4037	2.2411	2.9747
21	1.3997	2.2326	2.9609
22	1.3961	2.2249	2.9486
23	1.3928	2.2179	2.9374
24	1.3898	2.2116	2.9271
25	1.3870	2.2057	2.9177
26	1.3845	2.2004	2.9091
27	1.3822	2.1954	2.9012
28	1.3800	2.1908	2.8939
29	1.3780	2.1866	2.8871
30	1.3761	2.1826	2.8807
40	1.3626	2.1542	2.8354
60	1.3493	2.1264	2.7914
120	1.3362	2.0990	2.7478
∞	1.3233	2.0722	2.7055

TABLE A.4. (*Continued*)

b. Percentiles for P = 0.05, 0.01, and 0.005

Example: The 95th percentile (P = 0.05) for v_1 = 4 and v_2 = 15 is 3.0556.

$P = 0.05$

v_2 \ v_1	1	2	3	4	5	6	7	8	9
1	161·45	199·50	215·71	224·58	230·16	233·99	236·77	238·88	240·54
2	18·513	19·000	19·164	19·247	19·296	19·330	19·353	19·371	19·385
3	10·128	9·5521	9·2766	9·1172	9·0135	8·9406	8·8868	8·8452	8·8123
4	7·7086	6·9443	6·5914	6·3883	6·2560	6·1631	6·0942	6·0410	5·9988
5	6·6079	5·7861	5·4095	5·1922	5·0503	4·9503	4·8759	4·8183	4·7725
6	5·9874	5·1433	4·7571	4·5337	4·3874	4·2839	4·2066	4·1468	4·0990
7	5·5914	4·7374	4·3468	4·1203	3·9715	3·8660	3·7870	3·7257	3·6767
8	5·3177	4·4590	4·0662	3·8378	3·6875	3·5806	3·5005	3·4381	3·3881
9	5·1174	4·2565	3·8626	3·6331	3·4817	3·3738	3·2927	3·2296	3·1789
10	4·9646	4·1028	3·7083	3·4780	3·3258	3·2172	3·1355	3·0717	3·0204
11	4·8443	3·9823	3·5874	3·3567	3·2039	3·0946	3·0123	2·9480	2·8962
12	4·7472	3·8853	3·4903	3·2592	3·1059	2·9961	2·9134	2·8486	2·7964
13	4·6672	3·8056	3·4105	3·1791	3·0254	2·9153	2·8321	2·7669	2·7144
14	4·6001	3·7389	3·3439	3·1122	2·9582	2·8477	2·7642	2·6987	2·6458
15	4·5431	3·6823	3·2874	3·0556	2·9013	2·7905	2·7066	2·6408	2·5876
16	4·4940	3·6337	3·2389	3·0069	2·8524	2·7413	2·6572	2·5911	2·5377
17	4·4513	3·5915	3·1968	2·9647	2·8100	2·6987	2·6143	2·5480	2·4943
18	4·4139	3·5546	3·1599	2·9277	2·7729	2·6613	2·5767	2·5102	2·4563
19	4·3808	3·5219	3·1274	2·8951	2·7401	2·6283	2·5435	2·4768	2·4227
20	4·3513	3·4928	3·0984	2·8661	2·7109	2·5990	2·5140	2·4471	2·3928
21	4·3248	3·4668	3·0725	2·8401	2·6848	2·5727	2·4876	2·4205	2·3661
22	4·3009	3·4434	3·0491	2·8167	2·6613	2·5491	2·4638	2·3965	2·3419
23	4·2793	3·4221	3·0280	2·7955	2·6400	2·5277	2·4422	2·3748	2·3201
24	4·2597	3·4028	3·0088	2·7763	2·6207	2·5082	2·4226	2·3551	2·3002
25	4·2417	3·3852	2·9912	2·7587	2·6030	2·4904	2·4047	2·3371	2·2821
26	4·2252	3·3690	2·9751	2·7426	2·5868	2·4741	2·3883	2·3205	2·2655
27	4·2100	3·3541	2·9604	2·7278	2·5719	2·4591	2·3732	2·3053	2·2501
28	4·1960	3·3404	2·9467	2·7141	2·5581	2·4453	2·3593	2·2913	2·2360
29	4·1830	3·3277	2·9340	2·7014	2·5454	2·4324	2·3463	2·2782	2·2229
30	4·1709	3·3158	2·9223	2·6896	2·5336	2·4205	2·3343	2·2662	2·2107
40	4·0848	3·2317	2·8387	2·6060	2·4495	2·3359	2·2490	2·1802	2·1240
60	4·0012	3·1504	2·7581	2·5252	2·3683	2·2540	2·1665	2·0970	2·0401
120	3·9201	3·0718	2·6802	2·4472	2·2900	2·1750	2·0867	2·0164	1·9588
∞	3·8415	2·9957	2·6049	2·3719	2·2141	2·0986	2·0096	1·9384	1·8799

TABLE A.4. (*Continued*)

$$P = 0.05$$

v_1 / v_2	10	12	15	20	24	30	40	60	120	∞
1	241·88	243·91	245·95	248·01	249·05	250·09	251·14	252·20	253·25	254·32
2	19·396	19·413	19·429	19·446	19·454	19·462	19·471	19·479	19·487	19·496
3	8·7855	8·7446	8·7029	8·6602	8·6385	8·6166	8·5944	8·5720	8·5494	8·5265
4	5·9644	5·9117	5·8578	5·8025	5·7744	5·7459	5·7170	5·6878	5·6581	5·6281
5	4·7351	4·6777	4·6188	4·5581	4·5272	4·4957	4·4638	4·4314	4·3984	4·3650
6	4·0600	3·9999	3·9381	3·8742	3·8415	3·8082	3·7743	3·7398	3·7047	3·6688
7	3·6365	3·5747	3·5108	3·4445	3·4105	3·3758	3·3404	3·3043	3·2674	3·2298
8	3·3472	3·2840	3·2184	3·1503	3·1152	3·0794	3·0428	3·0053	2·9669	2·9276
9	3·1373	3·0729	3·0061	2·9365	2·9005	2·8637	2·8259	2·7872	2·7475	2·7067
10	2·9782	2·9130	2·8450	2·7740	2·7372	2·6996	2·6609	2·6211	2·5801	2·5379
11	2·8536	2·7876	2·7186	2·6464	2·6090	2·5705	2·5309	2·4901	2·4480	2·4045
12	2·7534	2·6866	2·6169	2·5436	2·5055	2·4663	2·4259	2·3842	2·3410	2·2962
13	2·6710	2·6037	2·5331	2·4589	2·4202	2·3803	2·3392	2·2966	2·2524	2·2064
14	2·6021	2·5342	2·4630	2·3879	2·3487	2·3082	2·2664	2·2230	2·1778	2·1307
15	2·5437	2·4753	2·4035	2·3275	2·2878	2·2468	2·2043	2·1601	2·1141	2·0658
16	2·4935	2·4247	2·3522	2·2756	2·2354	2·1938	2·1507	2·1058	2·0589	2·0096
17	2·4499	2·3807	2·3077	2·2304	2·1898	2·1477	2·1040	2·0584	2·0107	1·9604
18	2·4117	2·3421	2·2686	2·1906	2·1497	2·1071	2·0629	2·0166	1·9681	1·9168
19	2·3779	2·3080	2·2341	2·1555	2·1141	2·0712	2·0264	1·9796	1·9302	1·8780
20	2·3479	2·2776	2·2033	2·1242	2·0825	2·0391	1·9938	1·9464	1·8963	1·8432
21	2·3210	2·2504	2·1757	2·0960	2·0540	2·0102	1·9645	1·9165	1·8657	1·8117
22	2·2967	2·2258	2·1508	2·0707	2·0283	1·9842	1·9380	1·8895	1·8380	1·7831
23	2·2747	2·2036	2·1282	2·0476	2·0050	1·9605	1·9139	1·8649	1·8128	1·7570
24	2·2547	2·1834	2·1077	2·0267	1·9838	1·9390	1·8920	1·8424	1·7897	1·7331
25	2·2365	2·1649	2·0889	2·0075	1·9643	1·9192	1·8718	1·8217	1·7684	1·7110
26	2·2197	2·1479	2·0716	1·9898	1·9464	1·9010	1·8533	1·8027	1·7488	1·6906
27	2·2043	2·1323	2·0558	1·9736	1·9299	1·8842	1·8361	1·7851	1·7307	1·6717
28	2·1900	2·1179	2·0411	1·9586	1·9147	1·8687	1·8203	1·7689	1·7138	1·6541
29	2·1768	2·1045	2·0275	1·9446	1·9005	1·8543	1·8055	1·7537	1·6981	1·6377
30	2·1646	2·0921	2·0148	1·9317	1·8874	1·8409	1·7918	1·7396	1·6835	1·6223
40	2·0772	2·0035	1·9245	1·8389	1·7929	1·7444	1·6928	1·6373	1·5766	1·5089
60	1·9926	1·9174	1·8364	1·7480	1·7001	1·6491	1·5943	1·5343	1·4673	1·3893
120	1·9105	1·8337	1·7505	1·6587	1·6084	1·5543	1·4952	1·4290	1·3519	1·2539
∞	1·8307	1·7522	1·6664	1·5705	1·5173	1·4591	1·3940	1·3180	1·2214	1·0000

TABLE A.4. (*Continued*)

$$P = 0.01$$

ν_1 / ν_2	1	2	3	4	5	6	7	8	9
1	4052·2	4999·5	5403·3	5624·6	5763·7	5859·0	5928·3	5981·6	6022·5
2	98·503	99·000	99·166	99·249	99·299	99·332	99·356	99·374	99·388
3	34·116	30·817	29·457	28·710	28·237	27·911	27·672	27·489	27·345
4	21·198	18·000	16·694	15·977	15·522	15·207	14·976	14·799	14·659
5	16·258	13·274	12·060	11·392	10·967	10·672	10·456	10·289	10·158
6	13·745	10·925	9·7795	9·1483	8·7459	8·4661	8·2600	8·1016	7·9761
7	12·246	9·5466	8·4513	7·8467	7·4604	7·1914	6·9928	6·8401	6·7188
8	11·259	8·6491	7·5910	7·0060	6·6318	6·3707	6·1776	6·0289	5·9106
9	10·561	8·0215	6·9919	6·4221	6·0569	5·8018	5·6129	5·4671	5·3511
10	10·044	7·5594	6·5523	5·9943	5·6363	5·3858	5·2001	5·0567	4·9424
11	9·6460	7·2057	6·2167	5·6683	5·3160	5·0692	4·8861	4·7445	4·6315
12	9·3302	6·9266	5·9526	5·4119	5·0643	4·8206	4·6395	4·4994	4·3875
13	9·0738	6·7010	5·7394	5·2053	4·8616	4·6204	4·4410	4·3021	4·1911
14	8·8616	6·5149	5·5639	5·0354	4·6950	4·4558	4·2779	4·1399	4·0297
15	8·6831	6·3589	5·4170	4·8932	4·5556	4·3183	4·1415	4·0045	3·8948
16	8·5310	6·2262	5·2922	4·7726	4·4374	4·2016	4·0259	3·8896	3·7804
17	8·3997	6·1121	5·1850	4·6690	4·3359	4·1015	3·9267	3·7910	3·6822
18	8·2854	6·0129	5·0919	4·5790	4·2479	4·0146	3·8406	3·7054	3·5971
19	8·1850	5·9259	5·0103	4·5003	4·1708	3·9386	3·7653	3·6305	3·5225
20	8·0960	5·8489	4·9382	4·4307	4·1027	3·8714	3·6987	3·5644	3·4567
21	8·0166	5·7804	4·8740	4·3688	4·0421	3·8117	3·6396	3·5056	3·3981
22	7·9454	5·7190	4·8166	4·3134	3·9880	3·7583	3·5867	3·4530	3·3458
23	7·8811	5·6637	4·7649	4·2635	3·9392	3·7102	3·5390	3·4057	3·2986
24	7·8229	5·6136	4·7181	4·2184	3·8951	3·6667	3·4959	3·3629	3·2560
25	7·7698	5·5680	4·6755	4·1774	3·8550	3·6272	3·4568	3;3239	3·2172
26	7·7213	5·5263	4·6366	4·1400	3·8183	3·5911	3·4210	3·2884	3·1818
27	7·6767	5·4881	4·6009	4·1056	3·7848	3·5580	3·3882	3·2558	3·1494
28	7·6356	5·4529	4·5681	4·0740	3·7539	3·5276	3·3581	3·2259	3·1195
29	7·5976	5·4205	4·5378	4·0449	3·7254	3·4995	3·3302	3·1982	3·0920
30	7·5625	5·3904	4·5097	4·0179	3·6990	3·4735	3·3045	3·1726	3·0665
40	7·3141	5·1785	4·3126	3·8283	3·5138	3·2910	3·1238	2·9930	2·8876
60	7·0771	4·9774	4·1259	3·6491	3·3389	3·1187	2·9530	2·8233	2·7185
120	6·8510	4·7865	3·9493	3·4796	3·1735	2·9559	2·7918	2·6629	2·5586
∞	6·6349	4·6052	3·7816	3·3192	3·0173	2·8020	2·6393	2·5113	2·4073

TABLE A.4. (*Continued*)

P = 0.01

ν_1 / ν_2	10	12	15	20	24	30	40	60	120	∞
1	6055·8	6106·3	6157·3	6208·7	6234·6	6260·7	6286·8	6313·0	6339·4	6366·0
2	99·399	99·416	99·432	99·449	99·458	99·466	99·474	99·483	99·491	99·501
3	27·229	27·052	26·872	26·690	26·598	26·505	26·411	26·316	26·221	26·125
4	14·546	14·374	14·198	14·020	13·929	13·838	13·745	13·652	13·558	13·463
5	10·051	9·8883	9·7222	9·5527	9·4665	9·3793	9·2912	9·2020	9·1118	9·0204
6	7·8741	7·7183	7·5590	7·3958	7·3127	7·2285	7·1432	7·0568	6·9690	6·8801
7	6·6201	6·4691	6·3143	6·1554	6·0743	5·9921	5·9084	5·8236	5·7372	5·6495
8	5·8143	5·6668	5·5151	5·3591	5·2793	5·1981	5·1156	5·0316	4·9460	4·8588
9	5·2565	5·1114	4·9621	4·8080	4·7290	4·6486	4·5667	4·4831	4·3978	4·3105
10	4·8492	4·7059	4·5582	4·4054	4·3269	4·2469	4·1653	4·0819	3·9965	3·9090
11	4·5393	4·3974	4·2509	4·0990	4·0209	3·9411	3·8596	3·7761	3·6904	3·6025
12	4·2961	4·1553	4·0096	3·8584	3·7805	3·7008	3·6192	3·5355	3·4494	3·3608
13	4·1003	3·9603	3·8154	3·6646	3·5868	3·5070	3·4253	3·3413	3·2548	3·1654
14	3·9394	3·8001	3·6557	3·5052	3·4274	3·3476	3·2656	3·1813	3·0942	3·0040
15	3·8049	3·6662	3·5222	3·3719	3·2940	3·2141	3·1319	3·0471	2·9595	2·8684
16	3·6909	3·5527	3·4089	3·2588	3·1808	3·1007	3·0182	2·9330	2·8447	2·7528
17	3·5931	3·4552	3·3117	3·1615	3·0835	3·0032	2·9205	2·8348	2·7459	2·6530
18	3·5082	3·3706	3·2273	3·0771	2·9990	2·9185	2·8354	2·7493	2·6597	2·5660
19	3·4338	3·2965	3·1533	3·0031	2·9249	2·8442	2·7608	2·6742	2·5839	2·4893
20	3·3682	3·2311	3·0880	2·9377	2·8594	2·7785	2·6947	2·6077	2·5168	2·4212
21	3·3098	3·1729	3·0299	2·8796	2·8011	2·7200	2·6359	2·5484	2·4568	2·3603
22	3·2576	3·1209	2·9780	2·8274	2·7488	2·6675	2·5831	2·4951	2·4029	2·3055
23	3·2106	3·0740	2·9311	2·7805	2·7017	2·6202	2·5355	2·4471	2·3542	2·2559
24	3·1681	3·0316	2·8887	2·7380	2·6591	2·5773	2·4923	2·4035	2·3099	2·2107
25	3·1294	2·9931	2·8502	2·6993	2·6203	2·5383	2·4530	2·3637	2·2695	2·1694
26	3·0941	2·9579	2·8150	2·6640	2·5848	2·5026	2·4170	2·3273	2·2325	2·1315
27	3·0618	2·9256	2·7827	2·6316	2·5522	2·4699	2·3840	2·2938	2·1984	2·0965
28	3·0320	2·8959	2·7530	2·6017	2·5223	2·4397	2·3535	2·2629	2·1670	2·0642
29	3·0045	2·8685	2·7256	2·5742	2·4946	2·4118	2·3253	2·2344	2·1378	2·0342
30	2·9791	2·8431	2·7002	2·5487	2·4689	2·3860	2·2992	2·2079	2·1107	2·0062
40	2·8005	2·6648	2·5216	2·3689	2·2880	2·2034	2·1142	2·0194	1·9172	1·8047
60	2·6318	2·4961	2·3523	2·1978	2·1154	2·0285	1·9360	1·8363	1·7263	1·6006
120	2·4721	2·3363	2·1915	2·0346	1·9500	1·8600	1·7628	1·6557	1·5330	1·3805
∞	2·3209	2·1848	2·0385	1·8783	1·7908	1·6964	1·5923	1·4730	1·3246	1·0000

TABLE A.4. (*Continued*)

$$P = 0.005$$

ν_1 / ν_2	1	2	3	4	5	6	7	8	9
1	16211	20000	21615	22500	23056	23437	23715	23925	24091
2	198·50	199·00	199·17	199·25	199·30	199·33	199·36	199·37	199·39
3	55·552	49·799	47·467	46·195	45·392	44·838	44·434	44·126	43·882
4	31·333	26·284	24·259	23·155	22·456	21·975	21·622	21·352	21·139
5	22·785	18·314	16·530	15·556	14·940	14·513	14·200	13·961	13·772
6	18·635	14·544	12·917	12·028	11·464	11·073	10·786	10·566	10·391
7	16·236	12·404	10·882	10·050	9·5221	9·1554	8·8854	8·6781	8·5138
8	14·688	11·042	9·5965	8·8051	8·3018	7·9520	7·6942	7·4960	7·3386
9	13·614	10·107	8·7171	7·9559	7·4711	7·1338	6·8849	6·6933	6·5411
10	12·826	9·4270	8·0807	7·3428	6·8723	6·5446	6·3025	6·1159	5·9676
11	12·226	8·9122	7·6004	6·8809	6·4217	6·1015	5·8648	5·6821	5·5368
12	11·754	8·5096	7·2258	6·5211	6·0711	5·7570	5·5245	5·3451	5·2021
13	11·374	8·1865	6·9257	6·2335	5·7910	5·4819	5·2529	5·0761	4·9351
14	11·060	7·9217	6·6803	5·9984	5·5623	5·2574	5·0313	4·8566	4·7173
15	10·798	7·7008	6·4760	5·8029	5·3721	5·0708	4·8473	4·6743	4·5364
16	10·575	7·5138	6·3034	5·6378	5·2117	4·9134	4·6920	4·5207	4·3838
17	10·384	7·3536	6·1556	5·4967	5·0746	4·7789	4·5594	4·3893	4·2535
18	10·218	7·2148	6·0277	5·3746	4·9560	4·6627	4·4448	4·2759	4·1410
19	10·073	7·0935	5·9161	5·2681	4·8526	4·5614	4·3448	4·1770	4·0428
20	9·9439	6·9865	5·8177	5·1743	4·7616	4·4721	4·2569	4·0900	3·9564
21	9·8295	6·8914	5·7304	5·0911	4·6808	4·3931	4·1789	4·0128	3·8799
22	9·7271	6·8064	5·6524	5·0168	4·6088	4·3225	4·1094	3·9440	3·8116
23	9·6348	6·7300	5·5823	4·9500	4·5441	4·2591	4·0469	3·8822	3·7502
24	9·5513	6·6610	5·5190	4·8898	4·4857	4·2019	3·9905	3·8264	3·6949
25	9·4753	6·5982	5·4615	4·8351	4·4327	4·1500	3·9394	3·7758	3·6447
26	9·4059	6·5409	5·4091	4·7852	4·3844	4·1027	3·8928	3·7297	3·5989
27	9·3423	6·4885	5·3611	4·7396	4·3402	4·0594	3·8501	3·6875	3·5571
28	9·2838	6·4403	5·3170	4·6977	4·2996	4·0197	3·8110	3·6487	3·5186
29	9·2297	6·3958	5·2764	4·6591	4·2622	3·9830	3·7749	3·6130	3·4832
30	9·1797	6·3547	5·2388	4·6233	4·2276	3·9492	3·7416	3·5801	3·4505
40	8·8278	6·0664	4·9759	4·3738	3·9860	3·7129	3·5088	3·3498	3·2220
60	8·4946	5·7950	4·7290	4·1399	3·7600	3·4918	3·2911	3·1344	3·0083
120	8·1790	5·5393	4·4973	3·9207	3·5482	3·2849	3·0874	2·9330	2·8083
∞	7·8794	5·2983	4·2794	3·7151	3·3499	3·0913	2·8968	2·7444	2·6210

TABLE A.4. (*Concluded*)

P = 0.005

ν_1 / ν_2	10	12	15	20	24	30	40	60	120	∞
1	24224	24426	24630	24836	24940	25044	25148	25253	25359	25465
2	199·40	199·42	199·43	199·45	199·46	199·47	199·47	199·48	199·49	199·51
3	43·686	43·387	43·085	42·778	42·622	42·466	42·308	42·149	41·989	41·829
4	20·967	20·705	20·438	20·167	20·030	19·892	19·752	19·611	19·468	19·325
5	13·618	13·384	13·146	12·903	12·780	12·656	12·530	12·402	12·274	12·144
6	10·250	10·034	9·8140	9·5888	9·4741	9·3583	9·2408	9·1219	9·0015	8·8793
7	8·3803	8·1764	7·9678	7·7540	7·6450	7·5345	7·4225	7·3088	7·1933	7·0760
8	7·2107	7·0149	6·8143	6·6082	6·5029	6·3961	6·2875	6·1772	6·0649	5·9505
9	6·4171	6·2274	6·0325	5·8318	5·7292	5·6248	5·5186	5·4104	5·3001	5·1875
10	5·8467	5·6613	5·4707	5·2740	5·1732	5·0705	4·9659	4·8592	4·7501	4·6385
11	5·4182	5·2363	5·0489	4·8552	4·7557	4·6543	4·5508	4·4450	4·3367	4·2256
12	5·0855	4·9063	4·7214	4·5299	4·4315	4·3309	4·2282	4·1229	4·0149	3·9039
13	4·8199	4·6429	4·4600	4·2703	4·1726	4·0727	3·9704	3·8655	3·7577	3·6465
14	4·6034	4·4281	4·2468	4·0585	3·9614	3·8619	3·7600	3·6553	3·5473	3·4359
15	4·4236	4·2498	4·0698	3·8826	3·7859	3·6867	3·5850	3·4803	3·3722	3·2602
16	4·2719	4·0994	3·9205	3·7342	3·6378	3·5388	3·4372	3·3324	3·2240	3·1115
17	4·1423	3·9709	3·7929	3·6073	3·5112	3·4124	3·3107	3·2058	3·0971	2·9839
18	4·0305	3·8599	3·6827	3·4977	3·4017	3·3030	3·2014	3·0962	2·9871	2·8732
19	3·9329	3·7631	3·5866	3·4020	3·3062	3·2075	3·1058	3·0004	2·8908	2·7762
20	3·8470	3·6779	3·5020	3·3178	3·2220	3·1234	3·0215	2·9159	2·8058	2·6904
21	3·7709	3·6024	3·4270	3·2431	3·1474	3·0488	2·9467	2·8408	2·7302	2·6140
22	3·7030	3·5350	3·3600	3·1764	3·0807	2·9821	2·8799	2·7736	2·6625	2·5455
23	3·6420	3·4745	3·2999	3·1165	3·0208	2·9221	2·8198	2·7132	2·6016	2·4837
24	3·5870	3·4199	3·2456	3·0624	2·9667	2·8679	2·7654	2·6585	2·5463	2·4276
25	3·5370	3·3704	3·1963	3·0133	2·9176	2·8187	2·7160	2·6088	2·4960	2·3765
26	3·4916	3·3252	3·1515	2·9685	2·8728	2·7738	2·6709	2·5633	2·4501	2·3297
27	3·4499	3·2839	3·1104	2·9275	2·8318	2·7327	2·6296	2·5217	2·4078	2·2867
28	3·4117	3·2460	3·0727	2·8899	2·7941	2·6949	2·5916	2·4834	2·3689	2·2469
29	3·3765	3·2111	3·0379	2·8551	2·7594	2·6601	2·5565	2·4479	2·3330	2·2102
30	3·3440	3·1787	3·0057	2·8230	2·7272	2·6278	2·5241	2·4151	2·2997	2·1760
40	3·1167	2·9531	2·7811	2·5984	2·5020	2·4015	2·2958	2·1838	2·0635	1·9318
60	2·9042	2·7419	2·5705	2·3872	2·2898	2·1874	2·0789	1·9622	1·8341	1·6885
120	2·7052	2·5439	2·3727	2·1881	2·0890	1·9839	1·8709	1·7469	1·6055	1·4311
∞	2·5188	2·3583	2·1868	1·9998	1·8983	1·7891	1·6691	1·5325	1·3637	1·0000

APPENDIX B
LUNG FUNCTION
DATA

The lung function data set includes information on nonsmoking families from the UCLA study of chronic obstructive respiratory disease (CORD). In the CORD study persons seven years old and older from four areas (Burbank, Lancaster, Long Beach, and Glendora) were sampled, and information was obtained from them at two time periods. The data set presented here is a subset including families with both a mother and father, and one, two, or three children between the ages of 7 and 17 who answered the questionnaire and took the lung function tests at the first time period. The purpose of the CORD study was to determine the effects of different types of air pollutants on respiratory function, but numerous other types of studies have been performed on this data set. Further information concerning the CORD study can be found in the Bibliography at the end of this appendix.

The code book in Table B.1 summarizes the information in this subset of data. After the first two variables, note that the same set of data is given for fathers (F), mothers (M), oldest child (OC), middle child (MC), and youngest child (YC).

The format is F3.0, F1.0, 5(F1.0, 2F2.0, F3.0, 2F3.2). Note that in the listing of the data in Table B.2 columns of blanks have been added to isolate the data for different members of the family. The format statement and the code book assume no blanks. Some families have only one or two children between the ages of 7 and 17. If there is only one child, it is listed as the oldest child. Thus there are numerous missing values in the data for the middle and youngest child.

Forced vital capacity (FVC) is the volume of air, in liters, that can be expelled by the participant after having breathed in as deeply as possible,

TABLE B.1. Code Book for Lung Function Data Set

Variable Number	Variable Location	Variable Name	Description
1	1–3	ID	1–150
2	4	AREA	1 = Burbank
			2 = Lancaster
			3 = Long Beach
			4 = Glendora
3	5	FSEX	1 = male
4	6–7	FAGE	Age, father
5	8–9	FHEIGHT	Height (in), father
6	10–12	FWEIGHT	Weight (lb), father
7	13–15	FFVC	FVC father
8	16–18	FFEV1	FEV1 father
9	19	MSEX	2 = female
10	20–21	MAGE	Age, mother
11	22–23	MHEIGHT	Height (in), mother
12	24–26	MWEIGHT	Weight (lb), mother
13	27–29	MFVC	FVC mother
14	30–32	MFEV1	FEV1 mother
15	33	OCSEX	Sex, oldest child
			1 = male
			2 = female
16	34–35	OCAGE	Age, oldest child
17	36–37	OCHEIGHT	Height, oldest child
18	38–40	OCWEIGHT	Weight, oldest child
19	41–43	OCFVC	FVC oldest child
20	44–46	OCFEV1	FEV1 oldest child
21	47	MCSEX	Sex, middle child
22	48–49	MCAGE	Age, middle child
23	50–51	MCHEIGHT	Height, middle child
24	52–54	MCWEIGHT	Weight, middle child
25	55–57	MCFVC	FVC middle child
26	58–60	MCFEV1	FEV1 middle child
27	61	YCSEX	Sex, youngest child
28	62–63	YCAGE	Age, youngest child
29	64–65	YCHEIGHT	Height, youngest child
30	66–68	YCWEIGHT	Weight, youngest child
31	69–71	YCFVC	FVC youngest child
32	72–74	YCFEV1	FEV1 youngest child

that is, full expiration regardless of how long it takes. The measurement of FVC is affected by the ability of the participant to understand the instructions and by the amount of effort made to breathe in deeply and to expel as much air as possible. For this reason seven years is the practical lower age limit at which valid and reliable data can be obtained. Variable FEV1 is a measure of the volume of the air expelled in the first second after the start of expiration. Variable FEV1 or the ratio FEV1/FVC has been used as an outcome variable in many studies of the effects of smoking on lung function.

The authors wish to thank Dean Roger Detels, principal investigator of the CORD project, for the use of this data set and Miss Cathleen Reems for assembling it.

TABLE B.2. Lung Function Data Set

columns 123	4	1 5678901234567 8 (father)	2 9012345678901 2 (mother)	3 · 4 3456789012345 6 (child #1)	5 7890123456 7890 (child #2)	6 · 7 1234567890 1234 (child #3)
1	1	15361161391323	24362136370331	21259115296279		
2	1	14072198441395	23866160411347	11056 66323239		
3	1	12669210445347	22759114309265	1 850 59114111		
4	1	13468187433374	23658123265206	21157106256185	1 949 56159130	
5	1	14661121354290	23962128245233	11661 88260247	21260 85268234	21050 53154143
6	1	14472153610491	23666125349306	11567100389355	11357 87276237	21055 72195169
7	1	13564145345339	22768206492425	21154 70218163		
8	1	14569166484419	24563115342271	11567153460388	2 954 81193187	
9	1	14568180489429	24168144357313	21465144289272	11262108257235	
10	1	13066166550449	22667156364345	2 949 52192142		
11	1	14670188473390	24465136348243	11768145501381		
12	1	15068179391334	24864179232186	21150 54152144	1 750 61236153	
13	1	13174210526398	23066143399365	2 852 61148132		
14	1	13767195468401	23267164508357	21055 61224205	1 854 87241165	
15	1	14070190522456	23559112303257	21156 92244206		
16	1	13271170501411	23268138448354	2 955 61169149	2 749 50118108	
17	1	13774198595414	23562145360292	11361103296250	21058 85249218	
18	1	15472223361285	24765122363275	11159 84216213	2 954 58202184	
19	1	13466176467383	23057110260204	2 849 52171142		
20	1	14069178618479	23763131350287	21664117371353	11257 85254214	
21	1	13369176651472	23165142428319	21162101297258	2 850 52134131	
22	1	13172163512463	23066147417355	1 749 55164157		
23	1	15173215499416	24963122299294	11770180496436	11567148391287	11264105326299
24	1	13071163518470	22965125206202	1 850 50158144		
25	2	13967151458377	23464176418351	11771164567449	21361145266242	

TABLE B.2. (Continued)

columns 123	4	1 56789012345678	2 90123456789012	3	4 34567890123456	5 78901234567890	6 78901234567890	7 12345678901234
26	2	13668151488425	23363124319286		11360104313299		2 954 61195169	
27	2	13366178421375	22966170397325		11055 73233190			
28	2	13473209544470	23263108344285		1359 83235215		21053 55114 90	
29	2	15268182365318	24464128356298		21463115303291			
30	2	13968209417362	23863133339291		11463103330253	213611022226198		21156 68192168
31	2	14167155381250	23864132365334		21561102374235		1135910526 1219	
32	2	14370196470413	24467145411328		21766139366282	21661411356305		1146613047 3405
33	2	13270194592501	23160168215175		21054104238199		2 850 55174136	
34	2	13371178586466	23264123310297		21259 93226218		21054 61185169	1 746 44194112
35	2	13674191598496	23664133301257		1 956 82240203			
36	2	14766150465385	24665131394303		21359 85336233	2 750 51149126		2 750 51131127
37	2	12666158473377	22667151343302		2 951 57184166			
38	2	13569184400350	24265179362305		11768154482441			
39	2	15273187611472	25064140380304		11770170503439	212601092266227		11053 65179162
40	2	13566169302251	23461 90233194		11361 90250191		11257 85314173	
41	2	15070146510429	24867159317308		21669138468424			
42	2	14172234508409	23965241298254		21664147381303	212642030305262		11267191355273
43	2	13465159477401	23061150362297		21154 85217183		1 850 60194139	
44	2	15469176519424	25261125303258		11460 96361309			
45	2	14667186449405	24661231288248		21763174321287			
46	2	14866172437320	24765200252248		11463133403312		11054 72207193	
47	2	14268152391329	24262116281256		21661138294288			
48	2	13870214462408	23665129407294		11672141527431	21562113348293		
49	2	13873196606497	23666130422366		21566128353314	11368124413383		
50	2	14575186666585	24366151280227		11675171689545	21365114278264		

TABLE B.2. (Continued)

columns 123	4	1	2 3	4	5 6	7
		56789012345678	90123456789012	34567890123456	78901234567890	12345678901234
		(father)	(mother)	(child #1)	(child #2)	(child #3)
51	2	15171232448350	24862150345289	11769185680499	21567141371288	
52	2	14066187452397	24264140423369	11769161581544	11466131459419	
53	2	14367150438356	23866138275262	2 954 58128120		
54	2	14971172398325	23961150309268	21663111262245	11157 76217199	
55	2	13567158447404	23164123351311	11256 75236204	21055 68189182	
56	2	15973180512410	25667138418321	21769131463429	11674136439381	
57	2	13878172504449	23565146835309	11054 60153147	1 849 49136122	
58	2	14671194600489	24567128843283	21666105338310	11363 9431823	
59	2	14071176563476	24065128337300	11671148488453	21565106264212	11161 91337269
60	2	13571187566500	23665142356307	11056 69201180		
61	2	14065144475409	23863108302256	11256 90226189	21153 68175155	
62	2	15067170383306	24168145421305	11159122280226		
63	2	14368200496385	23863137349277	21767122379349		
64	2	15068205373309	25362160275225	11567186410334		
65	2	14270160518450	23962143412362	11665125325311	21357 86296253	
66	2	13966188488390	23262124257206	11049 50107 98		
67	2	13270235537430	23262181862316	11367140434356	11259108301256	1 956 82225213
68	2	14069213462395	23962134386285	11768151490456	21763111317304	11467126438372
69	2	15469179510384	25262121336318	11661183392380	11154 64202179	
70	2	13666173421352	23363113292278	21361104293269	21160 96264255	
71	2	14473178602450	24265161316295	21160 78149133		
72	2	14568173413339	24066125874307	21467116368336	21159 77208176	
73	2	14269220411372	24061164283256	11664125325272	21360109292258	11056 87280238
74	3	15173158657499	24766116404333	11772163604521	21162 98322269	
75	3	15466155340305	25264152296245	21663 95242238		

431

TABLE B.2. (*Continued*)

columns 123	4	1	2	3	4	5	6	7
		567890 12345678	901234 56789012	901234 56789012	345678 90123456	789012 34567890	789012 34567890	123456 78901234
76	3	14669192407343	24867205325283	21667142326301		21466136294275		11360134257209
77	3	12872172638545	22862124330328	1 952 64150140				
78	3	15370194553443	25662183282224	11465117376261				
79	3	12869194495407	22762128302259	2 848 53128108				
80	3	13967226319293	23462126245222	21462113289262				
81	3	14374216600490	24164141316266	21060 94335216				
82	3	13569164526402	23462144349305	11259100269240				
83	3	13870171569451	23564145432356	11264111406331		1 849 55172144		
84	3	13768138624537	23667128327307	11470156563474				
85	3	14271169407329	24169145447331	11259 80201197		1 957 84230206		
86	3	13367183511445	23363124287218	2 749 55164138				
87	3	14564153559471	24660150309252	11058 80240225		2 856 99272238		
88	3	13367175540399	23257108279221	21355 72152110				
89	3	14173219434362	24067170370305	21364108314289				
90	3	14169188466415	24268159243215	11160 98288228				
91	3	13474200585486	23168145567426	2 854 64193183		2 850 50189172		
92	3	13886143441384	24067130355308	2 846 45163127				
93	4	14270197568458	23768147398296	11260108321274		11061 91296233		1 955 74205183
94	4	14367171410327	24166260349289	2 952 66185160				
95	4	14070164606389	24065163333270	11665112391315		11362 95309251		2 954 70197144
96	4	13671205587398	23464195388321	2 952 58192155				
97	4	14473145413345	23763132384337	21459 86274247				
98	4	14270177509422	24266155359331	21364117334311		21057 82236212		2 750 65175142
99	4	13071156620525	22962143362299	1 748 55155138				
100	4	14669207375280	24461140352294	21660111279277				

TABLE B.2. (*Continued*)

columns 123	4	1 / 56789012345678 (father)	2 / 90123456789012 (mother)	4 / 34567890123456 (child #1)	5 / 78901234567890 (child #2)	7 / 12345678901234 (child #3)
101	4	14474202479385	23165189420380	2 851 62119116		
102	4	14371198652501	24065122402324	21566162392362	21364145399351	
103	4	14266162409309	23863120403343	1 953 60194155	2 748 55153136	
104	4	13370180532441	23061186357331	1 848 55167142		
105	4	14076245524455	23661185496460	11572172521462	11364115389310	21058 81194156
106	4	13371168477321	23226138407325	21466115361297	11259 91276218	
107	4	14369200447399	23869234447349	11670207498409	11572176454402	
108	4	14469151513438	23467140373325	21259 73299268	1 854 58156154	
109	4	13170183509441	23065132399375	2 853 66202182		
110	4	14070228650560	23465140301300	2 849 47147136		
111	4	14469205509428	24266168420377	21766172448402		
112	4	13969191468366	23761171371313	11568150474367	21361139274250	11260 95257216
113	4	13367174591497	23161108357318	1 954 83331203		
114	4	13768206496409	22862138405345	2 850 70148119		
115	4	13669184536423	23567220438380	11261 76319280		
116	4	13269190580518	23162118242241	21157 83264220		
117	4	13070210586480	22967211368295	1 850 59150139		
118	4	13472193642498	23262 98340327	2 851 64172155		
119	4	13871191466369	23666139423332	21363110310267	11365138335264	
120	4	13572209519426	23269184402307	21365104229209		
121	4	14771178521445	24563146319256	21766104376324	21665113385369	21565124460393
122	4	15268193512439	24963135290219	11666128464383		
123	4	14669169536462	24064146348264	11670206477389		
124	4	14268167427333	23565186333279	11461 98284233	21464123280251	11258 76222191
125	4	14468209452380	23562124304242	2 953 64186172		

TABLE B.2. (*Concluded*)

columns 123	4	1 56789012345678	2 90123456789012	3 34567890123456	4 34567890123456	5 78901234567890	6 78901234567890	7 12345678901234
126	4	13472198563440	23266152306263	2 952 58218137				
127	4	14068131487430	23762175357301	21463117321320		21262 85316254		
128	4	13971169562468	23762175315262	11366122435368		2 954 60168156		1 954 66224191
129	4	13770225497412	23068194442375	11157 75234220		11055 69224205		1 954 70220197
130	4	12975222640558	22969130334296	1 750 50157155				
131	4	13765167410329	23862110331290	11769144498453		2126110125244		
132	4	13964162409364	24062141289267	21560104288277				21156 73249210
133	4	13270167592485	23365135425389	21154 69191175		2 748 51144142		
134	4	13771180516451	23666131408328	11771170566446				
135	4	13672198452400	23565182328295	11263117267263		2 856 72201190		
136	4	13068166525481	23064124371329	2 851 53157146				
137	4	14569164461398	24262160305273	21766115329296				
138	4	13368132426372	23065176386338	21058 72198177				
139	4	15268195463399	24965152392306	11769122440335				
140	4	14574196555434	24464 93333299	21665100308292		11156 72186164		
141	4	12868198534451	22764220406348	11055 91259202				
142	4	14467149401347	23964143306253	2 951 60118114				
143	4	13670191440359	23367142378332	2 852 67160139				
144	4	14471220462376	23664267301296	11769136452407				
145	4	14073214537448	23561130307263	11463117346279			2 852 68182145	
146	4	15369162441340	25065147279245	21766127360347				
147	4	13772195473418	23764145404346	11155 66223169				
148	4	13967181549450	23363132260313	11156 96232211			1 849 55181165	
149	4	13666129495374	22960110380325	1 746 49161136				
150	4	14864170351292	24463150358289	21561115353285				

BIBLIOGRAPHY

Detels, R.; Rokaw, S. N.; Coulson, A. H.; Tashkin, D. P.; Sayre, J. W.; Massey, F. J. The UCLA population studies of chronic obstructive respiratory disease. I. Methodology and comparison of lung function in areas of high and low pollution. *American Journal of Epidemiology* 109 : 33–58, 1979.

Detels, R.; Sayre, J. W.; Coulson, A. H.; Rokaw, S. N.; Massey, F. J.; Tashkin, D. P.; Wu, M. The UCLA population studies of chronic obstructive respiratory disease. IV. Respiratory effects of long-term exposure to photochemical oxidants, nitrogen dioxide, and sulfates on current and never smokers. *American Review of Respiratory Disease* 124 : 673–680, 1981.

Rokaw, S. N.; Detels, R.; Coulson, A. H.; Sayre, J. W.; Tashkin, D. P.; Allwright, S. S.; Massey, F. J. The UCLA population studies of chronic respiratory disease. 3. Comparison of pulmonary function in three communities exposed to photochemical oxidants, multiple primary pollutants, or minimal pollutants. *Chest* 78 : 252–262, 1980.

Tashkin, D. P.; Detels, R.; Coulson, A. H.; Rokaw, S. N.; Sayre, J. W. The UCLA population studies of chronic obstructive respiratory disease. II. Determination of reliability and estimation of sensitivity and specificity. *Environmental Research* 20 : 403–424, 1979.

COMBINED BIBLIOGRAPHY

Ordered by first author, with cross-reference entries for other authors. For example, "Doe, J. See Smith (2); Jones." refers to two citations with Smith and one with Jones.

Aach, R. D. See Hollinger.

Abelson, R. P., and Tukey, J. W. Efficient conversion of nonmetric information into metric information. *Proceedings of the Social Statistics Section, American Statistical Association* 226–230, 1959.

———. Efficient utilization of non-numerical information in quantitative analysis: General theory and the case of simple order. *Annals of Mathematical Statistics* 34: 1347–69, 1963.

Acton, F. S. *The analysis of straight-line data*. New York: Wiley, 1959.

Afifi, A. A. See Bendel; Costanza; Hollinger.

Afifi, A. A., and Azen, S. P. *Statistical analysis: A computer oriented approach*. 2nd ed. New York: Academic Press, 1979.

Afifi, A. A., and Elashoff, R. M. Missing observations in multivariate statistics. I: Review of the literature. *Journal of the American Statistical Association* 61: 595–604, 1966.

———. Missing observations in multivariate statistics. III: Large sample analysis of simple linear regression. *Journal of the American Statistical Association*. 64: 337–358, 1969a.

———. Missing observations in multivariate statistics. IV: A note on simple linear regression. *Journal of the American Statistical Association* 64: 359–365, 1969b.

Allen, D. M., and Cady, F. B. *Analyzing experimental data by regression*. Belmont, Calif.: Lifetime Learning, 1982.

Allwright, S. S. See Rokaw.

Anderberg, M. R. *Cluster analysis for applications*. New York: Academic Press, 1973.

Anderson, T. W. *An introduction to multivariate statistical analysis*. New York: Wiley, 1958.

Andreasen, N. C., and Grove, W. M. The classification of depression: Traditional versus mathematical approaches. *American Journal of Psychiatry* 139:45–52, 1982.

Andrews, F. M.; Klem, L.; Davidson, T. N.; O'Malley, P. M.; and Rodgers, W. L. *A guide for selecting statistical techniques for analyzing social sciences data*. 2nd ed. Ann Arbor: Institute for Social Research, University of Michigan, 1981.

Andrews, R.; Morgan, J.; and Sonquist, J. *Multiple classification analysis*. Ann Arbor: Institute for Social Research, University of Michigan, 1969.

Aneshensel, C. S., and Frerichs, R. R. Stress, support, and depression: A longitudinal causal model. *Journal of Community Psychology* 10:363–376, 1982.

Aneshensel, C. S.; Frerichs, R. R.; and Clark, V. A. Family roles and sex differences in depression. *Journal of Health and Social Behavior* 22:379–393, 1981.

Aneshensel, C. S. See Clark; Frerichs (2).

Anscombe, F. J., and Tukey, F. W. The examination and analysis of residuals. *Technometrics* 5:141–160, 1963.

Augenstein, M. J., and Tenenbaum, A. M. *Data structures and PL/1 programming*. Englewood Cliffs, N.J.: Prentice-Hall, 1979.

Azen, S. P. See Afifi.

Bard, Y. *Nonlinear parameter estimation*. New York: Academic Press, 1974.

Barnett, V., and Lewis, T. *Outliers in statistical data*. New York: Wiley, 1978.

Bartlett, M. S. The statistical significance of canonical correlations. *Biometrika* 32:29–38, 1941.

――――. The use of transformation. *Biometrics* 3:39–52, 1947.

Beale, E. M. L.; Kendall, M. G.; and Mann, D. W. The discarding of variables in multivariate analysis. *Biometrika* 54:357–366, 1967.

Beale, E. M. L., and Little, R. J. A. Missing values in multivariate analysis. *Journal of the Royal Statistical Society* 37:129–145, 1975.

Belsley, D. A.; Kuh, E.; and Welsch, R. E. *Regression diagnostics: Identifying influential data and sources of collinearity*. New York: Wiley, 1980.

Bendel, R. B., and Afifi, A. A. Comparison of stopping rules in forward stepwise regression. *Journal of the American Statistical Association* 72:46–53, 1977.

Bennett, C. A., and Franklin, N. L. *Statistical analysis in chemistry and the chemical industry*. New York: Wiley, 1954.

Bent, D. H. See Nie.

Berglund, K. See Goldsmith.

Berk, K. W., and Francis, I. S. A review of the manuals for BMDP and SPSS. *Journal of the American Statistical Association* 73:65–71, 1978.

Berkson, J. Estimation of a linear function for a calibration line. *Technometrics* 11:649–660, 1969.

Bickel, P. J., and Doksum, K. A. An analysis of transformations revisited. *Journal of the American Statistical Association* 76:296–311, 1981.

Bishop, Y. M. M.; Fienberg, S. E.; and Holland, P. W. *Discrete multivariate analysis: Theory and practise.* Cambridge, Mass.: MIT Press, 1975.

Blackwelder, W. C. See Halperin.

Boggs, W. E. See Wilson.

Box, G. E. P. Use and abuse of regression. *Technometrics* 8:625–629, 1966.

Box, G. E. P., and Cox, D. R. Analysis of transformations. *Journal of the Royal Statistical Society*, Series B, 26:211-252, 1964.

Box, G. E. P.; Hunter, W. G.; and Hunter, J. S. *Statistics for experimenters.* New York: Wiley, 1978.

Box, G. E. P. and Jenkins, G. M. *Time series analysis; forecasting and control.* San Francisco: G. P. Holden-Day, 1976.

Breslow, N. E., and Day, N. E. *Statistical methods in cancer research.* IARC Scientific Publications No. 32. Lyon, France: WHO, 1980.

Briggs, W. R. See Wilson.

Brigham, E. F. *Fundamentals of financial management.* Hinsdale, Ill.: Dryden Press, 1978.

Brittain, E. *Probability of developing coronary heart disease.* Technical Report 54. Stanford, Calif: Division of Biostatistics, Stanford University, 1980.

Brownlee, K. A. *Statistical theory and methodology in science and engineering.* 2nd ed. New York: Wiley, 1965.

Cady, F. B. See Allen.

Capron, H. L., and Williams, B. K. *Computers and data processing.* Menlo Park, Calif.: Benjamin/Cummings, 1982.

Chatterjee, S., and Price, B. *Regression analysis by example.* New York: Wiley, 1977.

Churchman, C. W., and Ratoosh, P., eds., *Measurement: Definition and theories.* New York: Wiley, 1959.

Clark, V. A. See Aneshensel; Dunn; Frerichs (2).

Clark, V. A.; Aneshensel, C. S.; Frerichs, R. R.; and Morgan, T. M. Analysis of effects of sex and age in response to items on the CES-D scale. *Psychiatry Research* 5:171–181, 1981.

Cleary, J. P. See Levenbach (2).

Comstock, G.W., and Helsing, K. J. Symptoms of depression in two communities. *Psychological Medicine* 6:551–563, 1976.

Cook, R. D. Detection of influential observations in linear regression. *Technometrics* 19:15–18, 1977.

―――. Influential observations in linear regression. *Journal of the American Statistical Association* 74:169–174, 1979.

Cook, R. D., and Weisberg, S. Characterization of an empirical influence function for detecting influential cases in regression. *Technometrics* 22:495–508, 1980.

Cooley, W. W., and Lohnes, P. R. *Multivariate data analysis*. New York: Wiley, 1971.

Coombs, C. H. *A theory of data*. New York: Wiley, 1964.

Cormack. R. M. A review of classification. *Journal of the Royal Statistical Society, Series A* 134:321–367, 1971.

Cornfield, J. See Truett.

Costanza, M. C., and Afifi, A. A. Comparison of stopping rules for forward stepwise discriminant analysis. *Journal of the American Statistical Association* 74:777–785, 1979.

Coulson, A. H. See Detels (3); Rokaw; Tashkin.

Council, K. A. See Helwig.

Cox, D. R. *Analysis of binary data*. London: Methuen, 1970.

―――. See Box.

Dallal, G. E. See Wilkinson.

D'Amico, R. See Fennessey.

Daniel, C., and Wood, F.S. *Fitting equations to data*. New York: Wiley, 1980.

Davidson, T. N. See Andrews.

Day, N. E. See Breslow.

Detels, R. See Rokaw; Tashkin.

Detels, R.; Coulson, A.; Tashkin, D.; and Rokaw, S. Reliability of plethysmography, the single breath test, and spirometry in population studies. *Bullitin de Physiopathologie Respiratoire* 11:9–30, 1975.

Detels, R.; Rokaw, S. N.; Coulson, A. H.; Tashkin, D. P.; Sayre, J. W.; and Massey, F. J. The UCLA population studies of chronic obstructive respiratory disease. I. Methodology and comparison of lung function in areas of high and low pollution. *American Journal of Epidemiology* 109:33–58, 1979.

Detels, R.; Sayre, J. W.; Coulson, A. H.; Rokaw, S. N.; Massey, F. J.; Tashkin, D. P.; and Wu, M. The UCLA population studies of chronic obstructive respiratory disease. IV. Respiratory effects of long-term exposure to photochemical oxidants, nitrogen dioxide, and sulfates on current and never smokers. *American Review of Respiratory Disease* 124:673–680, 1981.

Devlin, S. J.; Gnanadesekan, R.; and Kettenring, J. R. Robust estimation of dispersion matrices and principal components. *Journal of the American Statistical Association*. 76:354–362, 1981.

Dickerson, R. E. See Wilson.

Dixon, W. J., ed. *BMDP statistical software 1983*. Berkeley: University of California Press, 1983.

Dixon, W. J., and Massey, F. J. *Introduction to statistical analysis*. 4th ed. New York: McGraw-Hill, 1983.

Doksum, K. A. See Bickel.

Draper, N. R., and Hunter, W. G. Transformations: Some examples revisited. *Technometrics* 11:23–40, 1969.

Draper, N. R., and Smith, H. *Applied regression analysis*. 2nd ed. New York: Wiley, 1981.

Duncan, O. D. Path analysis in sociological examples. *American Journal of Sociology* 72:1–16, 1966.

———. *Introduction to structural equation models*. New York: Academic Press, 1975.

Dunn, O. J., and Clark, V. A. *Applied statistics: Analysis of variance and regression*. New York: Wiley, 1974.

Eastment, H. T., and Krzanowski, W. J. Cross-validory choice of the number of components from a principal component analysis. *Technometrics* 24:73–77, 1982.

Efron, B. The efficiency of logistic regression compared to normal discriminant analysis. *Journal of the American Statistical Association* 70:892–898, 1975.

Efroymson, M. A. Multiple regression analysis. In *Mathematical methods for digital computers*, ed. Ralston and Wilf. New York: Wiley, 1960.

Eisner, T. See Wilson.

Elashoff, R. M. See Afifi (3).

Ellis, B. *Basic concepts of measurement*. London: Cambridge University Press, 1966.

Engleman, L. See Forsythe.

Everitt, B. *Cluster analysis*. London: Heinemann Educational Books, 1980.

Fennessey, J., and D'Amico, R. Collinearity, ridge regression, and investigator judgement. *Social Methods and Research* 8:309–340, 1980.

Fienberg, S. E. See Bishop.

Fisher, R. A. The use of multiple measurement in taxonomic problems. *Annals of Eugenics* 7:179–188, 1936.

Fleiss, J. L. *Statistical methods for rates and proportions*. New York: Wiley, 1981.

Flores, I. *Data structure and management*. 2nd ed. Englewood Cliffs, N.J.: Prentice-Hall, 1977.

Forsythe, A. B.; Engleman, L.; Jennrich, R. I.; and May, P. R. A. Stopping rules for variable selection in multiple regression. *Journal of the American Statistical Association* 68:75–77, 1973.

Francis, I. S. See Berk.

Franklin, N. L. See Bennett.

Frerichs, R. R. See Aneshensel (2); Clark.

Frerichs, R. R.; Aneshensel, C. S.; and Clark, V. A. Prevalence of depression in Los Angeles County. *American Journal of Epidemiology* 113:691–699, 1981*a*.

Frerichs, R.; Aneshensel, C. S.; Clark, V. A.; and Yokopenic, P. Smoking and depression: A community survey. *American Journal of Public Health* 71:637–640, 1981*b*.

Furnival, G. M., and Wilson, R. W. Regressions by leaps and bounds. *Technometrics* 16:499–512, 1974.

Gage, N. L., ed. *Handbook of research on teaching.* Chicago: Rand McNally, 1963.

Gallant, A. R. Nonlinear regression. *American Statistician* 29:73–81, 1975.

Gnanadesekan, R. See Devlin.

Goldsmith, J. R., and Berglund, K. Epidemiological approach to multiple factor interactions in pulmonary disease: The potential usefulness of path analysis. *Annals of the New York Academy of Sciences* 221:361–375, 1974.

Goldstein, H. *The design and analysis of longitudinal studies.* New York: Academic Press, 1979.

Gonzalez, R. C. See Ton.

Gordon, A. D. *Classification.* London: Chapman and Hall, 1981.

Gorman, J. W., and Toman, R. J. Selection of variables for fitting equations to data. *Technometrics* 8:27–51, 1966.

Gorsuch, R. L. *Factor analysis.* Philadelphia: Saunders, 1974.

Gower, J. C. A general coefficient of similarity and some of its properties. *Biometrics* 27:857–874, 1971.

Graybill, F. A. *Theory and application of the linear model.* N. Scituate Mass.: Duxbury Press, 1976.

Grove, W. M. See Andreasen.

Gunst, R. F., and Mason, R. L. *Regression analysis and its application.* New York: Dekker, 1980.

Hald, A. *Statistical theory with engineering applications.* New York: Wiley, 1952.

Halperin, M.; Blackwelder, W. C.; and Verter, J. I. Estimation of the multivariate logistic risk function: A comparison of the discriminant function and maximum likelihood approach. *Journal of Chronic Diseases* 24:125–158, 1977.

Hand, D. J. *Discrimination and classification.* New York: Wiley, 1981.

Hanushek, E. A., and Jackson, J. E. *Statistical methods for social scientists.* New York: Academic Press, 1977.

Harmon, H. H. *Modern factor analysis.* Chicago: University of Chicago Press, 1976.

Harris, R. J. *A primer of multivariate statistics.* New York: Academic Press, 1975.

Hartigan, J. A. *Clustering algorithms.* New York: Wiley, 1975.

Hartley, A. O. Modified Gauss-Newton method for fitting non-linear regression functions. *Technometrics* 3:269–280, 1961.

Hartley, H. O., and Hocking, R. R. The analysis of incomplete data. *Biometrics* 27:783–824, 1971.

Hartz, S. C. See O'Hara.

Hay, R. A. See McCleary.

Hearne, F. T. See Jackson.

Helsing, K. J. See Comstock.

Helwig, J. T., and Council, K. A., eds., User's SAS guide. Raleigh, N. C.: SAS Institute Inc., P.O. Box 10066, 1979.

Hildebrand, D. K.; Laing, J. D.; and Rosenthal, H. *Analysis of ordinal data*. Beverly Hills: Sage, 1977.

Hill, M. A. *BMDP user's digest*. 2nd ed. Los Angeles: BMDP Statistical Software, Inc., P.O. Box 24 A 26, 1982.

Hocking, R. R. Criteria for selection of a subset regression: Which one should be used. *Technometrics* 14:967–970, 1972.

———. The analysis and selection of variables in linear regression. *Biometrics* 32:1–50, 1976.

———. Developments in linear regression methodology. *Technometrics* 25:219–229, 1983.

———. See Hartley.

Hoerl, A. E., and Kennard, R. W. Ridge regression: Application to non-orthogonal problems. *Technometrics* 12:55–82, 1970.

Holford, T. R.; White, C.; and Kelsey, J. L. Multivariate analyses for matched case-control studies. *American Journal of Epidemiology* 107:245–256, 1978.

Holland, P. W. See Bishop.

Hollinger, F. B.; Mosley, J. W.; Szmuness, W.; Aach, R. D.; Melnick, J. L.; Afifi, A.; Stevens, C. E.; and Kahn, R. A. Non-A, non-B, hepatitis following blood transfusions: Risk factors associated with donor characteristics. In *Viral hepatitis: 1981 international symposium*, ed. W. Szmuness, H. J. Alter, and J. E. Maynard. Philadelphia: Franklin Institute Press, 1982.

Hopkins, C. E. Statistical analysis by canonical correlation: A computer application. *Health Services Research* 4: 304–312, Winter 1969.

Hosmer, D. W. See Lemeshow; O'Hara.

Hotelling, H. Analysis of a complex of statistical variables into principal components. *Journal of Educational Psychiatry* 24:417–441, 1933.

———. Relations between two sets of variables. *Biometrika* 28:321–377, 1936.

Hull, C. H. See Nie.

Hull, C. H., and Nie, N. H. *SPSS update 7–9*. New York: McGraw-Hill, 1981.

Hunter, J. S. See Box.

Hunter, W. G. See Box; Draper.

Hunter, W. G., and Lamboy, W. F. A Bayesian analysis of the linear calibration problem. *Technometrics* 23:323–343, 1981.

Jackson, J. E. See Hanushek.

Jackson, J. E., and Hearne, F. T. Relationships among coefficients of vectors in principal components. *Technometrics* 15:601–610, 1973.

Jenkins, G. M. See Box (2).

Jenkins, J. G. See Nie.

Jennrich, R. I. See Forsythe (2); Ralston.

Johnson, L. A. See Montgomery.

Johnson, R. A., and Wichern, D. W. *Applied multivariate statistical analysis*. Englewood Cliffs, N. J.: Prentice-Hall, 1982.

Johnson, S. C. Hierarchical clustering schemes. *Psychometrika* 32:241–244, 1967.

Joreskog, K. G., and Sorbom, D. *LISREL IV, analysis of linear structural relationships by the method of maximum likelihood*. Chicago: National Resources, 1978.

Kachigan, S. K. *Multivariate statistical analysis: A conceptual approach*. New York: Radius Press, 1982.

Kahn, R. A. See Hollinger.

Kannell, W. See Truett.

Kelsey, J. L. See Holford.

Kendall, M. G. See Beale.

Kendall, M. G., and Stuart, A. *The advanced theory of statistics*. New York: Hafner, 1967.

Kennard, R. W. See Hoerl.

Kettenring, J. R. See Devlin.

Kim, J. O., and Mueller, C. W. *Introduction to factor analysis*. Beverly Hills: Sage, 1978.

———. *Factor analysis*. Beverly Hills: Sage, 1978.

Klecka, W. R. *Discriminant analysis*. Beverly Hills: Sage, 1980.

Klem, L. See Andrews.

Kowalik, J., and Osborne, M. R. *Methods for unconstrained optimization problems*. New York: Elsevier, 1968.

Kowalski, C. J. The performance of some rough tests for bivariate normality before and after coordinate transformations to normality. *Technometrics* 12:517–544, 1970.

Krutchkoff, R. G. Classical and inverse regression methods of calibration. *Technometrics* 9:425–440, 1967.

Krzanowski, W. J. See Eastment.

Kuh, E. See Belsley.

Lachenbruch, P. A. *Discriminant analysis*. New York: Hafner Press, 1975.

————. See Woolson.

Laing, J. D. See Hildebrand.

Lamboy, W. F. See Hunter.

Lawing, W. D. See Pyne.

Lawley, D. N. Tests of significance for the latent roots of covariance and correlation matrices. *Biometrika* 43:128–136, 1956.

————. Tests of significance in canonical analysis. *Biometrika* 46:59–66, 1959.

Lawley, D. N., and Maxwell, A. E. *Factor analysis as a statistical method*. London: Butterworths, 1963.

Lemeshow, S. See O'Hara.

Lemeshow, S., and Hosmer, D. W. A review of goodness-of-fit statistics for use in the development of logistic regression models. *American Journal of Epidemiology* 115:92–106, 1982.

Levenbach, H., and Cleary, J. P. *The beginning forecaster: The forecasting process through data analysis*. Belmont, Calif.: Lifetime Learning, 1981.

————. *The professional forecaster: The forecasting process through data analysis*. Belmont, Calif.: Lifetime Learning, 1982.

Levine, M. S. *Canonical analysis and factor comparison*. Sage University Paper. Beverly Hills: Sage, 1977.

Lewis, T. See Barnett.

Lewis-Beck, M. S. *Applied regression: An introduction*. Beverly Hills: Sage, 1980.

Li, C. C. *Path analysis—A primer*. Pacific Grove, Calif.: Boxwood Press, 1975.

Little, R. J. A. See Beale.

Little, R. J. A. Models for nonresponse in sample surveys. *Journal of the American Statistical Association*, 77:237-250, 1982.

Lohnes, P. R. See Cooley.

Lwin, T., and Maritz, J.S. An analysis of the linear-calibration controversy from the perspective of compound estimation. *Technometrics* 24:235–242, 1982.

McCleary, R., and Hay, R. A. *Applied time series analysis for the social sciences*. Beverly Hills: Sage, 1980.

MacQueen, J. B. Some methods for classification and analysis of multivariate observations. *Proceedings of the Fifth Berkeley Symposium on Mathematical Statistics and Probability* 1:281–297, 1967.

Mage, D. T. An objective graphical method for testing normal distribution assumptions using probability plots. *American Statistician* 36:116–120, 1982.

Mallows, C. L. Some comments on Cp. *Technometrics* 15:661–676, 1973.

Mann, D. W. See Beale.

Mantel, N. Why stepdown procedures in variable selection. *Technometrics* 12:621–626, 1970.

Maritz, J. S. See Lwin.

Marquardt, D. W. An algorithm for the least-squares estimation of non-linear parameters. *Journal of Social and Industrial Applied Mathematics* 11:431–441, 1963.

Marquardt, D. W., and Snee, R. D. Ridge regression in practice. *American Statistician* 29:3–20, 1975.

Mason, R. L. See Gunst.

Massey, F. J. See Detels (2); Dixon; Rokaw.

Maxwell, A. E. See Lawley.

May, P. R. A. See Forsythe.

Melnick, J. L. See Hollinger.

Meredith, W. Canonical correlation with fallible data. *Psychometrika* 29:55–65, 1964.

Metzenberg, R. L. See Wilson.

Minton, P. D. See Schucany.

Montgomery, D. C., and Johnson, L. A. *Forecasting and time series analysis*. New York: McGraw-Hill, 1976.

Morgan, J. See Andrews.

Morgan, T. M. See Clark.

Morrison, D. F. *Multivariate statistical methods*. New York: McGraw-Hill, 1976.

Mosley, J. W. See Hollinger.

Mosteller, F., and Tukey, J. W. *Data analysis and regression*. Reading, Mass.: Addison-Wesley, 1977.

Mueller, C. W. See Kim (2).

Mulaik, S. A. *The foundations of factor analysis*. New York: McGraw-Hill, 1972.

Muller, M. E. A review of the manuals for BMDP and SPSS, (followed by comments by several authors). *Journal of the American Statistical Association* 73:71–80 (80–98), 1978.

Naszodi, L. J. Elimination of bias in the course of calibration. *Technometrics* 20:201–206, 1978.

Neffendorf, H. Statistical packages for microcomputers: A listing. *American Statistician* 37:83–86, 1983.

Neter, J., and Wasserman, W. *Applied linear statistical models*. Homewood, Ill.: Irwin, 1974.

Nie, N. H. See Hull.

Nie, C. H.; Hull, C. H.; Jenkins, J. G.; Steinbrenner, K.; and Bent, D. H. *SPSS*. New York: McGraw-Hill, 1975.

O'Brien, R. D. See Wilson.

O'Hara, T. F.; Hosmer, D. W.; Lemeshow, S.; and Hartz, S. C. A comparison of discriminant function and maximum likelihood estimates of logistic coefficients for categorical data. University of Massachusetts, Amherst, Mass; 1982.

O'Malley, P. M. See Andrews.

Orchard, T., and Woodbury, M. A. A missing information principle: Theory and application. *Proceedings of the Sixth Berkeley Symposium on Mathematical Statistics and Probability* 1:697–715, 1972.

Osborne, M. R. See Kowalik.

Ostrom, C. W. *Time series analysis: Regression techniques.* Beverly Hills: Sage, 1978.

Pope, P. T., and Webster, J. T. The use of an *F*-statistic in stepwise regression procedures. *Technometrics* 14:327–340, 1972.

Pough, F. H. *Field guide to rocks and minerals.* 3rd ed. Boston: Houghton Mifflin, 1960.

Press, S. J., and Wilson, S. Choosing between logistic regression and discriminant analysis. *Journal of the American Statistical Association* 73:699–705, 1978.

Price, B. See Chatterjee.

Pyne, D. A., and Lawing, W. D. *A note on the use of the Cp statistic and its relation to stepwise variable selection procedures.* Technical Report No. 210, Johns Hopkins University, 1974.

Quandt, R. E. The estimation of the parameters of a linear regression system obeying two separate regimes. *Journal of the American Statistical Association* 53:873–880, 1958.

————. Tests of the hypothesis that a linear regression system obeys two separate regimes. *Journal of the American Statistical Association* 55:324–331, 1960.

————. New approaches to estimating switching regressions. *Journal of the American Statistical Society* 67:306–310, 1972.

Radloff, L. S. The CES–D scale: A self-report depression scale for research in the general population. *Applied Psychological Measurement* 1:385–401, 1977.

Ralston, M. L., and Jennrich, R. I. Dud, a derivative-free algorithm for non-linear least squares. *Technometrics* 20:7–13, 1978.

Rao, C. R. *Linear inference and its application.* New York: Wiley, 1965.

Ratoosh, P. See Churchman.

Ray, A. A., ed. *SAS user's guide: Basics.* Cary, N.C.: SAS Institute, Inc., Box 8000, 1982.

————. *SAS user's guide: Statistics.* Cary, N.C.: SAS Institute, Inc., Box 8000, 1982.

Reinhardt, P. S. *SAS supplemental library user's guide, 1980.* Cary, N.C.: SAS Institute, Inc., Box 8000, 1980.

Reynolds, H. T. *Analysis of nominal data.* Beverly Hills: Sage, 1977.

Rodgers, W. L. See Andrews.

Rokaw, S. N. See Detels (3); Tashkin.

Rokaw, S. N.; Detels, R.; Coulson, A. H.; Sayre, J. W.; Tashkin, D. P.; Allwright, S.

S.; and Massey, F. J. The UCLA population studies of chronic respiratory disease. 3. Comparison of pulmonary function in three communities exposed to photochemical oxidants, multiple primary pollutants, or minimal pollutants. *Chest* 78:252–262, 1980.

Rosenthal, H. See Hildebrand.

SAS. SAS Institute Technical Report, SAS 79.5 Changes and Enhancements. Cary, N.C.: SAS Institute, Inc., Box 8000, 1981.

Sayre, J. W. See Detels (2); Rokaw; Tashkin.

Schlesselman, J. J. *Case-control studies*. New York: Oxford University Press, 1982.

Schucany, W. R.; Shannon, B. S.; and Minton, P. D. A survey of statistical packages. *Computer Surveys* 4:65–79, 1972.

Seal, H. *Multivariate statistical analysis for biologists*. New York: Wiley, 1964.

Shannon, B. S. See Schucany.

Shukla, G. K. On the problem of calibration. *Technometrics* 14:547–554, 1972.

Smith, H. See Draper.

Sneath, P. H., and Sokal, R. R. *Numerical taxonomy*. San Francisco: Freeman, 1973.

Snee, R. D. See Marquardt.

Sokal, R. R. See Sneath.

Sonquist, J. See Andrews.

Sorbom, D. See Joreskog.

Sprent, P. *Models in regression and related topics*. London: Methuen, 1969.

SPSS-X user's guide. New York: McGraw-Hill, 1983.

Steinbrenner, K. See Nie.

Stevens, C. E. See Hollinger.

Stevens, S. S. *Handbook of experimental psychology*. New York: Wiley, 1951.

Stuart, A. See Kendall.

Susman, M. See Wilson.

Szmuness, W. See Hollinger.

Tashkin, D, P. See Detels (3); Rokaw.

Tashkin, D. P.; Detels, R.; Coulson, A. H.; Rokaw, S. N.; and Sayre, J. W. The UCLA population studies of chronic obstructive respiratory disease. II. Determination of reliability and estimation of sensitivity and specificity. *Environmental Research* 20:403–424, 1979.

Tatsuoka, M. M. *Multivariate analysis: Techniques for educational and psychological research*. New York: Wiley, 1971.

Tchira, A. A. Stepwise regression applied to a residential income valuation system. *Assessors Journal* 8:23–35, 1973.

Tenenbaum, A. M. See Augenstein.

Theil, H. *Principles of econometrics*. New York: Wiley, 1971.

Thorndike, R. M. *Correlational procedures for research*. New York: Gardner Press, 1978.

Tobin, J. Estimation of relationships for limited dependent variables. *Econometrica* 26:24–36, 1958.

Toman, R. J. See Gorman.

Ton, J. T., and Gonzalez, R. C. *Pattern recognition principles*. Reading, Mass.: Addison-Wesley, 1974.

Torgerson, W. S. *Theory and methods of scaling*. New York: Wiley, 1958.

Truett, J.; Cornfield, J.; and Kannell, W. Multivariate analysis of the risk of coronary heart disease in Framingham. *Journal of Chronic Diseases* 20:511–524, 1967.

Tukey, J. W. On the comparative anatomy of transformations. *Annals of Mathematical Statistics* 28:602–632, 1957.

———. *Exploratory data analysis*. Reading, Mass.: Addison-Wesley, 1977.

———. See Abelson (2); Anscombe; Mosteller.

Upton, G. J. G. *The analysis of cross-tabulated data*. New York: Wiley, 1978.

Verter, J. I. See Halperin.

Ward, J. H. Hierarchical grouping to optimize an objective function. *Journal of the American Statistical Association* 58:236–244, 1963.

Wasserman, W. See Neter.

Waugh, F. V. Regressions between sets of variables. *Econometrica* 10:290–310, 1942.

Webster, J. T. See Pope.

Weisberg, S. See Cook.

Welsch, R. E. See Belsley.

White, C. See Holford.

Wichern, D. W. See Johnson.

Wilkinson, L., and Dallal, G. E. Tests of significance in forward selection regression with an F-to-enter stopping rule. *Technometrics* 23:377–380, 1981.

Williams, B. K. See Capron.

Williams, E. J. *Regression analysis*. New York: Wiley, 1959.

Wilson, E. O.; Eisner, T.; Briggs, W. R.; Dickerson, R. E.; Metzenberg, R. L.; O'Brien, R. D.; Susman, M.; and Boggs, W. E. *Life on earth*. Sunderland, Mass.: Sinauer Associates, 1973.

Wilson, R. W. See Furnival.

Wilson, S. See Press.

Winer, B. J. *Statistical principles in experimental design*. 2nd ed. New York: McGraw-Hill, 1971.

Wonnacott, R. J., and Wonnacott, T. H. *Econometrics*. 2nd ed. New York: Wiley, 1979.

Wonnacott, T. H. See Wonnacott.

Wood, F. S. See Daniel.

Woodbury, M. A. See Orchard.

Woolson, R. F., and Lachenbruch, P. A. Regression analysis of matched case-control data. *American Journal of Epidemiology* 115:444–452, 1982.

Wright, S. The method of path coefficients. *Annals of Mathematical Statistics* 5:161–215,1934.

Wu, M. See Detels.

Yokopenic, P. See Frerichs.

Younger, M. S. *A handbook for linear regression*. N. Scituate, Mass.: Duxbury Press, 1979.

INDEX